Y0-BCI-841

International Textbooks in

MECHANICAL ENGINEERING

Consulting Editor

EDWARD F. OBERT

Professor of Mechanical Engineering

University of Wisconsin

Introduction to
Automatic Controls

Introduction to

Automatic Controls

second edition

HOWARD L. HARRISON   **JOHN G. BOLLINGER**

Professor   *Professor*

The University of Wisconsin

INTERNATIONAL TEXTBOOK COMPANY

Scranton, Pennsylvania

B-B-RC

Preface to the Second Edition

In recent years control theory and control engineering practice have expanded to encompass many new topics and techniques. In education, mechanical engineers have become more concerned with controls and have provided the student with greater depth in the preparation of the mathematical tools associated with feedback control system analysis. As the state of the art progresses and as the ability to absorb more advanced topics is generated, the class text must be expanded.

The preparation of the second edition of *Introduction to Automatic Controls* has focused on two objectives. First, to expand the scope of the text to include an introduction to many new theoretical topics and control practices. Second, to uphold the original purpose of the first edition by presenting these topics in a fundamental, simplified manner so that they may be easily grasped and applied.

With these objectives in mind, new material has been added to a number of the existing chapters of the first edition. Examples of added topics include the convolution integral, analysis of systems with dead time, signal flow graphs, feedforward compensation, and cascade control. Two new chapters have been added. The first, Chapter 15, presents such topics as state variables, Liapunov's stability criterion, and optional control. The second, Chapter 16, is devoted to digital computer control and includes an introduction to digital computer algorithm formulation and analysis with z-transforms. Because of the importance of automated equipment in all phases of engineering, a new appendix on sequencing control (Appendix G) has been added. This material introduces some of the commonly used electromechanical devices used in switching circuits and develops the techniques for solving sequence control problems. The presentation is generalized to include a discussion of logic functions and Boolean algebra.

The authors have experienced widespread acceptance of the new material in classes involving beginning graduate students and practicing engineers.

HOWARD L. HARRISON
JOHN G. BOLLINGER

Madison, Wisconsin
December, 1968

Preface to the First Edition

The purpose of this book is to present basic automatic control fundamentals so that they can easily be grasped and applied. This means that concepts are stressed throughout, while unnecessary complication, which might obscure the basic ideas, is held to a minimum. The authors believe that, for engineering students, it is not enough just to understand fundamentals; some experience in applying them is equally as important. For this reason, a wide variety of physical systems is introduced for the purpose of illustrating the application of theory and indicating some of the assumptions made in combining theory and practice.

The book is suitable for undergraduate mechanical engineers and other undergraduate engineering students. No previous control study is assumed. Certainly any senior student in mechanical engineering has the necessary prerequisites. A course in differential equations should prove helpful but is not a necessity because the mathematics required, beyond elementary calculus, is carefully presented.

The first five chapters of this book deal with the time domain, in which the underlying principles of dynamic system response are most easily understood. Because the handling of even a moderately complicated control system is unwieldly in the time domain, Chapter 6 introduces the Laplace transformation, which serves as a basis for more efficient methods for control system analysis and design. The remainder of the book is concerned with these techniques.

Electronic computers play an important role in control engineering. Chapter 4 and Appendix D are devoted to the analog computer, while Appendix A deals with the digital computer. However, computer material can be omitted without destroying continuity.

It should be pointed out that no attempt has been made to extend the frontiers of control knowledge. The purpose has been to present basic control theory in such a manner as to permit effective teaching and understanding. If this purpose is achieved, the authors will feel that the book is a success and well worth the effort required in writing it.

It is impossible to acknowledge all contributors to this book. Ideas drawn from many books, articles, persons in the control field, as well as from personal experience have in some manner been woven into the fabric of the book. However, specific contributors include Professor E. F. Obert, University of Wisconsin, whose numerous suggestions were of great help in preparing the manuscript; Mr. Norbert Nitka of the Oilgear Company, who proved to be an excellent source of information regarding fluid power components and systems; and Mr. J. R. Gartner, University of Wisconsin, who read a large portion of the manuscript and offered many helpful suggestions. Particular thanks are due to Mrs. J. R. Gartner, whose typing efforts were most necessary in completing the manuscript.

HOWARD L. HARRISON
JOHN G. BOLLINGER

Madison, Wisconsin
March, 1963

Contents

Notation

SYMBOL	MEANING	UNITS

Letter Symbols

A	area	in.^{2}
$A(s)$	input elements	
A, B, C	constants	
a, b, c	constants	
a	acceleration	in./sec^2
$B(s)$	feedback signal	
b	damping coefficient	lb-in.-sec/radian or oz-in.-sec/radian
C	capacitance	farad
C_d	coefficient of discharge	
$C_1, C_2, \ldots, C_n$	constants	
$C(s)$	controlled variable	
$\Delta C(s)$	unwanted deviation of controlled variable caused by a disturbance input	
c	damping coefficient	lb-sec/in.
D	dead time	sec
D_m	hydraulic motor displacement	in.^{3}/radian
d	distance	
E, e	voltage	volt
$E(s)$	actuating signal	
e	base of natural logarithms	
F, f	force	lb
$f(t)$	general function of time	
$f^*(t)$	sampled-data form of a time function	
$F(s)$	general function transformed	
$F^*(s)$	Laplace transform of the sampled time function	
G	torsional spring constant	lb-in./radian
$G(s)$	forward transfer function	
$G(z)$	z transfer function	
$G_1(s)$	control elements	
$G_2(s)$	system elements	
$G_c(s)$	compensating elements	
$G_N(s)$	normalized transfer function	
g	acceleration of gravity	in./sec^2

SYMBOL	MEANING	UNITS
$g(t)$	impulse response	
H, h	head	in.
$H(s)$	feedback elements	
I, i	current	ampere
I	mass moment of inertia	oz-in.-sec^2
K, k	gain of an element or system	
K	a constant	
K'	a constant proportional to system gain K	
K_b	ball-head rate	
K_b	motor constant	oz-in.-sec/radian
K_d	derivative controller gain	
K_e	motor voltage constant	volt-sec/radian
K_i	integral controller gain	
K_o	effective spring rate	
K_p	proportional controller gain	
K_t	motor torque constant	oz-in./volt
K_t'	motor torque constant	oz-in./amp
K_v	valve constant	in.^{2}/sec
k	spring constant	lb/in.
L	inductance	henrys
L	length	in.
L	leakage coefficient	in.^{5}/lb-sec
M	mass	lb-sec^2/in.
$M(s)$	manipulated variable	
$M(\omega)$	magnitude ratio	
M_m	maximum magnitude ratio	
m_G	gain margin	
N	describing function	
n	gear ratio, indexing subscript	
P, p	pressure	lb/in.^{2}
p	operator ($py = dy/dt$)	
PI	performance index	
Q, q	charge	coulombs
Q, q	volume flow rate	in.^{3}/sec
q	heat flow rate	Btu/sec
R	resistance	ohms
$R(s)$	reference input	
r	root of equation	
s	Laplace transform operator $s = \sigma + j\omega$	
T	torque	lb-in.
T	temperature	°F
t	time	sec
Δt	time increment	
$U(s)$	disturbance input	
V, v	voltage	volt
V	volume	in.^{3}

Symbol	Meaning	Units
$V(s)$	command	
v	velocity	in./sec
w	width of valve port	in.
X, x	displacement	in.
Y, y	displacement	in.
Z, z	displacement	in.
Z_f	feedback impedance	
Z_i	input impedance	
z	z transform operator	

Greek Letters

Symbol	Meaning	Units
α	angular acceleration	radians/sec^2
α	asymptote angle	degrees
α	phase margin	degrees
β	fluid bulk modulus	lb/in.^{2}
β	angle locating line of constant damping ratio	degrees
Δ, δ	an increment	
δ	static spring deflection	in.
ϵ	system error	
ϵ	a small value	
ζ	damping ratio	
Θ, θ	angular displacement	radians
θ	angle	degrees
μ	coefficient of friction	
ρ	mass density	lb-sec^2/in.^{4}
$\sum$	summation	
σ_a	asymptote starting point	
λ	characteristic time, Lagrangian multiplier	
τ	time constant	sec
ϕ	angle	degrees
$\phi(\omega)$	phase angle	degrees
Ω, ω	angular speed	radians/sec
ω	frequency	radians/sec
ω_d	damped natural frequency	radians/sec
ω_n	undamped natural frequency	radians/sec

Chapter **1**

Orientation to
Automatic Controls

1-1. Introduction

Automatic controls play an ever-increasing role in our way of life, from the simple controls required in an automatic bread toaster to the sophisticated control systems necessary for space explorations. For this reason, most engineers will have occasion to work with control systems in some manner, even if only through the use of them. Thus, a study of automatic controls is highly desirable for its own sake. However, there are other benefits to be derived. Automatic control systems are dynamic systems, and a knowledge of control theory provides a basis for understanding the behavior of other dynamic systems. For example, many control concepts can be used in attacking vibration problems. In this sense, automatic control theory is but a portion of a larger body of theory concerning the behavior of all dynamic systems.

Control systems frequently employ components of different types, e.g., mechanical, electrical, hydraulic, pneumatic, and combinations of these. An engineer who works with controls must then be familiar with the fundamental laws underlying these components. Of course, these fundamentals are covered in various engineering courses. However, in too many cases among engineers, the fundamentals appear to exist as isolated islands of knowledge with few bridges built between them. The study of automatic controls can be beneficial for integrating knowledge (building bridges) from the different fields of study by bringing the various fundamentals into a common control problem.

The study of automatic controls is important because automatic controls are important. This study also provides a basic understanding of the behavior of all dynamic systems and leads to a better appreciation and utilization of the fundamental laws of nature.

1-2. Feedback Control Systems

Control systems can generally be classified as having feedback or not having feedback. Many systems of each type are in operation today. How-

1

ever, this book is largely concerned with control systems having feedback. In a *feedback control system*, the *controlled variable* (also called the *output* or *response*) is compared with the *reference variable* (the *input* or *command*) and any difference between the two, the *error*, is used to reduce the difference. Stated in its simplest terms, a feedback control system compares what we are getting with what we are asking for and uses any difference to drive the output into close correspondence with the input. The important distinguishing feature of a feedback control system is that a *comparison* is made; it is this comparison that makes the system so effective for control purposes.

To illustrate the concept of feedback control systems, consider the human being as the controller. Driving an automobile is a classic example. The objective is to keep the automobile on the roadway. The driver constantly compares the position of the automobile on the pavement with his idea of a safe position. When the controlled (actual) position is not in close correspondence with the reference (safe) position, the driver observes this error and turns the wheel to minimize the error. In the process of driving, the human being acts as a controller in a feedback control system by making the required comparison and then initiating a corrective action when the error exceeds appropriate limits. Many other examples could be cited but that of driving will serve to illustrate the idea of a feedback control system and also one in which man is an integral part of the system.

Feedback control systems can be divided into two broad categories, *regulator systems* and *follow-up systems*. A *regulator system* is one whose prime function is to maintain the controlled variable essentially constant despite unwanted disturbances to the system. In these systems, the reference variable is changed infrequently. An example is the home-heating system shown in block-diagram form in Fig. 1-1. Here the thermostat is the controller; it makes the comparison between the thermostat setting (the desired temperature) and the actual room temperature. If the temperature is too low, the thermostat initiates a signal to the furnace calling for heat. The furnace then supplies heat to the house with a resulting change in air temperature. The diagram shows a feedback line indicating that room-temperature information is supplied to the thermostat so that a comparison can be achieved.

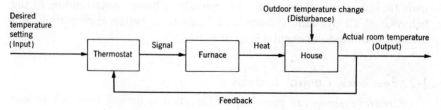

FIG. 1-1. Block diagram for home-heating system.

A change in outdoor temperature is a disturbance to the home-heating system. If the outside temperature falls, the room temperature will likewise tend to decrease. However, the room-temperature feedback to the thermostat provides it with the information required to initiate a longer furnace heating cycle.

The home-heating system is an example of a regulator system because it employs feedback and, although it will respond to a change in thermostat setting, its prime function is to maintain the desired room temperature despite changes in outdoor temperature.

A *follow-up system* is a feedback control system whose prime function is to keep the controlled variable in close correspondence with a reference variable which is frequently changed. The lathe tracer illustrated schematically in Fig. 1-2 is an example of such a system. The purpose of the tracer system is to provide a means for turning parts on the lathe with a contour corresponding to that of the template. Such a system permits rapid reproduction of the same part with a fair amount of flexibility—a new contour can be turned by making a new template.

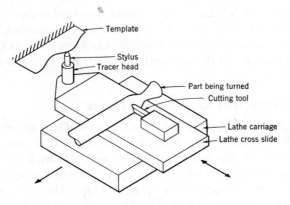

Fig. 1-2. Lathe tracer system.

Referring to Fig. 1-2, the template is mounted rigidly to the bed of the lathe. The tracer head, with its finger-like stylus which contacts the template, is fastened to the cross slide and therefore maintains a fixed position relative to the cutting tool. As long as the tracer head faithfully follows the template, the cutting tool will reproduce closely the desired contour. The in-and-out motion is controlled by the tracer system while the axial motion is obtained by engaging the lathe carriage feed.

Referring to the system block diagram, Fig. 1-3, the tracer head is the controller for the system. A particular point on the template corresponds to the desired position for the cutting tool. The tracer head can make the comparison between the desired position and the actual position because it

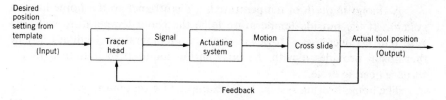

Fɪɢ. 1-3. Block diagram for lathe tracer system.

bears a fixed relation to the tool position. If there is an error, the tracer head initiates a signal to an actuating system which moves the cross slide and the cutting tool into the desired position. Feedback is also accomplished because the tracer head is mounted on the cross slide.

The tracing system is an example of a follow-up system because it is a feedback control system and, although it will correct for a disturbance such as a position variation caused by a change in tool loading, its prime function is to keep the cutting tool in correspondence with the ever-changing contour of the template.

Much has been written and said about *servomechanisms.* It is worthwhile to state how they relate to feedback control systems. A servomechanism is a feedback control system in which the controlled variable is mechanical position. Thus, servomechanisms are but one class of feedback control systems.

1-3. Closed-Loop and Open-Loop Systems

Feedback control systems are also called *closed-loop systems.* A reference to the block diagrams in Figs. 1-1 and 1-3 provides an insight into the name. When information about the output is fed back for comparison with the input, the diagram takes on the appearance of a complete loop and indeed the system itself becomes a complete loop. Thus, to provide feedback is to close the loop and so a closed-loop system is one with feedback.

There are also *open-loop systems*, that is, systems without feedback. An example might help to clarify the distinction. In the past, some apartment houses were built with heating systems controlled from one thermostat located in one of the apartments. The apartment with the thermostat had a closed-loop heating system; the thermostat could compare the room temperature with the thermostat setting. Not so in the other apartments; room temperatures could assume any value without recourse to modifying the output of the furnace. These apartments operated open-loop to the frequent consternation of the occupants.

It should not be concluded that open-loop control systems are bad and are to be avoided at all costs. Many open-loop systems exist and do serve a useful function. Such systems, however, because they make no comparison, must be carefully calibrated and designed to maintain that calibration.

The automatic washing machine will serve as a convenient example of a device with open-loop controls. The machine will go through a timed sequence of operations without regard for the cleanliness of the clothes being washed. Of course, when used by the housewife, the over-all system becomes a feedback control system with the housewife closing the loop. Nevertheless, the automatic washing machine itself is a device employing open-loop control and has proved to be a blessing to the housewife.

1-4. Requirements for the Control System

Stability, accuracy, and *speed of response* are requirements demanded of every control system. Certainly, a system must be stable. This means that the response to an input, be it a reference change or a disturbance, must reach and maintain some useful value within a reasonable period of time. An unstable control system will, by way of comparison, produce persistent, or even violent, oscillations of the output or will drive the output to some extreme limiting value. Any type of response characteristic of an unstable control system is obviously unsatisfactory.

A control system must be accurate within specified limits. This means that the system must be capable of reducing any error to some tolerable value. Note that no control system is able to maintain zero error at all. times because an error is required to initiate the corrective action. Even systems which are mathematically capable of reducing the system error to zero (under ideal conditions) do not actually accomplish this because of slight imperfections inherent in the components making up the system. Fortunately, many control applications do not require extreme accuracy. For example, a home-heating system which could keep the controlled temperature within ± 0.1 F would be of little value because the average human being, whose comfort is being considered, is probably incapable of sensing variations of ten times this amount. Thus, accuracy is a relative matter with limits based upon the particular application. The limits should be made as wide as possible because, in general, the cost of a control system increases rapidly if increased accuracy is demanded. In any event, all control systems must provide the demanded degree of accuracy.

A control system must complete its response to some input within an acceptable period of time. Although a system may be stable and exhibit the required accuracy, it has no real value if the time required to respond fully to some input is far greater than the time interval between inputs. In this case, the system may never "catch up."

The control engineer must design his system so that the three general requirements of stability, accuracy, and speed of response are met. This may not always be easy because the requirements tend to be incompatible; compromises between them must be made. As an example, consider the problem of increasing the accuracy of a system. Accuracy can be improved by making the controller in the system more sensitive, that is, the controller will then provide the same increment of correction for a reduced increment of error. This permits the system to respond to a smaller error with a resulting improvement in system accuracy. However, this change in the controller has an adverse effect on stability; the controller now provides a greater corrective action for the same magnitude of error. Carried to an extreme, the system may become unstable. Consider a system where a slight error exists. With a very sensitive controller, a large corrective action will be initiated. The correction might be sufficient to cause a response creating a greater error in the opposite sense, that is, the system will "overshoot" the proper value. The system will again attempt to reduce the error but now the error is greater in magnitude; a larger overshoot will occur. Thus, oscillations will increase in amplitude until either the system destroys itself or the amplitude of oscillation is limited by the physical nature of the system. In this example, a compromise must be made between accuracy and stability.

An ideal control system would be stable, would provide absolute accuracy (maintain zero error despite disturbances), and would respond instantaneously to a change in the reference variable. Such a system cannot, of course, be produced. However, the study of automatic control theory will provide the insight necessary to make the most effective compromises so that the engineer can design the best possible system.

1-5. Closure

The orientation to automatic controls is now complete, and attention can be turned to a preview of what is to follow. The mathematics underlying feedback control system behavior is concerned largely with differential equations. Therefore, Chapter 2 introduces the concept of obtaining these equations for a wide variety of physical systems. Throughout the greater portion of this book, we will work with linear differential equations with constant coefficients. Chapter 3 discusses one method for solving such equations. Chapter 4 introduces the analog computer, a device which can be used to solve linear equations but which is extremely effective for solving nonlinear equations. The analog computer is used widely to analyze and design control systems. (In addition, Appendix A explains the use of digital computers in some aspects of control work.) To illustrate the use of the material already covered, Chapter 5 treats the response of physical systems to various inputs. The mathematical concept of stability is also presented. Chapter 5 concludes with the analysis of two illustrative control systems.

Although the concepts advanced in the first five chapters will permit a reasonable analysis of many control systems, other methods and techniques are required to reduce the work involved in analyzing systems and to simplify making design changes. An alternative means for solving differential equations, the Laplace transform method, is introduced in Chapter 6. Chapter 7 discusses the way in which information contained in differential equations can be represented in block-diagram form. Block diagrams are useful in visualizing the control system and are an aid in simplifying the mathematical solution to the problem. Classifying control system performance based on the type of controller employed and the type of system being controlled is the subject of Chapter 8.

Chapters 9, 10, and 11 present a widely used technique known as *frequency response* while Chapter 12 introduces an alternative approach to control system analysis, the *root-locus method.* Each approach offers unique benefits. System compensation, the methods by which a system can be modified to obtain the desired performance, is the subject of Chapter 13.

One common complaint among engineers who have studied linear control theory (theory which assumes that the control system can be represented by linear differential equations) is that some of the components in most practical systems are distinctly nonlinear. There is some justification for this complaint. For this reason, a number of nonlinearities which frequently occur are considered in Chapter 14.

The general availability of the high-speed digital computer has had a pronounced influence on the automatic control field. The use of the digital computer for problem solution has prompted a change in outlook and brought about new theoretical concepts. An introduction to some of these more recent ideas is presented in Chapter 15. The digital computer is also used as a part of the control system. Chapter 16 discusses digital computer control.

The reader who masters the concepts presented in this book will have received an understanding of basic control theory, an insight into how control systems work, and will have experienced the use of both analog and digital computers in solving control problems.

Chapter 2

Differential Equations
for Physical Systems

2-1. Introduction

Automatic control systems, while physical in nature, have differential equations as mathematical models. This chapter is, therefore, concerned with writing differential equations which represent physical systems. For the most part, we will deal with the ordinary differential equation which is one involving a function and its derivatives. Basically, the process of obtaining differential equations is predicated more on engineering ability than on mathematical ability. The concept is simple: differential equations are written from an understanding of the basic laws underlying the system being considered. Newton's second law ($\Sigma F = Ma$) is an example of such a law. Systems of various types will be considered in this chapter. The solution of differential equations is the subject of Chapter 3.

2-2. Mechanical Systems

Our study of writing differential equations begins by considering a number of spring-mass-damper systems. These systems are also encountered in vibration work. The fundamental law used is

$$\Sigma F = Ma \qquad (2\text{-}1)$$

in which F is force, M is mass, and a is acceleration. Force may be expressed in pounds, mass in lb-sec^2/in. with the corresponding units of acceleration being in./sec^2. Of course, any other consistent units are valid.

Spring-Mass Systems (Damping Assumed Zero). Figure 2-1a shows a mass M resting on a frictionless surface and connected to a rigid wall through a spring with rate k(lb/in.). The displacement of the mass is y. The problem is to write an equation (the differential equation for the system) which, when solved, will yield an expression for the displacement of the mass as a function of time.

To solve the problem, the displacement of the mass will be considered zero with the system at equilibrium, that is, with the spring in neither tension nor compression and therefore exerting no force on the mass.

8

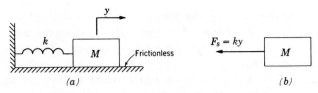

FIG. 2-1. Spring-mass system. Free-body diagram with mass displaced to the right (positive direction).

Further, displacements in the direction of the arrow, to the right in this case, will be taken as positive displacements while those to the left will be taken as negative displacements. Thus a displacement y can be zero, positive, or negative. Once a sign convention has been adopted, it must not be changed during the course of the problem. Also, and this statement is very important, the sign convention applies not only to displacement but to force, velocity, and acceleration as well. To illustrate, a velocity to the right is a positive velocity while a force acting on the mass and urging it to the left is a negative force.

Acceleration is, of course, the second derivative of displacement with respect to time. For this problem, Eq. 2-1 can therefore be written as

$$M \frac{d^2y}{dt^2} = \Sigma F$$

The only force acting on the mass is the spring force which has a magnitude ky (spring rate multiplied by displacement); however, the correct sign must be associated with this magnitude. Assume that the mass is displaced in the positive direction (to the right); the spring force F_s will act to restore the mass to the equilibrium position as shown in Fig. 2-1b. Following the sign convention established, this force is negative. Therefore,

$$F_s = -ky$$

It is instructive to study this expression and to be convinced that it will yield the correct spring force for any possible displacement. If y is zero, F_s is zero; that is correct because the displacement was initially taken to be zero with the spring relaxed. If y is negative, that is, the mass is displaced to the left, F_s is positive. This is correct because the spring will be placed in compression and will act to move the mass in the positive direction. Thus the expression for the spring force is correct for any possible displacement of the mass. Because the spring force is the only force acting on the mass, the differential equation for the system can be written as

$$M \frac{d^2y}{dt^2} = \Sigma F = F_s = -ky$$

or

$$M \frac{d^2y}{dt^2} + ky = 0 \tag{2-2}$$

Certainly the behavior of any physical system is independent of the sign convention one might adopt. For this reason, the reader may wish to rework the problem with the sign convention reversed. The same differential equation will result.

The weight of the mass was not a factor in the system just analyzed. However, weight is a factor in the spring-mass system shown in Fig. 2-2*a*.

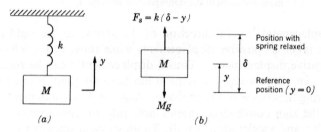

FIG. 2-2. System in which weight is a factor.

Consider the displacement y to be zero, i.e. choose a reference point, with the system at rest with the spring extended by some amount δ sufficient to support the weight of the mass. This, as will be shown, is a convenient reference point. The direction of the arrow indicates that the upward direction will be considered positive.

Figure 2-2*b* is a free-body diagram of the mass displaced upward from the equilibrium position. Two forces exist, the weight Mg (g is the acceleration of gravity) acting downward and the spring force F_s acting upward. The spring force is equal to the spring rate k multiplied by the extension of the spring which is $(\delta - y)$. From Eq. 2-1

$$M \frac{d^2y}{dt^2} = \Sigma F = k (\delta - y) - Mg = k\delta - ky - Mg$$

But $\delta = Mg/k$ = static deflection. Therefore,

$$M \frac{d^2y}{dt^2} = k \left(\frac{Mg}{k} \right) - ky - Mg = - ky$$

and
$$M \frac{d^2y}{dt^2} + ky = 0 \qquad (2\text{-}3)$$

Two points should be noted. First, the sign convention adopted initially was followed in introducing the forces into the first equation—an upward force is positive; a downward force, negative. Secondly, by choosing the reference at the equilibrium position, the weight of the mass had no effect on the differential equation which turned out to be the same as that of the previous system in which weight was not a factor. In the future, the reference will always be chosen at the equilibrium position so that the weight of the mass will not enter the problem.

Viscous Friction and the Dashpot. The nature of *viscous friction* (damping) is illustrated in Fig. 2-3. The curve shows that a device exhibiting viscous friction will offer a restraining force proportional to velocity:

$$F_f = cv$$

where F_f is the friction force (lb), v is the velocity (in./sec), and c is the damping coefficient (lb-sec/in.) and is the slope of the line shown in the figure.

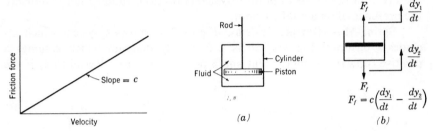

$$F_f = c\left(\frac{dy_1}{dt} - \frac{dy_2}{dt}\right)$$

FIG. 2-3. Nature of viscous friction.

(a)

(b)

FIG. 2-4. The dashpot.

By contrast, *coulomb friction* produces a constant friction force independent of velocity. A system exhibiting coulomb friction is described by a differential equation more difficult to solve than a system with viscous friction. Therefore in many cases, as a first approximation, it is assumed that coulomb friction is either negligible or can be represented by an equivalent viscous friction.

The dashpot is a frequently used example of a device providing viscous friction, or damping. A schematic drawing of a dashpot is shown in Fig. 2-4a. The device consists of a piston inside a fluid-filled cylinder. The piston rod extends from the cylinder through a suitable seal. Relative motion between the piston rod and the cylinder is resisted by the fluid because the fluid must move from one side of the piston to the other through orifices provided in or around the piston. The greater the relative velocity, the greater must be the flow rate of fluid past the piston and a pressure difference must exist across the piston to cause the fluid flow. It is this pressure difference which produces the restraining force. If the rate of fluid flow through the orifices is small, the flow is laminar (i.e., the flow rate is proportional to the pressure drop) and the force is proportional to the relative velocity (viscous damping). If, however, the rate of fluid flow is high, flow is turbulent (i.e., the flow rate is proportional to the square root of the pressure drop) and the force is proportional to the square of the velocity. Both these statements are made neglecting other friction effects such as that existing at the rod seal. (The student should verify the statements made by analyzing a dashpot under the two given flow conditions.)

In the work which follows, the assumption of an ideal (viscous friction) dashpot is made. A dashpot is usually represented as shown in Fig. 2-4*b*. The equation defining the behavior of such a device is given as

$$F_f = c \left(\frac{dy_1}{dt} - \frac{dy_2}{dt} \right) \tag{2-4}$$

where dy_1/dt and dy_2/dt are the velocities of the piston rod and cylinder, respectively. It should be recognized that the dashpot symbol is frequently used to denote the assumption of viscous friction, rather than to indicate the presence of an actual dashpot.

Spring-Mass-Damper Systems. Figure 2-5*a* shows a system which includes a dashpot; Figure 2-5*b* is a free-body diagram which assumes a positive (upward) displacement and a positive velocity. Recall that, if the

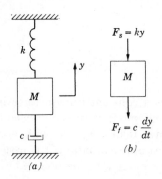
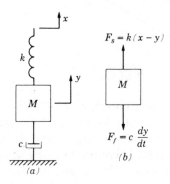

FIG. 2-5. Spring-mass-damper system. Free-body diagram assuming positive displacement and velocity.

FIG. 2-6. System with an input. Free-body diagram of mass responding to a positive displacement at *x*.

reference ($y = 0$) is chosen at the equilibrium position, the mass can be considered weightless. Two forces then exist on the mass. The spring force F_s, which equals ky, acts downward because of the "compression" in the spring. The dashpot, or damper, force from Eq. 2-4 is $c\,dy/dt$ (one end of the dashpot being fixed) and also acts downward in resisting the upward velocity. Therefore, both forces are negative in sign. From Eq. 2-1

$$M \frac{d^2y}{dt^2} = \Sigma F = -ky - c \frac{dy}{dt}$$

or

$$M \frac{d^2y}{dt^2} + c \frac{dy}{dt} + ky = 0 \tag{2-5}$$

Equation 2-5 is the differential equation of motion for the system. In the first equation, the dashpot force term, $-c\,dy/dt$, can be examined to insure that it is valid for all possible velocities. If dy/dt is zero, the force is zero and this is correct for a device exhibiting viscous friction. If dy/dt is negative, the force is positive. Looking at the system, this is correct because if the mass has a downward velocity, the dashpot will provide an upward restraining force. Thus the dashpot force term is valid for any velocity even though it was obtained by assuming a positive velocity.

A variation of the previous system is shown in Fig. 2-6a; here a provision is made for an input displacement x at the top of the spring. Figure 2-6b shows a free-body diagram of the mass responding to a positive input at x. It is assumed that x is displaced upward and that the mass is responding with a positive displacement y (less than x) and a positive velocity. With this assumption, the spring force is acting upward with a magnitude $k(x-y)$, and the dashpot force is acting downward with a magnitude $c\, dy/dt$. Therefore,

$$M\frac{d^2y}{dt^2} = k(x-y) - c\frac{dy}{dt}$$

or

$$M\frac{d^2y}{dt^2} + c\frac{dy}{dt} + ky = kx \qquad (2\text{-}6)$$

which is the differential equation for the system.

Torsional System. The principles already presented can be extended easily to set up the differential equation for the rotating system shown in Fig. 2-7a. The rotating mass I (lb-in.-sec^2), with angular position θ_o (radians), is driven through a torsion spring with spring rate G (lb-in./radians). Rotation is resisted by a rotary dashpot with a damping coefficient b (lb-in.-sec/radian). The input displacement is θ_i. The arrows indicate the direction considered as positive.

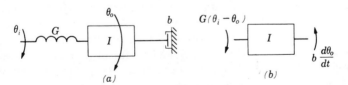

(a) (b)

FIG. 2-7. Torsional system.   Free-body diagram of mass responding to a positive displacement at θ_i.

The fundamental law to be applied is

$$\Sigma T = I\alpha \qquad (2\text{-}7)$$

where T is torque (lb-in.) and α is angular acceleration (radians/sec^2). Figure 2-7b shows a free-body diagram of the mass assuming that θ_i is displaced in the positive direction and that the mass is responding with a positive displacement θ_o (less than θ_i) and a positive velocity. With this assumption, the spring torque is acting in the positive direction with a magnitude $G(\theta_i-\theta_o)$ and the dashpot torque is acting in the negative direction with a magnitude $b \, d\theta_o/dt$. Therefore, from Eq. 2-7

$$I\alpha = I \frac{d^2\theta_o}{dt^2} = \Sigma T = G(\theta_i - \theta_o) - b \frac{d\theta_o}{dt}$$

or

$$I \frac{d^2\theta_o}{dt^2} + b \frac{d\theta_o}{dt} + G\theta_o = G\theta_i \qquad (2\text{-}8)$$

which is the differential equation for the system. Note the similarity between this equation and Eq. 2-6.

Systems with More Than One Mass. A system involving two masses is shown in Fig. 2-8*a*. The assumption is made that friction effects can be approximated by equivalent viscous friction with damping coefficients c_1 and c_2.

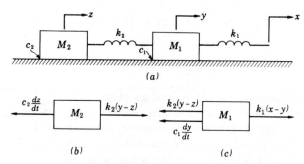

FIG. 2-8. System with two masses. Free-body diagram of each mass with $x > y > z > 0$, $dy/dt > 0$, and $dz/dt > 0$.

Two defining differential equations are required. They can be obtained from free-body diagrams of the two masses. It is convenient to base these diagrams on the assumption of a positive displacement at x. The system is then assumed to respond with positive displacements y and z with $x > y > z$. Also, positive velocities for both masses are assumed. The resulting free-body diagrams are given in Fig. 2-8*b* and *c*, from which the defining equations are written as

$$M_2 \frac{d^2z}{dt^2} = k_2(y - z) - c_2 \frac{dz}{dt}$$

or

$$M_2 \frac{d^2z}{dt^2} + c_2 \frac{dz}{dt} + k_2z = k_2y \qquad (2\text{-}9)$$

and

$$M_1 \frac{d^2y}{dt^2} = k_1(x - y) - k_2(y - z) - c_1 \frac{dy}{dt}$$

or

$$M_1 \frac{d^2y}{dt^2} + c_1 \frac{dy}{dt} + (k_1 + k_2)y = k_1x + k_2z \qquad (2\text{-}10)$$

Systems with more than two masses can be handled in a similar manner.

Systems with Gearing.   Numerous mechanical systems involve gearing.   Frequently it is convenient to represent a geared system by an equivalent one without gearing.   Techniques for making this transformation will now be discussed.

Figure 2-9a shows a rotating inertia *I* driven through a pair of mass-

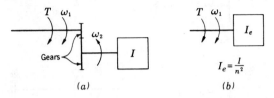

(a)   (b)

FIG. 2-9. A geared system and its equivalent.

less gears.   The problem is to determine an equivalent inertia I_e for the gearless system given in Fig. 2-9b.   The gear ratio *n* is defined as

$$n = \frac{\text{speed of driving shaft}}{\text{speed of driven shaft}} \qquad (2\text{-}11)$$

Thus, for a speed reducer, the gear ratio is greater than one.   The gear ratio in Fig. 2-9a is

$$n = \frac{\omega_1}{\omega_2} \qquad (2\text{-}12)$$

For the two systems to be equivalent, the kinetic energies must be equal. Therefore,

$$\frac{1}{2}I\omega_2^2 = \frac{1}{2}I_e\omega_1^2$$

With Eq. 2-12,

$$I_e = \frac{I\omega_2^2}{\omega_1^2} = \frac{I}{n^2} \qquad (2\text{-}13)$$

Note that the equivalent inertia at the drive shaft is smaller when a speed reducer is used.   Gear inertias can, of course, be included by adding them to other inertia values on their respective shafts.

A gear-driven spring system and its equivalent are shown in Fig. 2-10a and b.   For equal potential energies,

$$\frac{1}{2} G \theta_2^2 = \frac{1}{2} G_e \theta_1^2$$

Because $n = \theta_1/\theta_2$,

$$G_e = \frac{G \theta_2^2}{\theta_1^2} = \frac{G}{n^2} \qquad (2\text{-}14)$$

Equal power dissipation is required for the dashpot systems shown in Fig. 2-10c and d. Power P can be expressed as $P = T\omega$. For a dash-

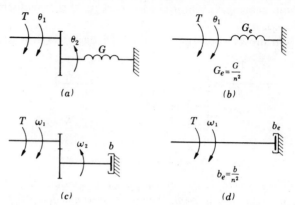

(a)

(b)

(c)

(d)

FIG. 2-10. Equivalent spring rate and damping coefficient.

pot, $T = b\omega$. Therefore, $P = b\omega^2$. In terms of the systems being considered,

$$b\omega_2^2 = b_e \omega_1^2$$

and
$$b_e = \frac{b}{n^2} \tag{2-15}$$

The use of Eqs. 2-13, 2-14, and 2-15 can be illustrated by converting the geared system given in Fig. 2-11a to its equivalent, Fig. 2-11b. Gear

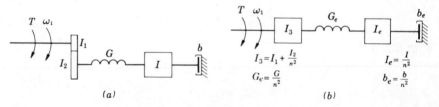

(a)

(b)

FIG. 2-11. Equivalent systems.

inertias I_1 and I_2 are to be included. Direct application of the equations yields I_e, G_e, and b_e. Inertia I_3 is obtained as the sum of inertia I_1 (on the drive shaft) and the equivalent value of I_2.

2-3. Electrical Systems

Attention is now turned to electrical systems, the second system type to be considered in this chapter. Several circuits and a d-c motor with load will be introduced.

Circuits. A summary of the equations for the common electrical elements—resistance, capacitance, and inductance—is given in Table 2-1.

TABLE 2-1
ELECTRICAL ELEMENTS

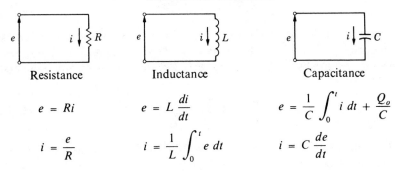

Resistance	Inductance	Capacitance
$e = Ri$	$e = L \dfrac{di}{dt}$	$e = \dfrac{1}{C} \displaystyle\int_0^t i\, dt + \dfrac{Q_o}{C}$
$i = \dfrac{e}{R}$	$i = \dfrac{1}{L} \displaystyle\int_0^t e\, dt$	$i = C \dfrac{de}{dt}$

Electrical units:
  t time, seconds R resistance, ohms
  e voltage, volts L inductance, henrys
  i current, amperes C capacitance, farads
                 Q_o initial charge, coulombs

This information will be used in writing the differential equations for the two circuits shown in Fig. 2-12. The fundamental law to be applied is Kirchhoff's first law which states that the algebraic sum of the voltages around a closed loop is zero:

$$\Sigma e = 0 \qquad (2\text{-}16)$$

The circuit shown in Fig. 2-12*a* is composed of a resistance and inductance in series with an impressed voltage *e*. Using the equations for the

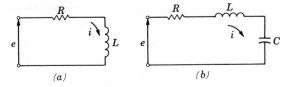

(a)   *(b)*

FIG. 2-12. Electrical circuits.

voltage drop across the elements from Table 2-1 and Eq. 2-16 yields

$$e - Ri - L \frac{di}{dt} = 0$$

or

$$L \frac{di}{dt} + Ri = e \qquad (2\text{-}17)$$

The latter equation is the required differential equation for the circuit.

The circuit shown in Fig. 2-12b is handled in a similar manner. Assuming the capacitor initially discharged, that is $Q_o = 0$, the result is

$$L \frac{di}{dt} + Ri + \frac{1}{C} \int_0^t i \, dt = e \tag{2-18}$$

This equation can be written in an alternate form by recalling that current is the time rate of flow of charge, that is, $i = dq/dt$ where q is charge expressed in coulombs. Substituting this relationship into the previous equation yields

$$L \frac{d^2q}{dt^2} + R \frac{dq}{dt} + \frac{q}{C} = e \tag{2-19}$$

The simple network shown in Fig. 2-13 implies that the relationship between the output voltage e_{out} and the input voltage e_{in} is desired. In this case, two equations are required. The use of Eq. 2-16 yields

$$Ri + \frac{1}{C} \int_0^t i \, dt = e_{in} \tag{2-20}$$

which is the first equation. The second equation can be written by observing that the output voltage is simply the voltage drop across the capacitor,

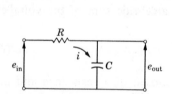

FIG. 2-13. Electrical network.

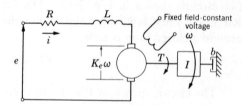

FIG. 2-14. D-c motor with load.

or

$$e_{out} = \frac{1}{C} \int_0^t i \, dt \tag{2-21}$$

Again the assumption was made that the capacitor was initially discharged. If the current i were eliminated between the two equations, a single equation would result relating the output and input voltages. A method for doing this will be presented later in the book. For the present, however, the emphasis is placed on writing equations, not on solving them.

D-C Motor with Load. A schematic drawing of a d-c motor with a load involving inertia and damping is given in Fig. 2-14. A desirable system equation would be one which relates the output speed ω (radians/sec) with the input voltage e to the motor armature. Note that the motor field is supplied from another voltage source which is independent of the armature voltage e. Three equations are required to define the behavior of this system.

The first equation is based on the algebraic sum of the voltages around the armature loop. In addition to the voltage drops across the armature resistance R and inductance L, the voltage induced by the armature winding cutting the lines of field flux must be considered. Recall that a voltage, proportional to velocity, is induced in a conductor which cuts lines of magnetic flux providing, of course, that the flux field is constant as is true in this case. Thus, the induced voltage in the armature winding is proportional to rotating speed and can be expressed as $K_e\omega$ where K_e is the voltage constant for the motor. This induced voltage opposes the impressed voltage e. Therefore, from Eq. 2-16

$$e = Ri + L\frac{di}{dt} + K_e\omega \qquad (2\text{-}22)$$

which is the first of the required equations.

The second equation relates the output torque T with the armature current i. Recall that a magnetic field surrounds a conductor carrying a current and that the intensity of the field is proportional to the current. Further, if the current carrying conductor is suitably located in a magnetic field, the interaction between the fields results in a force on the conductor. These phenomena are utilized in a d-c motor (with a fixed field) to provide an output torque proportional to the armature current, or

$$T = K_t i \qquad (2\text{-}23)$$

where K_t is the torque constant of the motor.

The final equation relates the output speed with the torque input to the load. This equation, based on Eq. 2-7, is

$$T = I\frac{d\omega}{dt} + b\omega \qquad (2\text{-}24)$$

The three equations can, by methods to be presented later, be combined to yield a single equation defining the behavior of the overall system, that is, $\omega = f(e, t)$.

2-4. The Beginnings of a Network Approach

A frequently used technique for writing and solving differential equations is to reduce the system under consideration to an equivalent electrical circuit or network. This method is known as the network approach. The approach, once the transformation (which may or may not be obvious) is accomplished, is quite powerful because electrical network analysis methods have been developed to a high degree. For example, an engineer with a complicated hydraulics system might better be able to effect a solution by converting it to an equivalent electrical system. The network approach, while beneficial to all engineers, finds particular favor with electrical engineers because of their extensive training in circuit analysis.

A complete presentation of the network approach is beyond the scope of this book. However, general definitions for resistance, capacitance, and inductance will be introduced. These definitions will serve a useful purpose here and can be helpful to the reader who wishes to pursue the network approach further. Because the reader is already familiar with electrical resistance, inductance, and capacitance, these will be discussed first following each definition to show that they are but specialized cases of the general definitions. Examples from other systems will then follow. The purpose of the examples is not to provide a tabulation of analogies but merely to illustrate the general definitions.

Resistance is that which opposes flow; it can be defined as the change of potential required to cause a unit change of flow rate. In an electrical system, potential is measured in volts while flow rate is measured in amperes (coulombs/sec). Therefore, from the definition, electrical resistance is

$$R = \frac{de}{di}$$

If the relationship between e and i is linear, the usual case for electrical resistance, the resistance is said to be linear and the previous equation becomes simply

$$R = \frac{e}{i}$$

The units are volts/(coulomb/sec) which has been given the name *ohm*. This illustration shows that electrical resistance is defined by the general concept of resistance.

Consider a thermal system; here the potential involved is temperature difference (degrees) and heat (Btu) is the quantity which flows. From the general definition, thermal resistance then has the units degree/(Btu/sec). The reader may not have encountered thermal resistance as such but probably has come in contact with its reciprocal, thermal conductance.

A capacitor is a storage element. *Capacitance* can be defined as the change in quantity contained per unit change in some reference variable. For example, an electrical capacitor contains charge q (coulombs) and the reference variable is volts. Linear electrical capacitance, the usual case, is then

$$C = \frac{q}{e}$$

The units are coulombs/volt which has been given the name *farad*.

A block of steel will exhibit thermal capacitance because it is capable of storing heat. The units for thermal capacitance are Btu/degree. Thermal capacitance is probably more commonly called *heat capacity* which is defined as the quantity of heat required to change the temperature of some body by one degree. The two definitions involved are, of course, the same.

Inductance is that which opposes acceleration; it can be defined as the potential change required to cause a unit change in acceleration. In an electrical system, the potential is voltage and, if the current i (dq/dt) is thought of as the velocity of charge, di/dt is the acceleration of charge. Therefore, from the general definition,

$$L = \frac{e}{d^2q/dt^2} = \frac{e}{di/dt}$$

which is a statement of the equation given in Table 2-1.

Consider a mechanical system involving a mass M, with displacement y, with a net force F acting on it. If force is thought of as the potential (that which drives the system), the inductance is, by definition,

$$L = \frac{F}{d^2y/dt^2} = \frac{F}{a} = M$$

Here, from Newton's law, the mechanical inductance is seen to be the mass M.

The purpose of this section has been to introduce the general definitions for resistance, capacitance, and inductance. Illustrations of the use of the definitions will be given in the next two sections of this chapter.

2-5. Hydraulic Systems

Two distinct types of hydraulic systems will be discussed in this section. The first has to do with the liquid level in a tank, an illustration closely related to control problems encountered in the chemical processing industry [10].[1] The second type of hydraulic system is related to fluid power applications which are common throughout industry and the military [2]. In this case, the fluid used is oil.

The fundamental relationship to be used for both types of systems is

$$q = Av \tag{2-25}$$

in which q is volume flow rate (in.^{3}/sec or similar units), A is cross-sectional area (in.^{2}), and v is velocity (in./sec). The equation states that the flow rate is equal to the cross-sectional area multiplied by the average velocity normal to the area.

Liquid-Level System. Figure 2-15 shows a tank with inflow q_{in}. The liquid level h (for head) provides a potential which causes an outflow q_{out}

[1]Numbers in brackets are those of references listed in the bibliography.

through the valve resistance R. The cross-sectional area A of the tank is constant throughout the height. Using Eq. 2-25, the net flow to the tank equals the area A multiplied by the velocity of the liquid level, or dh/dt. Therefore,

$$q_{in} - q_{out} = A \frac{dh}{dt}$$

Note that if a rise in level is considered positive, the equation is correct in sign for any inflow-outflow condition. For example, if the inflow exceeds the outflow, the level will surely rise; in the equation, substitution of the flow values will give a positive dh/dt, A being a positive quantity.

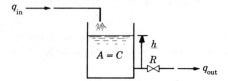

FIG. 2-15. Hydraulic system involving liquid level.

The previous equation can be modified by using the general definitions introduced in Sec. 2-4. Capacitance, in this case hydraulic capacitance, is the change in quantity contained (volume) per unit change in the reference variable (head). Thus, by definition

$$C = \frac{A \Delta h}{\Delta h} = A$$

in which Δh is the change of head. The equation shows that the capacitance is equal to the cross-sectional area of the tank.

Hydraulic resistance is the potential (head) change required to cause a unit change in flow rate (q_{out}). Therefore,

$$R = \frac{dh}{dq_{out}}$$

and if the resistance is linear (laminar, *not* turbulent, flow),

$$R = \frac{h}{q_{out}} \qquad \text{or} \qquad q_{out} = \frac{h}{R}$$

Substituting the expressions involving capacitance and resistance into the original equation yields

$$q_{in} - \frac{h}{R} = C \frac{dh}{dt}$$

or
$$RC \frac{dh}{dt} + h = Rq_{in} \qquad (2\text{-}26)$$

which is the differential equation for the system based on the assumption that the flow through the outlet valve is laminar. Turbulent flow would produce a nonlinear resistance; nonlinearities are the subject of Chapter 14.

It will be instructive to obtain the system differential equation in a different way. The conservation law requires that the outflow plus the rate at which liquid is stored must equal the inflow. Consider the question of liquid storage; the hydraulic capacitance is

$$C = \frac{V}{h} \quad \text{and} \quad V = Ch$$

where V is volume. Differentiating with respect to time provides an expression for the rate at which liquid is stored, q_s, which is

$$q_s = \frac{dV}{dt} = C \frac{dh}{dt}$$

The conservation law, as it applies in this case, can be written as

$$q_{out} + q_s = q_{in}$$

Substituting for q_s,

$$q_{out} + C \frac{dh}{dt} = q_{in} = C \frac{dh}{dt} + \frac{h}{R}$$

which is the system equation.

The analogy-minded reader might remark that too much effort was expended in obtaining the expression for q_s when it could have been written, by analogy, from the similar equation for electrical capacitance ($i = C \, de/dt$) given in Table 2-1. While analogies are useful, the reader is advised to work first with the laws governing the system to insure obtaining the proper equations; analogies can then be used as a check. Analogies contain no magic and are not a substitute for knowledge about a particular system. Indiscriminate use of analogies will surely lead to bloodshed, mainly that of the indiscriminate user.

Fluid Power Systems. Figure 2-16 shows a schematic drawing of a valve-controlled hydraulic actuator, or cylinder. Oil under pressure is supplied to the pressure port of the valve. In an actual valve, the two return ports shown are joined internally, and a single return port is connected to the pump tank. The remaining two valve ports are connected to the cylinder. The valve spool is shown in the centered position blocking flow to and from all ports. With the spool centered, the piston will remain stationary. Displacing the valve spool to the right will permit oil flow to the cylinder causing the piston to move to the right with the excess oil in the right end of the cylinder being expelled through the return port. A leftward motion of the piston and its rod is obtained, of course, by displacing the valve spool to the left from the centered position.

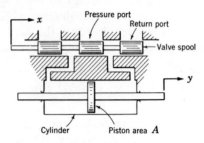

FIG. 2-16. Valve-controlled
hydraulic actuator.

To obtain the differential equation for the system, references are chosen. The spool displacement x is considered zero with the spool centered, and displacements to the right will be taken as positive. The piston rod displacement y will be considered zero in its last position before the start of the problem.

The equation commonly used for determining the flow-rate through a spool-type hydraulic valve is that for a plane orifice. This equation is

$$q = C_d A v$$

where q is the flow rate (in.^{3}/sec), A is the orifice area (in.^{2}), v is the theoretical velocity at the vena contracta of the jet assuming no losses (in./sec), and finally, C_d is the coefficient of discharge which is the unitless factor which corrects for the contraction of the jet and for the losses in available energy actually involved. In a spool-type valve, the orifice area A is the product of the port width w and the spool displacement x. The theoretical velocity v, frequently expressed as $\sqrt{2gh}$, is more conveniently used as $\sqrt{2p/\rho}$ in which p is the pressure drop across the orifice (lb/in.^{2}) and ρ is the mass density of the fluid (lb-sec^2/in.^{4}). Substituting the expressions for A and v into the previous equation yields

$$q = C_d \, wx \, \sqrt{2p/\rho} \qquad (2\text{-}27)$$

which can be used to approximate the flow rate through a spool-type valve. Values for C_d generally range from 0.6 to 0.8.

Equation (2-27) can be simplified considerably if the pressure drop across the valve is constant or reasonably so; the factors C_d, w, and ρ are either constant or nearly so. In this case, Eq. 2-27 becomes

$$q = K_v x \qquad (2\text{-}28)$$

in which the valve constant $K_v = C_d w \sqrt{2p/\rho}$.

Returning to the system shown in Fig. 2-16 and making the assumption of a constant pressure drop across the valve, the valve flow rate from Eq. 2-28 is

$$q = K_v x$$

This flow produces piston motion; the relationship, from Eq. 2-25, is

$$q = Av = A\,\frac{dy}{dt}$$

The last two equations combine to give

$$\frac{dy}{dt} = \frac{K_v}{A}\,x \qquad (2\text{-}29)$$

which is the differential equation for the system as approximated.

A schematic drawing of a more sophisticated system is shown in Fig. 2-17. It is actually a closed-loop positioning system, or servomechanism. The new features shown are the feedback lever connecting the piston rod with the spring-biased sliding sleeve which surrounds the valve spool. In operation, if the spool is displaced to the right, the piston and rod will move to the right. In so doing, however, the feedback lever will permit the spring to force the sliding sleeve to the right. Piston motion will cease

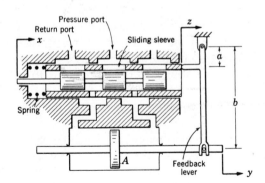

FIG. 2-17. Hydromechanical servomechanism.

when the sleeve overtakes the spool shutting off flow to the cylinder. In the previous example, a fixed spool displacement produced a piston velocity; in this example, the spool displacement causes a proportional piston rod displacement.

References are chosen with x (the spool displacement), y (the piston-rod displacement), and z (the sliding-sleeve displacement) being assumed zero with the system at rest. Displacements to the right are positive. The flow rate through the valve, with the assumption of a constant pressure drop, can be obtained from Eq. 2-28 with one modification. The valve opening in this case is not x but rather $(x\text{-}z)$. Therefore,

$$q = K_v (x\text{-}z)$$

This flow produces piston motion and, as in the last illustration,

$$q = A \frac{dy}{dt}$$

Finally, a study of the feedback lever shows the relationship between the sliding-sleeve and piston-rod displacements to be approximately

$$z = (a/b)y$$

Combining the three equations provides some exercise in algebra and the result

$$A \frac{dy}{dt} + \frac{aK_v}{b} y = K_v x \qquad (2\text{-}30)$$

which is the differential equation for the system as approximated.

2-6. A Thermal System

A single example will be presented to illustrate setting up differential equations for thermal systems. Consider an oven with provision for supplying heat at the rate of q_{in} Btu/sec. (Note that q_{in} is now being used to denote heat flow rather than fluid flow as in the previous section; however, since the two types of systems are distinct, no confusion should result.) The oven is loaded with a number of small parts, suitably separated, to be heat-treated. Heat is then applied. The problem is to write an equation which, when solved, will yield an expression for the oven temperature T as a function of time. This equation is, of course, the differential equation for the system.

The ambient (surrounding) temperature will be chosen as a reference and therefore equal to zero. Thus, the difference between oven temperature and ambient temperature $(T_{oven}\text{-}T_{amb})$ can be expressed as the oven temperature and simply denoted by T. If the heat loss through the walls of the oven is denoted as q_{out} (Btu/sec) and, the rate at which heat is stored by the parts in the oven by q_s (Btu/sec), then the conservation law requires that

$$q_{out} + q_s = q_{in}$$

This equation points out the fact that this system is the thermal equivalent of the liquid-level system introduced in the previous section.

Thermal resistance, by general definition, is the change of potential required to cause a unit change of flow rate. Thermal resistance is considered linear and so

$$R = \frac{T_{\text{oven}} - 0}{q_{\text{out}}} = \frac{T}{q_{\text{out}}}$$

This equation suggests that the thermal resistance of the oven could be obtained by supplying a known rate of heat input and determining the final temperature rise; at equilibrium, $q_{\text{out}} = q_{\text{in}}$.

Upon reviewing the equations involving both electrical and hydraulic capacitance, one notes that the flow rate in each case is equal to the capacitance multiplied by the time rate of change of potential. For the thermal system, then, by analogy

$$q_s = C \frac{dT}{dt} = [\text{Btu/degree}][\text{degree/sec}] = [\text{Btu/sec}]$$

The units do check, and although this does not prove we are right, even the most skeptical reader must admit that our result is far more satisfying than had the units check failed. The value for the thermal capacitance would be obtained by multiplying the specific heat of the material by the weight of the parts.

The differential equation for the system is obtained by substituting the expressions for thermal resistance and capacitance into that expression which satisfies the continuity law. The result is

$$\frac{T}{R} + C \frac{dT}{dt} = q_{\text{in}}$$

or

$$RC \frac{dT}{dt} + T = Rq_{\text{in}} \qquad (2\text{-}31)$$

Note that the parts in the oven were assumed to exhibit pure capacitance with no resistance to heat flow. Further, the capacitance of the oven walls was neglected. With these assumptions, the differential equation was simplified although neither assumption is strictly correct. Assumptions of this nature are frequently made to obtain a "first approximation" of system behavior.

2-7. A Biological System

From an engineering viewpoint, the human body could be regarded as being composed of a number of dynamic systems. The purpose of this section is to consider one such system and to illustrate an approach for obtaining the defining differential equation.

The problem is to determine the dynamic relationship between body weight and calorie intake. Diet experts state that a moderately active person burns about 20 calories per pound of body weight during each day. Further, 4,000 calories is roughly equivalent to one pound of body weight;

that is, an excess of 4,000 calories will result in a one-pound weight increase, while one pound will be lost if the body burns 4,000 calories over and above those obtained from food intake.

A convenient approach to formulating the defining equation is to consider what takes place in an increment of time dt. The units of time will be days. In time dt, the number of calories taken in, C_i, can be expressed as

$$C_i = q_i dt$$

where q_i is the rate of calorie intake (calories/day). The number of calories burned, C_b, is

$$C_b = 20 W dt$$

where W is body weight (pounds). The change of weight can be written as

$$dW = \frac{C_i - C_b}{4000}$$

Substitution of the expressions for C_i and C_b yields

$$4000 \, dW = q_i dt - 20 W dt$$

The resulting differential equation is

$$4000 \frac{dW}{dt} + 20 W = q_i \tag{2-32}$$

The basic approach used in this section (that of formulating an equation on the basis of a time increment dt) is applicable to a wide variety of systems and is a very commonly used technique.

2-8. Closure

Several points need to be discussed at the end of this chapter. The first has to do with a matter of notation. The differential equations (Eqs. 2-6 and 2-19 for the spring-mass-damper system shown in Fig. 2-6a and the electrical circuit shown in Fig. 2-12b can be written as

$$M \frac{d^2 y(t)}{dt^2} + c \frac{dy(t)}{dt} + ky(t) = kx(t)$$

and
$$L \frac{d^2 q(t)}{dt^2} + R \frac{dq(t)}{dt} + \frac{q(t)}{C} = e(t)$$

Comparison of these equations with those originally derived shows that they are identical except for notation; here $y(t)$ replaces y; $x(t)$ replaces x, and so forth. Of course, the notation merely indicates that the quantities are functions of time. This notation is not used solely to provide a more elegant equation but to emphasize the nature of the equation.

The two equations can be solved for displacement and charge. Designating these quantities as $y(t)$ and $q(t)$ is a reminder that the solutions will be functions of time and that we are dealing with dynamic, not static, systems. Our concern is with defining the behavior of a system for all values of time rather than just for some ultimate steady-state condition. Expressing the inputs as $x(t)$ and $e(t)$ is a reminder that the differential equations are general in nature and therefore are independent of the type of input. The input can be any time-varying function one wishes to introduce into the equation. With a particular input, the solution of the equation yields the appropriate system response.

The notation discussed is widely used. The reader may use it or not as he sees fit. However, this book will use the $f(t)$ notation only when its use will better serve to illustrate a point.

A second point worthy of note is that the differential equations obtained for the various systems discussed are remarkably similar. This fact can be verified by thumbing through the chapter and observing the form of the equations. As a specific example, compare the two equations given in this section. Mathematically speaking, these equations are identical because constants are constants and functions are functions no matter what name we give them. For this reason, one would expect the solutions of the equations, and therefore the responses of the systems, to be the same. Such is the case. This illustrates the fact that a knowledge of the solution of differential equations can provide considerable insight into the dynamic response of systems of many types. Also, an intuitive feel for one type of system can be extended to other types once the differential equations are known to be similar.

A third point relates to the assumptions made in writing differential equations. In obtaining the equations in this chapter, some assumptions were stated while others were only implied. In all cases, however, the assumptions made were for the purpose of simplifying the final equation. The best equation, from a practical viewpoint, is the one most easily solved provided it is reasonably descriptive of the system under consideration.

In closing, the reader who feels that he understands the concepts presented in this chapter is cautioned not to think that setting up differential equations for all physical systems is a simple matter. The examples chosen have been rather obvious compared with the problems usually encountered in practice. Nevertheless, the examples presented do illustrate the basic concept of writing differential equations, namely, differential equations are written from an understanding of the laws of nature underlying the system being considered.

Problems

2-1. Obtain the differential equation for each of the following:

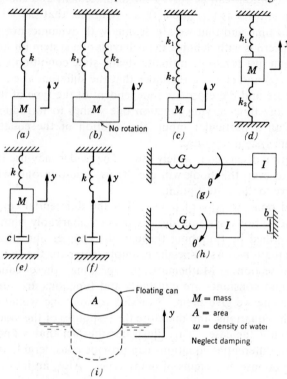

PROB. 2-1

2-2. Obtain the appropriate differential equation for each of the following. In (a) through (d), x is input and y is response; in (e), T is input and ω is response; in (f), force f is the input and y is the response.

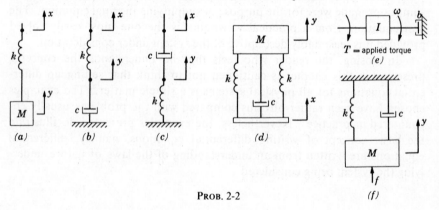

PROB. 2-2

2-3. a) Obtain the two differential equations necessary to define the system shown in the accompanying figure. Displacement θ_i is the input. b) Represent the geared system shown by an equivalent system without gears. c) The first part of the accompanying illustration shows a mass M which can be translated by rotation of a lead-screw. The screw has lead L, i.e., the mass advances L units of length for one revolution of the screw. Replace the mass by an equivalent rotating inertia I_e, as shown in the second part of the figure, so that the driving source would experience the same load in either case. d) Shown in the accompanying illustration is a schematic drawing of a mass M driven through a gear and rack. The gear shaft is supported in frictionless bearings and thus can only rotate. Assume that gear and rack inertias are included in mass M. Obtain the system differential equation in which angular displacement θ_i is the input and translational displacement x is the response (or output).

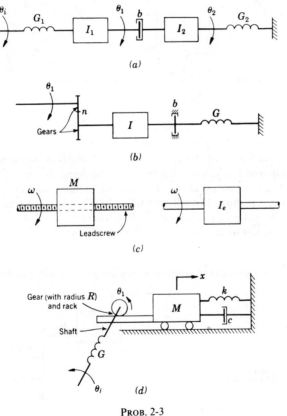

(a)

(b)

(c)

(d)

PROB. 2-3

2-4. Analyze the dashpot in Fig. 2-4a to show that if the rate of fluid flow across the piston is proportional to the pressure drop (laminar flow), the restraining force is proportional to velocity but if the flow rate is proportional to the square root of the pressure drop (turbulent flow), the restraining force is proportional to the square of the velocity. Neglect other friction effects.

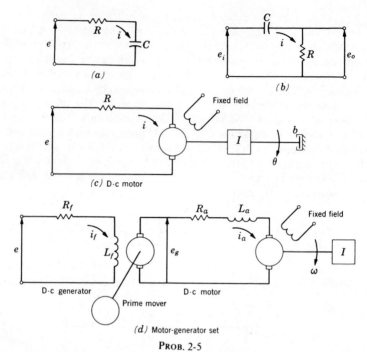

(a)

(b)

(c) D-c motor

(d) Motor-generator set

PROB. 2-5

2-5. For each part of the accompanying illustration, write as many equations as are required to relate the response with the input. In (a), e is input and i is response; in (b), e_i is input and e_o is response (assume capacitor initially discharged); in (c), e is input and θ is response. In (d), e is input and ω is response. Assume that the generated voltage $e_g = K_g i_f$ where K_g is the generator constant. The subscript f refers to the generator field while the subscript a refers to armature values.

2-6. For each part of the accompanying illustration, write as many equations

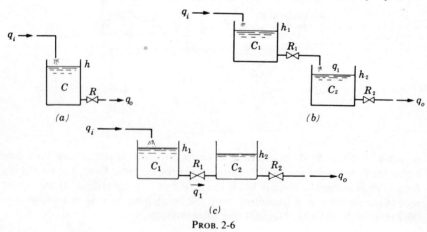

(a)

(b)

(c)

PROB. 2-6

as are required to relate the response with the input. In (*a*), q_i is input and q_o is response; in (*b*) and (*c*), q_i is input and h_2 is response.

2-7. Obtain the differential equation, in which *x* is the input and *y* is the output, for the value-controlled hydraulic actuator shown in the accompanying illustration. In this particular system, the piston rod is fixed while the cylinder is free to move. The valve body is rigidly attached to the cylinder and moves with it. Displacement of the valve spool opens the valve, causing cylinder and valve-body displacements which tend to close the valve.

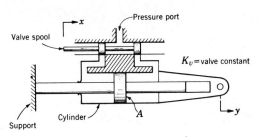

PROB. 2-7

2-8. A part, at an elevated temperature, is quenched in a water bath. The water temperature may be assumed to remain constant at T_w. Obtain the appropriate differential equation involving the part temperature *T*. Assume that the part exhibits pure capacitance *C* with no internal resistance to heat flow, i.e., the temperature of the part is uniform throughout. Thermal resistance *R* occurs only at the surface of the part.

2-9. Shown in the accompanying illustration is a schematic drawing of a perfectly insulated tank through which water is pumped at a constant volumetric flow rate *Q*. The tank contains volume *V* of water at temperature *T*, the assumption being made that perfect mixing occurs. Water density *w* and specific heat *c* are assumed to be constant. The thermal capacitance of the tank itself is assumed negligible. Obtain the differential equation which relates the response temperature *T* with a change in inlet temperature T_i. Assume that, before $t = 0$, the system is in equilibrium with inlet, tank, and outlet temperatures being the same. You may wish to consider this temperature zero as a reference.

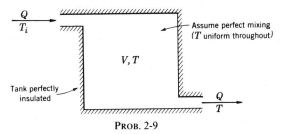

PROB. 2-9

2-10. The hopper, shown in figure, receives an input q_i (lb/hr) of material from the belt conveyor. The screw conveyor removes material from the hopper at a rate q_o. This rate is adjusted to meet plant needs and is in no way related to the weight W of material in the hopper. Obtain the differential equation which relates weight W with rates q_i and q_o.

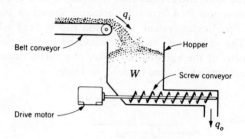

PROB. 2-10

2-11. For the system given in the accompanying illustration, obtain the differential equation which could be solved for the output voltage e_o from the

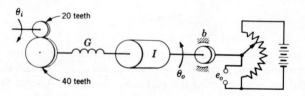

PROB. 2-11

potentiometer for an input displacement θ_i. Assume $e_o = K\theta_o$ and that gear inertias can be neglected.

Chapter 3

Solving Differential Equations

3-1. Introduction

This chapter deals with the solution of the ordinary linear differential equation with constant coefficients, the basic type of equation found in this book [33]. However, before discussing solutions, some preliminary remarks about differential equations will be made to provide a review and to put the material which follows into its proper perspective.

Suppose that $t, u, v, \ldots$ are independent variables and that $x, y, z, \ldots$ are dependent variables and are functions of $t, u, v, \ldots$. Then the functional dependence is shown as

$$x = x(t,u,v,\ldots)$$
$$y = y(t,u,v,\ldots)$$
$$z = z(t,u,v,\ldots)$$

A differential equation shows the relationship between the independent variables, the dependent variables, and the derivatives of the dependent variables with respect to the independent variables. For a simple case, this relationship can be expressed as

$$f\left(t,x,\frac{dx}{dt}\right) = 0$$

Consider the following examples:

$$\frac{d^2x(t)}{dt^2} + \frac{dx(t)}{dt} + 3\,x(t) = 5t$$

$$\frac{dy(u)}{du} + y(u) = \cos u$$

$$\frac{\partial^2 x}{\partial t^2} + \frac{\partial^2 x}{\partial u^2} + \frac{\partial^2 x}{\partial v^2} = \frac{\partial x}{\partial \lambda}$$

where $x = x(t,u,v,\lambda)$.

If there is *one* independent variable, as in the first two examples, the differential equation is *ordinary*. If there are several independent variables so that the derivatives are partial derivatives, as in the third example, the equation is a *partial* differential equation. Our concern is with ordinary differential equations.

The *order* of a differential equation is the order of the highest derivative. For example

$$\frac{d^2x}{dt^2} + x = 5$$

is a second-order equation while

$$\frac{d^5y}{dt^5} + \frac{d^3y}{dt^3} + y = \sin t$$

is a fifth-order equation. The *degree* of a differential equation is the power to which the highest derivative is raised. The equation

$$\left(\frac{dx}{dt}\right)^2 + x = 0$$

is an illustration of a second-degree equation.

Ordinary differential equations may be classified by the nature of the coefficients of the terms in the equation. These coefficients may be variable or constant. An example of an equation with *variable coefficients* is

$$t^3 \frac{d^2x}{dt^2} + t \frac{dx}{dt} + t^2x = 15 \sin 3t$$

while an equation with *constant coefficients* is

$$8 \frac{d^2y}{dt^2} + 4 \frac{dy}{dt} + 2y = 0$$

Differential equations may be further classified as *linear* or *nonlinear*. The two previous examples illustrate linear equations. These would become nonlinear if the variable coefficients were functions of the dependent variable or if any of the derivatives were raised to a power other than one. Examples of nonlinear equations are

$$x^3 \frac{dx}{dt} + x = 10$$

and
$$8 \left(\frac{d^2y}{dt^2}\right)^2 + 2y = 0$$

Linear equations are characterized by the fact that the principle of superposition holds. Consider the linear equation

$$\frac{d^2y}{dt^2} + \frac{dy}{dt} + y = f(t)$$

where $f(t)$ is called the *driving function*, or the *forcing function*. If the solution of the equation with a forcing function $f_1(t)$ is $y_1(t)$ and with $f_2(t)$ is $y_2(t)$ then the solution with a forcing function $f_1(t) + f_2(t)$ is $y_1(t) + y_2(t)$.

In the case of a linear physical system (one whose behavior can be defined by a linear differential equation), this means that the response of the system to the sum of two inputs is equal to the sum of the responses to the inputs taken one at a time. The principle of superposition *does not* hold for a nonlinear system.

This chapter presents a method for solving ordinary linear differential equations with constant coefficients. An operational method which will prove to be very useful in solving differential equations will be employed. An operator p (frequently given elsewhere as D or s) is defined such that

$$py \equiv \frac{dy}{dt} \qquad p^2y \equiv \frac{d^2y}{dt^2} \qquad p^ny \equiv \frac{d^ny}{dt^n} \tag{3-1}$$

and

$$\frac{1}{p}y \equiv \int_0^t y \, dt \tag{3-2}$$

Thus p denotes differentiation with respect to time while $1/p$ denotes integration with respect to time. The utility of the p-operator is that, for the most part, it can be treated as an algebraic quantity. This permits the replacement of differential equations with algebraic equations which are more easily handled.

3-2. Solution of Homogeneous Equations

The solution of ordinary linear differential equations with constant coefficients will be presented in three parts. The first part involves the solution of homogeneous equations. A homogeneous equation is one in which the forcing function is zero.

First-Order Equations. Consider the general equation

$$\frac{dy}{dt} + ky = 0$$

where k is a constant. Using Eq. 3-1,

$$py + ky = 0 = y(p + k)$$

Because y is the required non-trivial solution, and therefore is not zero, then

$$p + k = 0$$

This expression is called the *characteristic equation*. From it

$$p = -k$$

The solution of all first-order equations is of the form

$$y = C_1 e^{pt} \tag{3-3}$$

where C_1 is an arbitrary constant which is determined from a knowledge of the value of y when time is assumed to be zero. This value of y is called an *initial condition*. From Eq. 3-3, our solution is

$$y = C_1 e^{-kt}$$

But is it? We say we have a solution when, upon substitution, it reduces the original equation to an identity or, said in another way, it satisfies the equation. From our solution

$$\frac{dy}{dt} = -kC_1 e^{-kt}$$

Substituting the expressions for dy/dt and y into the original equation yields

$$-kC_1 e^{-kt} + kC_1 e^{-kt} = 0$$

which shows that the solution is correct.

EXAMPLE 3-1. Solve the equation

$$\frac{dy}{dt} + 2y = 0$$

Solution: $py + 2y = 0 = y(p + 2)$ and $p = -2$. From Eq. 3-3,

$$y = C_1 e^{-2t} \qquad\qquad Ans.$$

Second-Order Equations. Consider the general equation

$$\frac{d^2 y}{dt^2} + g\,\frac{dy}{dt} + ky = 0$$

where g and k are constants. Using Eq. 3-1

$$p^2 y + gpy + ky = 0 = y(p^2 + gp + k)$$

and the characteristic equation is

$$p^2 + gp + k = 0$$

Solving this equation leads to two values of p which are

$$p = \frac{-g \pm \sqrt{g^2 - 4k}}{2}$$

There are three different possibilities for the values of p depending on whether the quantity under the radical is positive, zero, or negative. These three cases will be discussed separately.

Case 1. If the quantity under the radical is positive, the roots of the characteristic equation, p_1 and p_2, will be real and unequal numbers. The solution is then

$$y = C_1 e^{p_1 t} + C_2 e^{p_2 t} \qquad\qquad (3\text{-}4a)$$

where C_1 and C_2 are arbitrary constants which are determined from initial conditions.

EXAMPLE 3-2. Solve the equation

$$\frac{d^2 y}{dt^2} + 3\,\frac{dy}{dt} + 2y = 0$$

Solution: $p^2 y + 3py + 2y = 0$ and $p^2 + 3p + 2 = 0$; $p_1 = -1$ and $p_2 = -2$. From Eq. 3-4a,

$$y = C_1 e^{-t} + C_2 e^{-2t} \qquad\qquad Ans.$$

The reader may verify this solution by substituting it into the original equation.

Case 2. If the quantity under the radical is zero, the roots of the characteristic equation will be real and equal numbers. In this special case of repeated roots, the solution is

$$y = C_1 e^{p_1 t} + C_2 t e^{p_2 t} \qquad\qquad (3\text{-}4b)$$

where $p_1 = p_2$.

EXAMPLE 3-3. Solve the equation

$$\frac{d^2 y}{dt^2} + 2\frac{dy}{dt} + y = 0$$

Solution: The characteristic equation is $p^2 + 2p + 1 = 0$ and $p_1 = p_2 = -1$. From Eq. 3-4b,

$$y = C_1 e^{-t} + C_2 t e^{-t} \qquad\qquad Ans.$$

This solution can be checked by substitution into the original equation.

Case 3. If the quantity under the radical is negative, the roots of the characteristic equation will be complex numbers, actually complex conjugates. In this case, the roots of the generalized characteristic equation $(p^2 + gp + k = 0)$ are

$$p = \frac{-g \pm j\sqrt{4k - g^2}}{2}$$

where $j = \sqrt{-1}$. Letting $a = -g/2$ and $b = (4k - g^2)^{1/2}/2$, the roots are $p_1 = a + jb$ and $p_2 = a - jb$, and the solution is

$$y = C_1 e^{p_1 t} + C_2 e^{p_2 t}$$

where C_1 and C_2 are also complex conjugates. This solution can be modified into

$$y = e^{at}(C_1 \sin bt + C_2 \cos bt) \qquad\qquad (3\text{-}4c)[1]$$

where C_1 and C_2 are real numbers.

EXAMPLE 3-4. Solve the equation

$$\frac{d^2 y}{dt^2} + 4\frac{dy}{dt} + 13 y = 0$$

---

[1] Equation 3-4c can also be expressed as

$$y = C_3 e^{at} \sin(bt + \phi)$$

in which

$$C_3 = (C_1^2 + C_2^2)^{1/2} \qquad \text{and} \qquad \phi = \tan^{-1} C_2/C_1$$

Solution: The characteristic equation is $p^2 + 4p + 13 = 0$.

$$p = \frac{-4 \pm (-36)^{\frac{1}{2}}}{2} = \frac{-4 \pm j6}{2} = -2 \pm j3$$

From Eq. 3-4c, $y = e^{-2t}(C_1 \sin 3t + C_2 \cos 3t)$ *Ans.*

Higher-Order Equations. A third-order differential equation will have a third-degree characteristic equation which can be solved for three roots. In general, an nth-degree characteristic equation will result from an nth-order differential equation and there will be n roots. If special provisions are made for repeated roots and if the roots and arbitrary constants are permitted to be real or complex numbers, the general solution for all ordinary linear differential equations with constant coefficients is

$$y = C_1 e^{p_1 t} + C_2 e^{p_2 t} + C_3 e^{p_3 t} + \cdots + C_n e^{p_n t} \tag{3-5}$$

The solution of an nth-order differential equation will involve n arbitrary constants which can be determined from n initial conditions.

To illustrate the use of Eq. 3-5 consider the following examples: If $p_1 = -1$, $p_2 = -2$, and $p_3 = -3$,

$$y = C_1 e^{-t} + C_2 e^{-2t} + C_3 e^{-3t}$$

If $p_1 = -2$, $p_2 = -3 + j2$, and $p_3 = -3 - j2$,

$$y = C_1 e^{-2t} + e^{-3t}(C_2 \sin 2t + C_3 \cos 2t)$$

This result is obtained by using Eq. 3-4c as the basis for handling the pair of complex roots. If $p_1 = 0$, $p_2 = -1$, $p_3 = -1$, and $p_4 = -2$,

$$y = C_1 + C_2 e^{-t} + C_3 t e^{-t} + C_4 e^{-2t}$$

This result is obtained by remembering that e^0 is unity and by using Eq. 3-4b as the basis for handling the repeated roots. Finally, if $p_1 = p_2 = p_3 = -2$,

$$y = C_1 e^{-2t} + C_2 t e^{-2t} + C_3 t^2 e^{-2t}$$

This last example illustrates that for m repeated roots, the solution is

$$y = C_1 e^{p_1 t} + C_2 t e^{p_2 t} + C_3 t^2 e^{p_3 t} + \cdots + C_m t^{m-1} e^{p_m t} \tag{3-6}$$

3-3. Consideration of Initial Conditions

The solutions for the differential equations presented in the previous section involved one or more arbitrary constants. These constants are determined from the initial conditions, i.e. the values of the dependent variable and its derivatives existing when time is assumed to be zero. The method for obtaining the constants is straightforward and will be illustrated by three examples involving first-order and second-order differential equations. The method is, of course, quite general and is applicable to equations of higher order.

EXAMPLE 3-5. Solve the equation

$$\frac{dy}{dt} + 2y = 0$$

with $y = 3$ when $t = 0$.

Solution: By the method of Sec. 3-2 (see Eq. 3-3)

$$y = C_1 e^{-2t}$$

This expression for y is valid for all positive values of t including $t = 0$. From the knowledge that $y = 3$ when $t = 0$, one writes

$$3 = C_1 e^0 = C_1$$

The arbitrary constant C_1 is now evaluated and the solution is

$$y = 3e^{-2t} \qquad\qquad Ans.$$

EXAMPLE 3-6. Solve the equation

$$\frac{d^2y}{dt^2} + 5\frac{dy}{dt} + 6y = 0$$

with $y = 3$ and $dy/dt = -7$ when $t = 0$.

Solution: By the method of Sec. 3-2 (see Eq. 3-4a)

$$y = C_1 e^{-3t} + C_2 e^{-2t}$$

Differentiation yields

$$\frac{dy}{dt} = -3\,C_1 e^{-3t} - 2\,C_2 e^{-2t}$$

By substituting the initial conditions, the result is

$$3 = C_1 + C_2 \quad\text{and}\quad -7 = -3C_1 - 2C_2$$

from which $C_1 = 1$ and $C_2 = 2$. Therefore, the required solution is

$$y = e^{-3t} + 2\,e^{-2t} \qquad\qquad Ans.$$

Note that the solution of the second-order equation involved two arbitrary constants and that two initial conditions were necessary for evaluation.

EXAMPLE 3-7. Solve the equation

$$\frac{d^2y}{dt^2} + 4y = 0$$

with $y = 0$ and $dy/dt = 8$ when $t = 0$.

Solution: By the method previously described (see Eq. 3-4c),

$$y = e^0(C_1 \sin 2t + C_2 \cos 2t) = C_1 \sin 2t + C_2 \cos 2t$$

Differentiation yields

$$\frac{dy}{dt} = 2C_1 \cos 2t - 2C_2 \sin 2t$$

Substituting the initial conditions:

$$0 = C_1(0) + C_2(1) \quad \text{and} \quad C_2 = 0$$
$$8 = 2C_1 - 0 \quad \text{and} \quad C_1 = 4$$

Therefore, the solution is

$$y = 4 \sin 2t \qquad\qquad Ans.$$

Check: A check is always desirable. The expressions for y and its derivatives are

$$y = 4 \sin 2t \qquad \frac{dy}{dt} = 8 \cos 2t \qquad \frac{d^2y}{dt^2} = -16 \sin 2t$$

Substituting these expressions into the original equation yields

$$-16 \sin 2t + 4(4 \sin 2t) = 0$$

which proves that the solution is correct.

3-4. Solution of Nonhomogeneous Equations

The nonhomogeneous differential equation is most common in control work because the determination of the response of a system to some input is the usual case. A nonhomogeneous equation is one in which the forcing function is not zero. The equation

$$\frac{d^2y}{dt^2} + g\frac{dy}{dt} + ky = f(t)$$

illustrates a second-order nonhomogeneous equation.

The solution for a nonhomogeneous equation is composed of the sum of two parts. The first part is the solution of the associated homogeneous equation [that is, let $f(t) = 0$] which is called the *complementary function*. The second part is a solution which satisfies the entire equation; this solution is called the *particular integral*. Letting y_c denote the complementary function and y_p the particular integral, the complete solution can be expressed as

$$y = y_c + y_p \tag{3-7}$$

The method for determining y_c, the solution for the homogeneous equation, has been discussed. All that remains is to learn how to obtain the particular integral y_p. *Note, and this is very important, that the arbitrary constants associated with y_c must not be determined until the complete solution y is obtained.*

The procedure for obtaining the particular integral y_p is to substitute an assumed solution into the entire equation for the purpose of evaluating the unknown constants. The appropriate assumed solutions for forcing functions commonly encountered in control work are given in Table 3-1. Notice that the assumed solution has the same form as the forcing function.

TABLE 3-1

ASSUMED SOLUTIONS FOR OBTAINING THE PARTICULAR
INTEGRAL y_p

($K, A, B, C,$ and ω are constants)

Forcing Function	Assumed Solution (y_p)
K	A
Kt	$At + B$
Kt^2	$At^2 + Bt + C$
$K \sin \omega t$	$A \sin \omega t + B \cos \omega t$

Three examples will now be given to illustrate the method for obtaining the particular integral for a nonhomogeneous differential equation.

EXAMPLE 3-8. Obtain the particular integral for the equation

$$\frac{d^2y}{dt^2} + 3\frac{dy}{dt} + 5y = 15$$

Solution: The forcing function is a constant, and from Table 3-1, one assumes $y_p = A$. This assumed solution must be substituted into the entire equation so that the constant A can be determined. The derivatives of y_p are

$$\frac{dy_p}{dt} = 0 \quad \text{and} \quad \frac{d^2y_p}{dt^2} = 0$$

Upon substitution into the original equation, the result is $0 + 0 + 5A = 15$; $A = 3$. The particular integral is therefore

$$y_p = A = 3 \qquad \qquad Ans.$$

Note that $y = 3$ satisfies the equation.

EXAMPLE 3-9. Obtain the particular integral for the equation

$$\frac{d^2y}{dt^2} + \frac{dy}{dt} + 2y = 4t$$

Solution: From Table 3-1, the assumed solution is $y_p = At + B$. The derivatives are $dy_p/dt = A$ and $d^2y_p/dt^2 = 0$. Substituting these values into the original equation yields

$$0 + A + 2At + 2B = 4t$$

Upon equating the coefficients of like terms (that is, the t terms and the constant terms), the results are $2A = 4$ and $A = 2$; $A + 2B = 0$ and $B = -1$. Therefore, the particular integral is

$$y_p = 2t - 1 \qquad \qquad Ans.$$

EXAMPLE 3-10. Obtain the particular integral for the equation

$$\frac{dy}{dt} + 5y = 10 \sin 5t$$

Solution: From Table 3-1, the assumed solution is $y_p = A \sin 5t + B \cos 5t$. The first derivative is $dy_p/dt = 5A \cos 5t - 5B \sin 5t$. Substituting these expressions into the original equation yields

$$5A \cos 5t - 5B \sin 5t + 5A \sin 5t + 5B \cos 5t = 10 \sin 5t$$

By equating coefficients of like terms, $5A + 5B = 0$ and $5A - 5B = 10$ from which $A = 1$ and $B = -1$. The particular integral is

$$y_p = \sin 5t - \cos 5t \qquad\qquad Ans.$$

A fourth example will be given to summarize the material presented thus far and to illustrate a minor difficulty which can occur in assuming the form for the particular integral.

EXAMPLE 3-11. Solve the equation

$$\frac{d^2y}{dt^2} + 2\frac{dy}{dt} = 4$$

with $y = 0$ and $dy/dt = 6$ when $t = 0$.

Solution: The complete solution is composed of the sum of the complementary function and the particular integral. To obtain the complementary function, the associated homogeneous equation is written as

$$\frac{d^2y}{dt^2} + 2\frac{dy}{dt} = 0$$

From Eq. 3-1,

$$p^2y + 2py = 0 = y(p^2 + 2p)$$

Since the solution y is not zero,

$$p^2 + 2p = p(p + 2) = 0$$

and this is the characteristic equation. Its roots are $p_1 = 0$ and $p_2 = -2$. These roots are real and unequal. Therefore, from Eq. 3-4a, the complementary function y_c is

$$y_c = C_1 + C_2 e^{-2t}$$

The arbitrary constants C_1 and C_2 must not be evaluated until the complete solution is obtained.

To obtain the particular integral, one notes that the forcing function is a constant, and from Table 3-1, assumes $y_p = A$. The derivatives are $dy_p/dt = d^2y_p/dt^2 = 0$. Upon substitution into the original equation, the result is

$$0 + 0 = 4$$

This is not an acceptable equation and the conclusion is that the assumed solution is at fault. However, an assumed solution whose first derivative is a constant would appear to work. Therefore, assume

$$y_p = At*$$

The derivatives are $dy_p/dt = A$ and $d^2y_p/dt^2 = 0$. Substituting these expressions into the original equation yields $0 + 2A = 4$; $A = 2$. Therefore,

$$y_p = 2t$$

From Eq. 3-7

$$y = y_c + y_p = C_1 + C_2e^{-2t} + 2t$$

This is the complete solution and now the arbitrary constants can be determined. The derivative of the solution is

$$\frac{dy}{dt} = -2C_2e^{-2t} + 2$$

Upon substitution of the initial conditions, the result is $0 = C_1 + C_2$ and $6 = -2C_2 + 2$ from which $C_1 = 2$ and $C_2 = -2$. Therefore,

$$y = 2 - 2e^{-2t} + 2t \qquad\qquad Ans.$$

3-5. Combining Simultaneous Equations

There are times when it is desirable to combine a number of simultaneous equations into one differential equation. Here the p operator is very useful, because the set of differential equations can be replaced by a set of algebraic equations. Once the algebraic equations have been solved, the desired differential equation can be recovered. The approach will be illustrated with an example.

EXAMPLE 3-12. Combine the two given equations into one which can be solved for z. The forcing function is to be x.

$$\frac{dy}{dt} + 2y = 5x$$

$$3\frac{dz}{dt} + z = 2y$$

Solution: Introduction of the p operator (Eq. 3-1) yields

$$y(p + 2) = 5x \qquad \text{and} \qquad y = \frac{5x}{p + 2}$$

$$z(3p + 1) = 2y = \frac{10x}{p + 2}$$

---

*This new assumption can also be based on the mathematical statement that, if the assumed particular integral duplicates any term in the complementary function, it must be multiplied by the lowest power of the independent variable sufficient to eliminate the duplication.

Multiplying both sides of the latter equation by $p + 2$,

$$z(3p + 1)(p + 2) = 10x$$
$$z(3p^2 + 7p + 2) = 10x$$

The required differential equation, then, is

$$3\frac{d^2z}{dt^2} + 7\frac{dz}{dt} + 2z = 10x \qquad\qquad Ans.$$

3-6. Closure

The concepts required for solving ordinary linear differential equations with constant coefficients have been developed. These are sufficient to permit us to continue our study of the dynamic behavior of physical systems.

Although primary attention, in this chapter, has been given to the solution of first- and second-order equations, the concepts can easily be extended to higher-order equations. The work involved, however, is greater. One troublesome aspect is the determination of the roots of the characteristic equation, but this can readily be done with the aid of a digital computer. The use of the digital computer for root solving is discussed in Appendix A. Appendix B briefly discusses alternative techniques if the roots must be obtained by hand calculations.

Problems

3-1. Solve the following differential equations:

a) $\dfrac{dy}{dt} + 3y = 0$

b) $\dfrac{dy}{dt} + \dfrac{y}{2} = 0$

c) $\dfrac{d^2y}{dt^2} + 4\dfrac{dy}{dt} + 3y = 0$

d) $\dfrac{d^2y}{dt^2} + 7\dfrac{dy}{dt} + 12y = 0$

e) $\dfrac{d^2y}{dt^2} + 4\dfrac{dy}{dt} + 4y = 0$

f) $\dfrac{d^2y}{dt^2} + 3\dfrac{dy}{dt} = 0$

g) $\dfrac{d^2y}{dt^2} + 9y = 0$

h) $\dfrac{d^2y}{dt^2} + 2\dfrac{dy}{dt} + 2y = 0$

i) $\dfrac{d^2y}{dt^2} + 6\dfrac{dy}{dt} + 13y = 0$

j) $\dfrac{d^3y}{dt^3} + 3\dfrac{d^2y}{dt^2} + 3\dfrac{dy}{dt} + y = 0$

k) $\dfrac{d^3y}{dt^3} + 5\dfrac{d^2y}{dt^2} + 11\dfrac{dy}{dt} + 15y = 0$

3-2. Solve the following differential equations and determine the values of the constants from the initial conditions specified:

a) $\dfrac{dy}{dt} + 5y = 0;$ at $t = 0$, $y = 2$.

b) $\dfrac{d^2y}{dt^2} + 3\dfrac{dy}{dt} + 2y = 0$; at $t = 0$, $y = 1$ and $dy/dt = 2$.

c) $\dfrac{d^2y}{dt^2} + 6\dfrac{dy}{dt} + 9y = 0$; at $t = 0$, $y = 0$ and $dy/dt = 2$.

d) $\dfrac{d^2y}{dt^2} + 8\dfrac{dy}{dt} + 25y = 0$; at $t = 0$, $y = 1$ and $dy/dt = 8$.

3-3. Obtain the particular integrals for the following differential equations:

a) $\dfrac{d^2y}{dt^2} + 6\dfrac{dy}{dt} + 3y = 6$

b) $\dfrac{d^2y}{dt^2} + 4\dfrac{dy}{dt} + 2y = 6t$

c) $\dfrac{dy}{dt} + 4y = 10\sin 2t$

3-4. Solve the following differential equations:

a) $\dfrac{d^2y}{dt^2} + 8\dfrac{dy}{dt} + 12y = 12t$; at $t = 0$, $y = dy/dt = 0$.

b) $\dfrac{d^2y}{dt^2} + 4\dfrac{dy}{dt} + 8y = 4$; at $t = 0$, $y = dy/dt = 0$.

3-5. Check the answers for Prob. 3-4 by substitution into the original equations.

3-6. a) Combine the following three equations into one in which i is the response and g is the input (forcing function).

$A\dfrac{df}{dt} + Bf = Cg$

$h = Df$

$E\dfrac{di}{dt} + i = Kh$

b) Combine the following three equations into one in which v is the response and r is the input.

$B\dfrac{ds}{dt} + s = Ar$

$\dfrac{du}{dt} = Cs$

$D\dfrac{du}{dt} + Eu = v$

3-7. Factor the following polynomials:

a) $p^3 + 4p^2 + 14p + 20 = 0$
b) $p^3 + 9.5p^2 + 31p + 35 = 0$

c) $p^4 + 8p^3 + 32p^2 + 80p + 100 = 0$

d) $p^4 + 8p^3 + 27p^2 + 50p + 50 = 0$

e) $p^5 + 8p^4 + 35p^3 + 96p^2 + 150p + 100 = 0$

Chapter **4**

Applying the
Analog Computer

4-1. Introduction

The general purpose analog computer (electronic differential analyzer) is a device that is naturally suited for studying automatic control systems. It is extremely useful in both the analysis of system elements and in the study of over-all control system response.

The analog computer is an electronic machine in which all quantities are represented by voltages—system inputs are represented by voltages as well as system outputs, or responses. System behavior may then be observed and data recorded using oscilloscopes and electro-mechanical recorders. The accuracy of analog computer results is naturally limited by the precision of the components which make up the computer and the ability to measure voltages accurately. Most modern commercial computers are, however, capable of providing results sufficiently accurate for engineering analysis and synthesis.

The analog computer is used as a means of control system analysis in basically two ways, first, as a device to solve a differential equation, and second, as a device to simulate the behavior of the actual components of the control system. As the first part of this book is concerned with the writing and solution of the differential equations of physical elements and systems, only the former application of the analog computer will be considered in this chapter. The second approach, which is widely employed, is more specialized and is thus reserved for discussion in Appendix D. The usefulness of the method of direct simulation can be appreciated only after completing Chapters 7 through 11.

This chapter first describes the mathematical operations that may be accomplished on the analog computer and then shows how these operations may be combined to obtain the solution to a differential equation. Next, the computer elements are discussed and the formulation of computer diagrams is illustrated. Finally, techniques which aid in making the most effective use of the computer are presented.

4-2. Generation of the Solution of a Differential Equation

The mathematical operations of modern analog computers include integration, algebraic summation, multiplication, and function generation as represented by the symbols of Fig. 4-1. The elements which perform these operations function in such a way that the analog computer is ideal for obtaining the continuous solution of a differential equation, either linear or nonlinear.

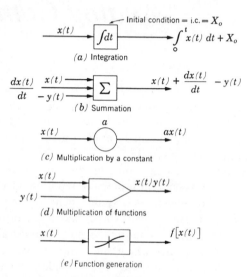

FIG. 4-1. Mathematical operations performed by an analog computer.

The underlying principle of solution of a differential equation on an analog computer is simple, but perhaps somewhat deceiving at first. The concept is best grasped by following a simple example.

EXAMPLE 4-1. Set up the computer block diagram required to solve the differential equation

$$\frac{d^2x}{dt^2} + 5\frac{dx}{dt} + 8x = 10t$$

where the initial conditions of the problem are specified as

$$x\,\big|_{t=0} = 5 \qquad \frac{dx}{dt}\bigg|_{t=0} = 1$$

Solution: The equation may be rearranged to give

$$\frac{d^2x}{dt^2} = -5\frac{dx}{dt} - 8x + 10t$$

which expresses the equation in terms of the highest derivative.

The first step is to assume that the highest derivative is available. It is then possible, with the use of the integrating box of Fig. 4-1a, to obtain the other terms of the equation. By cascading two integrators as shown in Fig. 4-2a, the terms dx/dt and x are obtained. The addition of constant multipliers provide the terms $-5dx/dt$ and $-8x$ (Fig. 4-2b), while the term $10t$ is obtained by integrating the constant 10 with respect to the independent variable t. Note that the terms at the input of the summation block provided are those required to give the second derivative which was initially assumed to be available.

Wherever an integrating box is seen, there must also be a symbol (i.c.) indicating the initial value of the output of the integrator at time equals zero. For example, when dx/dt is integrated to get x, then the integrating box must indicate the initial value of x. The complete diagram is seen in Fig. 4-2b.　　　　*Ans.*

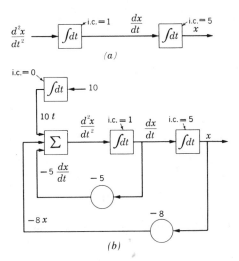

Fig. 4-2. Computer block diagram for
Example 4-1.

In the general case, consider the nth order differential equation expressed in terms of the highest derivative

$$\frac{d^n x}{dt^n} = -\frac{a_{n-1}}{a_n}\frac{d^{n-1}x}{dt^{n-1}} - \cdots - \frac{a_1}{a_n}\frac{dx}{dt} - \frac{a_0 x}{a_n} + \frac{f(t)}{a_n}$$

The forcing function is $f(t)$ and the initial conditions, for example, are specified as

$$x\big|_{t=0} = X_o \qquad \frac{dx}{dt}\bigg|_{t=0} = X_o^1 \qquad \frac{d^{n-1}x}{dt^{n-1}}\bigg|_{t=0} = X_o^{n-1}$$

The complete generation of this solution is given by the block diagram of Fig. 4-3. Observe that each output of the integrating boxes represents the succeeding lower derivative. All of the terms required are summed in the summing box to give the highest derivative that was originally assumed

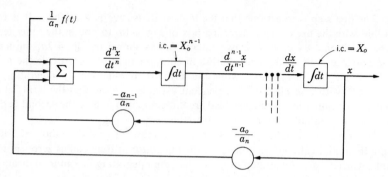

FIG. 4-3. Block diagram for the general nth-order equation.

available. The procedure may be further illustrated by another simple example.

EXAMPLE 4-2. Consider the problem of generating the function $f(t) = Ae^{-\alpha t}$ where A and α are constants.

Solution: This is not a differential equation; however, a differential equation can be devised which has as a solution this specific function. Differentiating $f(t)$ yields

$$\frac{df(t)}{dt} = -A\alpha e^{-\alpha t} = -\alpha f(t)$$

or

$$\frac{df(t)}{dt} + \alpha f(t) = 0$$

The initial condition in the problem is obtained by evaluating $f(t)$ at $t = 0$. Specifically,

$$f(t)\big|_{t=0} = A$$

This problem provides a simple method of obtaining an exponential function which may be used as a forcing function in a more complicated problem. The complete block diagram is given in Fig. 4-4.   *Ans.*

The solution of differential equations and the generation of functions are thus obtained by the use of operational building blocks. Although

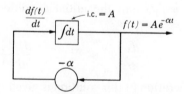

FIG. 4-4. Block   diagram
for Example 4-2.

operational blocks have been devised which are mechanical, hydraulic, or pneumatic, this chapter will be devoted to the principles of operation and programming of an electronic analog computer. The principles of operation of most commercial analog computers are the same, although there is some deviation from one manufacturer to another in the symbolism used in programming. The symbols adopted here for making computer diagrams are widely used.

4-3. Analog Computer Elements

The heart of the analog computer is the d-c operational amplifier. This element, with the proper input and feedback impedances, forms the basic electronic building block for summation and integration. To use an analog computer properly it is necessary to become familiar with the operation of the amplifier. However, one need not be concerned with the detailed circuitry or design of the amplifier itself.

The Operational Amplifier. The high-gain d-c operational amplifier is represented by the diagram of Fig. 4-5a. The amplifier is specified as high gain because it is desirable, as will be seen, to have the ratio of output to input signal as great as possible. In most commercial computers, the gain A ranges from 10^5 to 10^8. For computing purposes it is desirable to have a linear relationship between output and input over the entire range of output voltage. Further, it is desirable to have a flat response (the output of

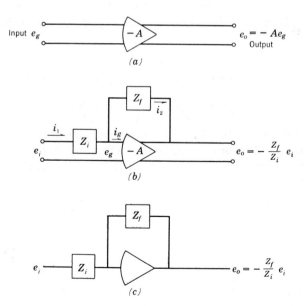

FIG. 4-5. The operational amplifier and impedance configurations.

the amplifier is not affected by the input frequency) over a wide range of frequency, commonly from d-c to several thousand cycles per second. Further requirements of a good operational amplifier include zero output voltage for zero input voltage, extremely high internal input impedance, and low noise generated within the amplifier. The amplifier will either have a reversal of polarity between output and input or a bipolar output.

Suppose that the amplifier of Fig. 4-5a is modified by the addition of an input impedance Z_i in series with the amplifier and a feedback impedance Z_f from output to the amplifier input, as shown in Fig. 4-5b. The relationship between the output and input voltages may be developed in the following manner. As the internal input impedance of the amplifier is high, there will be essentially no input current. Therefore,

$$i_g \approx 0$$

and by Kirchhoff's law

$$i_1 = i_2 \tag{4-1}$$

Noting that the current through an electrical impedance is equal to the voltage drop divided by the impedance gives

$$i_1 = \frac{e_i - e_g}{Z_i} \quad \text{and} \quad i_2 = \frac{e_g - e_o}{Z_f} \tag{4-2}$$

Substituting Eq. 4-2 into Eq. 4-1,

$$\frac{e_i}{Z_i} = \frac{e_g}{Z_i} + \frac{e_g}{Z_f} - \frac{e_o}{Z_f} \tag{4-3}$$

Noting that the ratio of the output voltage to grid voltage of the amplifier (Fig. 4-5a) is

$$e_o = -A e_g \tag{4-4}$$

Equation 4-3 becomes, after substitution of Eq. 4-4 to eliminate e_g,

$$\frac{e_i}{Z_i} = -\frac{e_o}{Z_f} - \frac{e_o}{A}\left(\frac{1}{Z_f} + \frac{1}{Z_i}\right) \tag{4-5}$$

In general, $|A| = 10^5$ to 10^8 and for many computers $|e_o| \leq 100$ volts. Also, Z_i and Z_f will be large impedances. Therefore, one may neglect the second term on the right-hand side of Eq. 4-5, thus obtaining the simple relationship

$$e_o = -\frac{Z_f}{Z_i} e_i \tag{4-6}$$

Equation 4-6 is the basic transfer relationship of the computer amplifier. The basic computer element is then represented as shown in Fig. 4-5c. On the computer, the ground is common to all amplifiers thus eliminating the necessity for showing two connections between amplifiers. This simplifies the drawing of computer programs and the common ground of Fig. 4-5c is understood.

Summation with the Operational Amplifier. The operation of algebraic summation of several variables is accomplished by using resistors as the input and feedback impedances of an operational amplifier. In the process of summation, using this element, each of the summed variables is multiplied by a constant equal to the respective ratio of feedback to input impedance in accordance with Eq. 4-6.

EXAMPLE 4-3. Consider the case where there is a single variable and the input and feedback impedances are resistors R_i and R_f, respectively, as shown in Fig. 4-6a.

Solution: If $R_f = 1$ megohm or 10^6 ohms and $R_i = 0.5$ megohm, or 500,000 ohms, then Fig. 4-6a provides the operation

$$e_o = -2e_i \qquad \textit{Ans.}$$

The general case of the summation of n input variables is obtained by placing several inputs into the amplifier, each through its own input impedance as shown in Fig. 4-6b. The general relationship for algebraic sum-

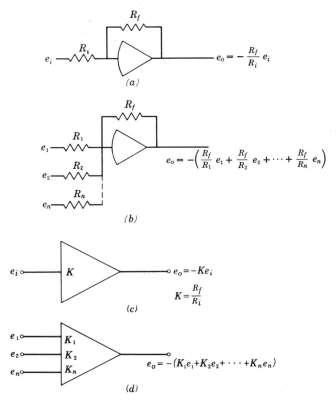

FIG. 4-6. Algebraic summation with the operational amplifier.

mation is then

$$e_o = -\left(\frac{R_f}{R_1} e_1 + \frac{R_f}{R_2} e_2 + \cdots + \frac{R_f}{R_n} e_n\right) \qquad (4\text{-}7)$$

There is an alternate form of the computer diagram which is particularly useful and often employed. Also, some computers have resistor gains built-in and the programmer must choose a particular value, say, 0.1, 1, or 10. The simpler alternate symbols for the summer are shown in Figs. 4-6c and d.

EXAMPLE 4-4. Consider the case where the following resistance values are chosen for the input and feedback resistors:

$$R_1 = 0.2 \text{ megohm} \qquad R_3 = 0.1 \text{ megohm}$$
$$R_2 = 1.0 \text{ megohm} \qquad R_f = 0.5 \text{ megohm}$$

Solution:

$$e_o = -(2.5e_1 + 0.5e_2 + 5e_3) \qquad\qquad Ans.$$

Integration with the Operational Amplifier. To build an integration device using the operational amplifier, it is necessary to choose a capacitor

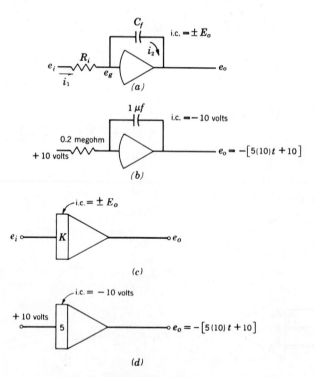

FIG. 4-7. Integration with the operational amplifier.

as the feedback impedance and a resistor for the input impedance, as illustrated in Fig. 4-7a. The simplified form of the integrator symbol is shown in Fig. 4-7c.

The basic transfer relationship for this case may be obtained by a derivation similar to Eqs. 4-1 through 4-5. As before, $i_1 = i_2$. Also,

$$i_1 = \frac{e_i - e_g}{R_i} \tag{4-8}$$

and

$$i_2 = C_f \frac{d(e_g - e_o)}{dt} \tag{4-9}$$

Therefore,

$$\frac{e_i - e_g}{R_i} = C_f \frac{d(e_g - e_o)}{dt} \tag{4-10}$$

Remembering that $e_o = -Ae_g$,

$$\frac{e_i}{R_i} + \frac{e_o/A}{R_i} = C_f \frac{d\left(\dfrac{-e_o}{A} - e_o\right)}{dt} \tag{4-11}$$

Neglecting the terms e_o/A, which will be exceedingly small, results in

$$\frac{e_i}{R_i} = -C_f \frac{de_o}{dt} \tag{4-12}$$

Integrating both sides with respect to t and rearranging yields

$$e_o = -\frac{1}{R_i C_f} \int_0^t e_i \, dt \tag{4-13}$$

which is the expression for the integrating element of the computer. In the course of using this element, it may be necessary to place an initial condition on the output variable (e_o has some initial value E_o at time zero). The use of initial conditions on the output of an integrator was illustrated in Examples 4-1 and 4-2. A convenient method of providing this initial condition is to place an initial charge on the feedback capacitor. Thus, when the zero time reference is taken, the output of the integrator is equal to the initial condition E_o.[1] The complete expression for the integrator is thus obtained by adding the constant E_o to Eq. 4-13 with the proper sign necessary to take into account the sign change in the amplifier.

$$e_o = -\frac{1}{R_i C_f} \int_0^t e_i \, dt \pm E_o \tag{4-14}$$

---

[1] In the case of an integration element, it is necessary to have a special switching device to disconnect the input of the integrator and place the initial condition on the capacitor. There are many mechanical and electronic devices for this switching process, thus, for simplicity in programming, one need only specify an initial condition (i.c.) at the output of each integrator. The actual nature of the necessary switching at time zero may be left for the programmer to study in the operation manual of the specific equipment employed.

EXAMPLE 4-5. Consider the integrator in Fig. 4-7b with a constant input of 10 volts. The input resistance is 0.2 megohm and the feedback capacitor is 1.0 μf. What is the output voltage in one second if the initial value of e_o is −10 volts? Fig. 4-7d illustrates the simplified computer diagram.

Solution: The output voltage, using Eq. 4-14, is

$$e_o = -\frac{1}{0.2} \int_0^1 10\,dt - 10 = -50 - 10 = -60 \text{ volts} \qquad Ans.$$

In the general case, where there is more than a single input into the integrator, the expression for the integration process is

$$e_o = -\left(\frac{1}{R_1 C_f} \int_0^t e_1\,dt + \cdots + \frac{1}{R_n C_f} \int_0^t e_n\,dt \pm E_o\right) \qquad (4\text{-}15)$$

Multiplication by a Constant. The analog computer has a simple method of multiplying by a quantity less than or equal to unity by the use of a "voltage divider," or potentiometer, as illustrated in Fig. 4-8a. The expression relating the output voltage to the input voltage is simply

$$e_o = \frac{R_p}{R_t} e_i$$

or where $R_p/R_t = a$, this expression becomes

$$e_o = ae_i \qquad (4\text{-}16)$$

where $a \leq 1$. The expression is commonly denoted by Fig. 4-8b.[2]

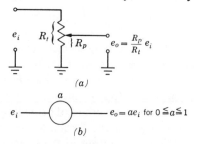

(a)

(b)

FIG. 4-8. The potentiometer.

Multiplication by a constant greater than unity may be accomplished by using a combination of a potentiometer and an impedance ratio at the amplifier. For example, the constant 2.312 may be obtained by a potentiometer setting of 0.2312 and an impedance ratio of 10, that is, 2.312 = 10(0.2312).

---

[2] It should be noted that Eq. 4-16 assumes that there is no loading effect of the next computing element at the output e_o, that is, no additional element with finite impedance from the output to ground. Unfortunately, this is not always true. However, most commercial computers provide a circuit for the accurate setting of potentiometers after a problem is completely programmed. Some more elaborate computing installations provide automatic setting of the many potentiometers by a rapid and accurate programming scheme.

Multiplication of Two Variables. The multiplication of two variables is a simple matter for the computer programmer. Common commercial function multipliers operate on a variety of basic techniques, each of which requires a special understanding of the principles involved. For a full treatment of the many multiplication methods the reader is referred to one of the references specifically concerned with analog computing [17, 18, 27]. If a particular commercial computer is available, the instruction manual will generally illuminate the operating procedure and the multiplier's limitations. For this introduction to analog computer techniques it is sufficient to recognize that the product of two functions may be represented by the symbol of Fig. 4-1*d*.

Generation of a Function of a Variable. In the solution of differential equations and in the direct simulation of physical devices (including automatic controls) it is often necessary to generate special functions. With the use of the operational amplifier and special function generation equipment it is possible to produce a wide variety of mathematical functions on the computer. The function generation process is represented by the block of Fig. 4-1*e*.

A common example of a nonlinear function which may be generated is the limited function. A good example is a hydraulic valve that has some limiting flow regardless of the spool position. The valve flow rate $Q(x)$ versus spool position x is represented by Fig. 4-9*a*. The flow is seen to increase with spool position until a maximum is reached. The flow is then limited and any further movement will result only in the same flow.

Another commonly used function is that of dead zone, as shown in

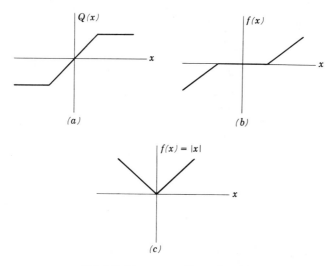

FIG. 4-9. Common nonlinear functions.

Fig. 4-9*b*. A computer analysis for a pair of gears with backlash would require the generation of such a function.

Another common function which may easily be generated is the absolute value of a function, as illustrated in Fig. 4-9*c*. All three of the examples of Fig. 4-9 are simple functions which may be generated using analog equipment. More elaborate functions such as powers and roots, trigonometric functions, exponential, and logarithmic functions may also be obtained. The reader is once again referred to the many complete texts on analog computing for a thorough treatment of the various methods and techniques.

4-4. Computer Solution Diagrams

The concept of the solution block diagrams of Sec. 4-2, using the operational building blocks, may now be expanded to build the wiring diagrams or computer solution diagrams with symbols for the actual computing elements discussed in Sec. 4-3. This development is best carried out by a series of simple examples.

EXAMPLE 4-6. The problem is to generate the function $f(t) = t^2$ for the interval of time 0 to 10 sec.[3]

Solution: The block diagram of the solution is given in Fig. 4-10*a*. Figure 4-10*b* is the actual computer solution diagram. The simplified version of the

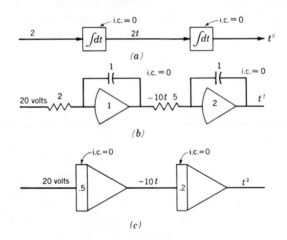

FIG. 4-10. Computer diagram for Example 4-6.

computer diagram is shown in Fig. 4-10*c*. The input is 20 volts d-c and the out-

---

[3]In this text an operational amplifier with a linear output range of ± 100 volts will be assumed. All resistors are given in megohms, and all capacitors are given in microfarads.

put of the first integrator (identified as amplifier 1 by the number within the amplifier figure) is given by use of Eq. 4-14 as

$$e_o = -\frac{1}{2} \int_0^t 20\,dt + 0 = -10t$$

In the same manner the output of amplifier 2 is given by

$$e_o = -\frac{1}{5} \int_0^t -10t\,dt + 0 = t^2$$

One should not fail to recognize that there is a sign change associated with each operational amplifier. Fortunately, the output of amplifier 2 will just reach, and not exceed, 100 volts in the time interval desired. If, however, one desired to generate t^2 for 20 sec, the problem must be scaled so that the ± 100 volt limit will not be exceeded. The process of scaling will be developed further in the following section.

Ans.

EXAMPLE 4-7. Let us reconsider the problem of generation of an exponential function, specifically, $f(t) = 50e^{-2t}$.

Solution: The solution block diagram was given in Fig. 4-4. The analog computer diagram is obtained by differentiating $f(t)$ to yield

$$\frac{df(t)}{dt} = -100e^{-2t} = -2f(t)$$

or

$$-\frac{1}{2}\frac{df(t)}{dt} = f(t)$$

The analog computer diagram is given in Fig. 4-11.   *Ans.*

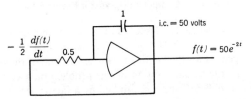

FIG. 4-11. Computer diagram for Example 4-7.

EXAMPLE 4-8. Draw the analog computer solution of the equation of Example 4-1. The equation is

$$\frac{d^2x}{dt^2} + 5\frac{dx}{dt} + 8x = 10t$$

with the initial conditions that

$$x\big|_{t=0} = 5 \quad \text{and} \quad \frac{dx}{dt}\bigg|_{t=0} = 1$$

Solution: For the actual solution of this problem it is necessary to specify a time range for the solution. (Why?) Consider the range 0 to 1 sec. Recall that the first step in the solution is the rearrangement of the equation to express the highest derivative in terms of the other quantities. Thus,

$$\frac{d^2x}{dt^2} = -\left(5\frac{dx}{dt} + 8x - 10t\right)$$

A computer diagram for this equation is illustrated in Fig. 4-12. Notice that the negative sign in front of the parentheses on the right-hand side is accounted for

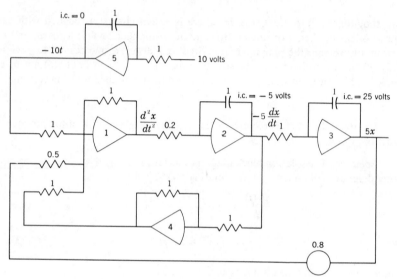

FIG. 4-12. Computer diagram for Example 4-8.

in the first summer (amplifier 1). Note also that the output of amplifier 3 is given as $5x$ rather than just x. If x is the quantity to be plotted versus time, then it is only necessary to account for the factor of 5 on the x scale of the data recorded. It may be observed, by the actual solution of this problem on a computer, that even with the output of amplifiers 2 and 3 multiplied by a factor of 5, none of the amplifiers exceed ± 100 volts in the course of the solution. This problem actually introduces a simple form of magnitude scaling which is discussed further in Sec. 4-5. *Ans.*

4-5. Formal Computer Scaling

Besides the formulation of correct computer diagrams, proper scaling is the most important aspect of successful analog computing. In particular, it is necessary to relate the variables of the problem to voltages and times in the computer so that the solutions are slow or fast enough to be accurately recorded and so that the voltages at any amplifier do not exceed the maximum linear range at any time. In addition, the voltages should not be too small as the solution is more susceptible to noise and error.

Magnitude Scaling. The first step of the two fundamental aspects of scaling is magnitude scaling. It is simple to understand that if a problem calls for a displacement of 50,000 ft for the initial condition, 1 volt cannot equal 1 ft. On the other hand, if it could be ascertained that 50,000 ft is the greatest height that will be encountered, then a natural choice would be to equate 50,000 ft to the maximum linear range of the amplifier, say 100 volts. Then, one must consider the scaling factor of 500 ft/volt relating the computer variable to the actual variable. Example 4-8 presented a form of magnitude scaling wherein the variables were multiplied by some convenient factor to increase the voltage range of each amplifier. It could have been assumed that 1 volt was equivalent to 1 unit of x. However, working in a lower voltage range on the computer is less accurate and the solution is not as good.

In computer scaling, as in many other types of engineering problems, it is extremely useful to keep careful account of units. The analog computer operates in the units of volts, regardless of the problem. It is the duty of the programmer to translate the results to the proper physical units with the use of the proper scale factors. The technique of magnitude scaling is best understood with the aid of an illustrative example.

EXAMPLE 4-9. Consider the problem of generating a sine function, specifically, $x = 40 \sin 2t$. The derivatives are then

$$\dot{x} = 80 \cos 2t$$
$$\ddot{x} = -160 \sin 2t = -4x$$

Solution: If one sets up a solution for the equation $\ddot{x} = -4x$ and equates 1 volt to 1 unit of x, then $x = 40 \sin 2t$ (volts), $\ddot{x} = -160 \sin 2t$ (volts), and it becomes obvious that the amplifier with output $\ddot{x}$ will exceed the ±100-volt limit. Specifically, the maximum value of this term is ±160 volts. In an effort to solve this problem on the computer, let the subscript c represent the computer variable and let X represent the scale factor. Then,

$$x_c = X x \quad \text{volts}$$
$$\dot{x}_c = \dot{X} \dot{x} \quad \text{volts}$$
$$\ddot{x}_c = \ddot{X} \ddot{x} \quad \text{volts}$$

The scale factors therefore have the following units:

$$X = \text{volts/in.}$$
$$\dot{X} = \text{volts/(in./sec)}$$
$$\ddot{X} = \text{volts/(in./sec}^2)$$

Substituting, to express the original function and its derivatives in terms of the computer variables,

$$x_c = X \, 40 \sin 2t$$
$$\dot{x}_c = \dot{X} \, 80 \cos 2t$$
$$\ddot{x}_c = -\ddot{X} \, 160 \sin 2t$$

Suppose that the scale factors X, $\dot{X}$, and $\ddot{X}$ are chosen to be equal. A convenient value here is $\frac{1}{4}$. Thus, the scaled computer terms are

$$x_c = 10 \sin 2t$$
$$\dot{x}_c = 20 \cos 2t$$
$$\ddot{x}_c = -40 \sin 2t$$

where the initial conditions will be $x_c = 0$ at $t = 0$ and $\dot{x}_c = 20$ volts at $t = 0$.

The complete computer diagram for this case is given in Fig. 4-13a. This solution would be entirely satisfactory, except that one may note that amplifier 1 oscillates between ± 40 volts, amplifier 2 between ± 20 volts, and amplifier 3 between

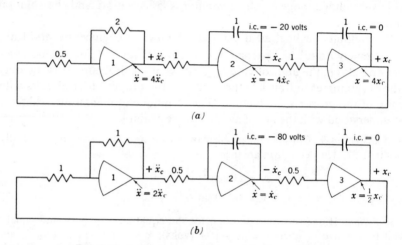

(a)

(b)

FIG. 4-13. Analog computer diagrams for Example 4-9.

± 10 volts. It would be far more desirable to have amplifiers 2 and 3 swing over a larger voltage range. The solution is more accurate when all amplifiers are using a large portion of their operating range. To this end, consider choosing the scale factors unequal. For example,

$$X = 2$$
$$\dot{X} = 1$$
$$\ddot{X} = \frac{1}{2}$$

There is no restriction that the scale factors must be chosen equal.

Substituting these scale factors into the expressions for the computer variables yields

$$x_c = 80 \sin 2t$$
$$\dot{x}_c = 80 \cos 2t$$
$$\ddot{x}_c = -80 \sin 2t$$

with the initial conditions that $x_c = 0$ and $\dot{x}_c = 80$ volts at time zero.

At first, the latter equations may appear contrary to the derivatives of $\sin 2t$, but it must be realized that the equations with the computer subscript c are "scaled" equations. On the computer solution, one must satisfy the relation $\ddot{x} = -4x$ and the missing factor of 2 in the computer equations must be accounted for at each integrator on the computer diagram. The final computer diagram is given in Fig. 4-13b. The mathematical relationship that must be satisfied at amplifier 2 in Fig. 4-13b is[4]

$$\dot{x}_c = \frac{1}{R_i C_f} \int_0^t \ddot{x}_c \, dt \pm \dot{x}_{co}$$

where $\dot{x}_{co}$ is the initial condition. Substituting to obtain the original problem variables yields

$$\dot{X}\dot{x} = \frac{1}{R_i C_f} \int_0^t \ddot{X}\ddot{x} \, dt + \dot{X}\dot{x}_o$$

or

$$\dot{x} = \frac{1}{R_i C_f} \frac{\ddot{X}}{\dot{X}} \int_0^t \ddot{x} \, dt + \dot{x}_o$$

For this expression to be valid the coefficient of the integral must be unity. Then, with the chosen scale factors,

$$\frac{1}{R_i C_f} = 2$$

A good choice of computer elements is then a 0.5 megohm resistor and a 1 μf capacitor. In a parallel fashion, one may verify the choice of the elements for amplifier 3.

The solution of Fig. 4-13b is far better than the previous solution, as all amplifiers are operating between ± 80 volts. Note that if one follows around the loop, the total gain of all the amplifiers in either Fig. 4-13a or Fig. 4-13b is -4. This satisfies the relationship $\ddot{x} = -4x$, and each output is proportional to the actual value by the chosen scale factor.

One general principle which may be derived from this example is that it is advantageous to distribute the total gain between several of the amplifiers rather than have all of the gain on a single amplifier. In this case, the total gain of 4 is on the first amplifier in Fig. 4-13a. In the latter solution of the problem the total gain is distributed as a gain of 2 on amplifier 2 and a gain of 2 on amplifier 3. *Ans.*

Another method of magnitude scaling is based on the approximation of the characteristic (or undamped oscillating) frequency of the equation. This method is extremely effective and after some exercise, it allows one to scale rapidly.

EXAMPLE 4-10. Develop the solution to a differential equation. The equation to be solved is that of a simple spring-mass-damper system

$$\frac{d^2x}{dt^2} + 3.725 \frac{dx}{dt} + 7.312x = 0$$

---

[4] Note that the sign change introduced by an integrator, as given in Eq. 4-13, is accounted for on the computer diagram—mathematically there is no sign reversal.

with the initial conditions

$$x\big|_{t=0} = 25 \text{ in.} \quad \text{and} \quad \dot{x}\big|_{t=0} = 0 \text{ in./sec}$$

Solution: The undamped resonant frequency of the system is

$$\omega_n = (7.312)^{1/2}$$

and for the purpose of scaling we may assume $\omega = 3$, approximately. Assuming a sinusoidal solution (we wish only to make a rough guess of the magnitudes of the variables) gives

$$x = 25 \cos 3t \text{ in.}$$
$$\dot{x} = -75 \sin 3t \text{ in./sec}$$
$$\ddot{x} = -225 \cos 3t \text{ in./sec}^2$$

The method of Example 4-9 may now be applied to choose scale factors for x and its derivatives. In an effort to choose convenient numbers so that none of the computing variables exceeds ± 100 volts, consider the following scale factors:

$$X = 2 \text{ volts/in.}$$
$$\dot{X} = 1 \text{ volt/(in./sec)}$$
$$\ddot{X} = \tfrac{1}{5} \text{ volt/(in./sec}^2)$$

Rearranging the original equation in terms of the highest derivative and substituting the computer variables with the subscript c yields

$$\frac{1}{\ddot{X}} \frac{d^2 x_c}{dt^2} = -\left[\frac{10(0.3725)}{\dot{X}} \frac{dx_c}{dt} + \frac{10(0.7312)}{X} x_c \right]$$

Substituting the chosen values of the scale factors gives the final equation to be solved as

$$\frac{d^2 x_c}{dt^2} = -\left[2(0.3725) \frac{dx_c}{dt} + (0.7312)x_c \right]$$

The computer diagram for this solution is given in Fig. 4-14.

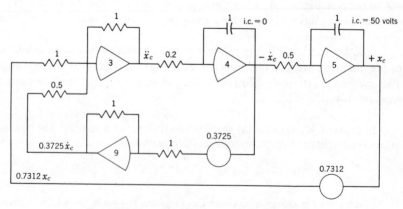

FIG. 4-14. Computer diagram for Example 4-10.

The conditions of the latter equation are satisfied at the summing element which is amplifier 3. With the introduction of the scale factors, however, the $1/RC$ gains of the integrators must be adjusted to give the proper integrals. The mathematical relationship for the integrator using amplifier 5 is

$$x = \int_0^t \dot{x}\, dt + x_o$$

Substituting the scaled computer variables yields

$$\frac{x_c}{2} = \int_0^t \frac{\dot{x}_c}{1}\, dt + \frac{50}{2}$$

or

$$x_c = 2 \int_0^t \dot{x}_c\, dt + 50$$

The latter equation is expressed in terms of the computer variable x_c. The coefficient of the integral is actually the $1/RC$ gain of the integrator. Thus, $1/R_5C_5 = 2$, and the initial condition on amplifier 5 is 50 volts. A good choice for R_5 is then 0.5 megohm, with C_5 being a 1 μf capacitor. One should note that in Example 4-9 the choice of $1/RC$ was based on looking at the integral expressed in terms of the actual variable, whereas in this example, the choice was made by looking at the integral in terms of the computer variable. Either approach leads to the correct conclusion and may be used with equal success.

The gain on integrator 4 may be obtained by considering the integral

$$\dot{x} = \int_0^t \ddot{x}\, dt + \dot{x}_o$$

or scaling

$$\dot{x}_c = 5 \int_0^t \ddot{x}_c\, dt + 0$$

This indicates that $1/R_4C_4 = 5$, and a good choice for these elements is $C_4 = 1\ \mu$f and $R_4 = 0.2$ megohm. The necessary scaling of the problem is complete.   *Ans.*

Time Scaling. Time scaling is the second half of the total problem of computer scaling. The purpose here is to relate the independent variable of the problem to the independent variable of the machine when other than a unity relationship is necessary. It was pointed out previously that the speed at which the machine may operate is often limited by the response of the electronic equipment or electromechanical recording equipment. We will adopt what is, in effect, a change of the independent variable.

The simplest form of time scaling is to speed or slow all of the time dependent components of the computer by the same factor. The basic integrator relationship expressed by Eq. 4-14 may be modified by recognizing that $1/R_iC_f$ is a constant and may be moved within the integral sign. Thus, Eq. 4-14 becomes

$$e_o(t) = -\int_0^t e_i(t)\, \frac{dt}{R_iC_f} \pm E_o \qquad (4\text{-}17)$$

Further, one may justify that

$$\frac{dt}{R_i C_f} = d\left(\frac{t}{R_i C_f}\right)$$

The resulting conclusion is that $1/R_i C_f$ may be used as a device to change the independent variable t. For example, let

$$t = T t_c \qquad (4\text{-}18)$$

where t is the real time in seconds, t_c is the computer time in seconds, and T is a constant, specifically,

$$T = R_i C_f$$

Substituting the change of variable of Eq. 4-18 into Eq. 4-17 yields

$$e_o(t_c) = -\int_0^{Tt_c} e_i(t_c)\, dt_c \pm E_o \qquad (4\text{-}19)$$

Let us consider an example to illustrate the change of the independent variable.

EXAMPLE 4-11. Consider the generation of $f(t) = t^2$, as illustrated in Example 4-6. Suppose that both capacitors of Fig. 4-10b are changed to 0.1 μf.

Solution: In this case we may consider the input resistors of each integrator as gain factors in front of the integral sign and take the factor of change of the capacitors ($T = 0.1$) within the integral. Then,

$$t = 0.1\, t_c$$

This expression states that 1 sec of real time is equivalent to 10 computer seconds. Thus, we must modify the length of time for the solution stated in Example 4-6. Rather than taking the solution from 0 to 10 sec. in real time we must consider the solution over 1 sec of real time. Clearly, this is equivalent to 10 sec on the computer. The solution has thus been speeded by a factor of 10.

Consider, further, the possibility of replacing all capacitors of Fig. 4-10 by 10 μf capacitors. All of the time controlling integrators are then related to real time by the relation

$$t = 10 t_c$$

Now, 1 sec of real time is equivalent to only 1/10 of a computer second. Thus events on the computer are occurring in ten times the length of real time seconds, or the solution has been slowed by a factor of 10.   *Ans.*

The concepts of time and magnitude scaling may be combined in a method of formal scaling that may be used to scale a wide variety of problems.

Consider the general third-order differential equation with constant coefficients,

$$a_0 \frac{d^3 x}{dt^3} + a_1 \frac{d^2 x}{dt^2} + a_2 \frac{dx}{dt} + a_3 x = f(t) \qquad (4\text{-}20)$$

One may choose for the change of the independent variable,

$$t = Tt_c \qquad (4\text{-}21)$$

and for the dependent variable,

$$x = Xx_c \qquad (4\text{-}22)$$

Equations 4-21 and 4-22 may be substituted in Eq. 4-20 to give a new equation expressed in volts (the units of the dependent variable) and computer seconds (the units of the independent variable).

$$\frac{d^3(Xx_c)}{d(Tt_c)^3} + \frac{a_1 d^2(Xx_c)}{a_0 d(Tt_c)^2} + \frac{a_2 d(Xx_c)}{a_0 d(Tt_c)} + \frac{a_3}{a_0}(Xx_c) = \frac{1}{a_0}f(Tt_c)$$

Since both T and X are constant scale factors, they may be brought out in front of each term to give

$$\frac{X}{T^3}\frac{d^3 x_c}{dt_c^3} + \frac{a_1}{a_0}\frac{X}{T^2}\frac{d^2 x_c}{dt_c^2} + \frac{a_2}{a_0}\frac{X}{T}\frac{dx_c}{dt_c} + \frac{a_3}{a_0}Xx_c = \frac{1}{a_0}f(Tt_c)$$

Dividing by X and multiplying by T^3 yields the scaled equation

$$\frac{d^3 x_c}{dt_c^3} + \frac{a_1}{a_0}T\frac{d^2 x_c}{dt_c^2} + \frac{a_2}{a_0}T^2\frac{dx_c}{dt_c} + \frac{a_3}{a_0}T^3 x_c = \frac{T^3}{a_0 X}f(Tt_c) \qquad (4\text{-}23)$$

If one then chooses a value for T such that the coefficients of all the derivatives are close to unity, or at least less than some maximum gain K, then the solution will generally function properly on the computer. In addition, it is necessary to choose a value of X such that a convenient magnitude of the forcing function is obtained that will drive all of the amplifiers over a large portion of their linear range, yet not cause any amplifier to be overloaded. In general, it is desirable to have K less than 10 and as close to unity as possible. An aid in choosing the value of T then involves a value of T near to and smaller than the smallest of the following terms:

$$K\frac{a_0}{a_1} \qquad \left(K\frac{a_0}{a_2}\right)^{1/2} \qquad \left(K\frac{a_0}{a_3}\right)^{1/3}$$

The remaining step is to choose a value of X such that the term $T^3 f(Tt_c)/a_0 X$ provides a forcing function of convenient magnitude.

EXAMPLE 4-12. The problem is to scale and draw a computer diagram for the solution of the differential equation

$$1000\frac{d^3 x}{dt^3} + 100\frac{d^2 x}{dt^2} + 20\frac{dx}{dt} + x = 500 \sin 0.1t$$

with the initial conditions

$$x|_{t=0} = 100 \qquad \frac{dx}{dt}\bigg|_{t=0} = 5 \qquad \frac{d^2 x}{dt^2}\bigg|_{t=0} = 0$$

Solution:

$$\frac{d^3x_c}{dt_c^3} + 0.1\,T\,\frac{d^2x_c}{dt_c^2} + 0.02\,T^2\,\frac{dx_c}{dt_c} + 0.001\,T^3x_c = \frac{0.5\,T^3\,\sin(0.1\,Tt_c)}{X}$$

If K is chosen as 2, then

$$K\,\frac{a_0}{a_1} = 20 \qquad \left(K\,\frac{a_0}{a_2}\right)^{1/2} = 10 \qquad \left(K\,\frac{a_0}{a_3}\right)^{1/3} > 10$$

Therefore, let $T = 10$. The equation then becomes

$$\frac{d^3x_c}{dt_c^3} + \frac{d^2x_c}{dt_c^2} + 2\,\frac{dx_c}{dt_c} + x_c = \frac{500}{X}\,\sin t_c$$

Thus, let $X = 10$ units/volt. The final scaled equation for computer solution is

$$\frac{d^3x_c}{dt_c^3} = -\left(\frac{d^2x_c}{dt_c^2} + 2\,\frac{dx_c}{dt_c} + x_c - 50\,\sin t_c\right)$$

The final computer diagram is given in Fig. 4-15. The initial conditions on the com-

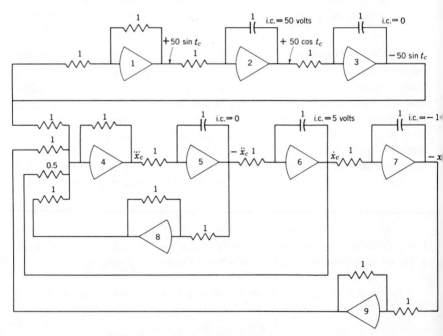

Fig. 4-15. Computer diagram for solution of Example 4-12.

puter are determined by applying the scale factors. Thus,

$$x = Xx_c = 100 \quad \text{at } t = 0$$

or

$$x_c = 10 \text{ volts} \quad \text{at } t = 0$$

The first derivative is

$$\frac{dx}{dt}\bigg|_{t=0} = \frac{X}{T}\frac{dx_c}{dt_c}\bigg|_{t=0} = 5$$

or

$$\frac{dx_c}{dt_c}\bigg|_{t=0} = 5 \text{ volts}$$

and the initial condition on the second derivative is zero.   *Ans.*

4-6. Closure

The analog computer is an electronic tool designed to solve differential equations, and it is particularly suited for the analysis of dynamic systems in all branches of engineering problems. Through the use of integration, summation, multiplication, and function generation equipment, it is possible to solve a wide variety of problems. Although the problems in this chapter are linear problems, the analog computer is well suited for solving nonlinear problems. In fact, it is occasionally found that the analog computer is the only practical means of obtaining a solution to some nonlinear differential equations.

There is, indeed, considerable art involved in good analog computing. No amount of formalism can replace a thorough understanding of the equipment involved, a comprehensive knowledge of the physical behavior of the system being studied, good judgment in the choice of scale factors, and a certain amount of plain cleverness in the programming. One does not undertake a difficult calculation the first time. Rather, the best approach is to digest the fundamentals through the working of some simple examples, perhaps with the instruction manual in one hand. Likewise, the analog computer requires the digesting of its operation procedures through the solution of many illustrative examples, the answers of which are available. Experience is, without question, one of the most useful teachers in analog computing. With some practice, most engineers should be fluent in obtaining reliable and accurate solutions to difficult problems in simulation and analysis. The analog computer is an engineering tool particularly suited for this purpose.

Problems

4-1. Draw the operational block diagrams for the solution of the following equations or sets of equations. (Do not overlook the initial conditions.)

a) Generation of functions of the independent variable t.

1. $f(t) = Kt^4$
2. $f(t) = e^{10t}$
3. $f(t) = 10 \sin 3t$

b)

$$\frac{d^5\theta}{dt^5} + 2\frac{d^4\theta}{dt^4} + 5\frac{d^3\theta}{dt^3} + 2\frac{d^2\theta}{dt^2} + 10\theta = 100$$

$$\left.\frac{d^4\theta}{dt^4}\right|_{t=0} = 0 \quad \left.\frac{d^3\theta}{dt^3}\right|_{t=0} = 10 \quad \left.\frac{d^2\theta}{dt^2}\right|_{t=0} = 1 \quad \left.\frac{d\theta}{dt}\right|_{t=0} = 5 \quad \theta|_{t=0} = 0$$

c)
$$\frac{d^2x}{dt^2} + 8\frac{dx}{dt} + 100(x - y) = t$$

$$\frac{dy}{dt} + xy = 0$$

$$x|_{t=0} = 0 \quad \left.\frac{dx}{dt}\right|_{t=0} = 0 \quad y|_{t=0} = 10$$

d)
$$\frac{d^2x}{dt^2} + K\frac{dx}{dt} + \left[x - \left(\frac{dx}{dt}\right)^2\right]x = 0$$

$$x|_{t=0} = X \quad \left.\frac{dx}{dt}\right|_{t=0} = 0$$

NOTE: It is suggested that the following computer components be used in the remaining problems:

Resistors, in megohms: 1, 2, 5, 10, 0.5, 0.2, 0.1

Capacitors, in microfarads: 0.01, 0.1, 1.0

4-2. Write equations for the accompanying computer configurations.

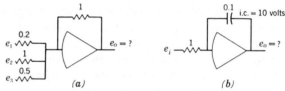

PROB. 4-2

4-3. Make a computer diagram to generate the following functions:

a) $f(t) = 10t$ for $t = 0$ to $t = 10\,\text{sec}$

b) $f(t) = 5e^{-2t}$

c) $f(t) = 40 \sin 3t$

4-4. Make a computer diagram with proper scaling to solve

$$\frac{d^2x}{dt^2} + 21.4\frac{dx}{dt} + 400x = 0$$

$$x|_{t=0} = 100 \quad \left.\frac{dx}{dt}\right|_{t=0} = 0$$

4-5. Make a computer diagram with proper scaling to solve

$$\frac{d^5x}{dt^5} + 4\frac{d^4x}{dt^4} + 20\frac{d^3x}{dt^3} + 5\frac{dx}{dt} + 40x = 0$$

$$x|_{t=0} = 100$$

All other initial conditions are zero.

4-6. Make a computer diagram for the solution of

$$3\frac{d^2y}{dt^2} + \frac{dy}{dt} + 9y = 0$$

$$y\big|_{t=0} = 0 \qquad \frac{dy}{dt}\bigg|_{t=0} = 10$$

4-7. Make a computer diagram for the solution of

$$\ddot{\theta} + 3\dot{\theta} + 67\theta = 30 \sin \omega t \qquad \omega = 5 \text{ radians/sec}$$

All initial conditions are zero.

4-8. The two equations which follow are to be solved simultaneously on the analog computer. Make a complete computer diagram.

$$\ddot{x}_1 + 4\dot{x}_1 + 10x_1 = +5x_2 + 10$$

$$20\ddot{x}_2 + 5x_2 - 5x_1 = 0$$

$$x_1\big|_{t=0} = 5 \qquad \dot{x}_1\big|_{t=0} = 0 \qquad x_2\big|_{t=0} = 3 \qquad \dot{x}_2\big|_{t=0} = 0$$

Chapter 5

System Response

5-1. Introduction

The response of dynamic systems to various inputs is of extreme importance in control work. For example, a control system constantly receives inputs of one sort or another, and its response to these inputs must be known in order to achieve a satisfactory system. The purpose of the present chapter is to provide an understanding of dynamic behavior by analyzing a number of systems including two simple feedback control systems. It should be noted that, although the viewpoint presented here is that of control, the principles apply equally well to any and all dynamic systems which are mathematically similar.

Prediction of the performance of dynamic systems requires the ability to write and solve differential equations. This chapter integrates the material of earlier chapters by dealing, in particular, with the response of first- and second-order systems (systems whose behavior is defined by first- and second-order differential equations). This does not imply that most control systems are simple enough to be represented by lower-order equations; such is not the case. However, first- and second-order systems do offer a convenient starting point and, in addition, the behavior of many of the elements which make up control systems can be defined by these lower-order equations.

Three very common inputs used in control work are those which yield forcing functions which are constants K, linear functions of time Kt, and sinusoidals $K \sin \omega t$. These inputs are plotted against time in Fig. 5-1. In all cases, the input is assumed to begin at $t = 0$. In Fig. 5-1a, the input is zero until $t = 0$ at which time it takes on a value K which remains constant for $t > 0$. Control engineers call this input a *step input* because of the appearance of the plot.[1] In Fig. 5-1b, the input increases with time in a linear manner; the appearance of the plot is that of a ramp, and this input is frequently called a *ramp input*. The sinusoidal input, shown in Fig. 5-1c, is generally called by that name—a *sinusoidal input*.

---

[1] The input shown in Fig. 5-1a is sometimes expressed as $Ku(t)$ where $u(t)$ is a notation indicating a step input. Actually $u(t)$ represents a unit step, that is, a step input with unity magnitude. Thus $Ku(t)$ indicates a step input with a magnitude of K.

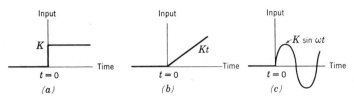

FIG. 5-1. Common system inputs.

In this chapter, various systems will be analyzed with step and/or ramp inputs. Treatment of responses to sinusoidal inputs will be deferred. The reason is that, although the approach is straightforward (solving a differential equation with a sinusoidal forcing function by the method already introduced), the work involved is tedious. A superior technique will be developed later.

5-2. First-Order System Response

A mechanical and an electrical system will be used to illustrate the response of first-order systems to step inputs. These two examples will demonstrate the similar response of two dissimilar physical systems; the idea will be extended to show that the response of all first-order systems is the same.

Mechanical System. The system of Fig. 5-2a is composed of a spring and dashpot; there is no mass. The displacement x is the input while the displacement y is the output, or response. At an equilibrium position (spring relaxed), $x = y = 0$. The differential equation is obtained from Newton's law:

$$M \frac{d^2y}{dt^2} = (0) \frac{d^2y}{dt^2} = \Sigma F = -c \frac{dy}{dt} + k(x - y)$$

or

$$c \frac{dy}{dt} + ky = kx(t)$$

The input is written as $x(t)$ in the latter equation to recall that the equation

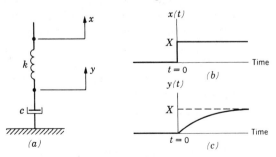

FIG. 5-2. Mechanical system subjected to step input X.

is general in nature and will yield a solution for any time-varying input. For a step input, $x(t) = X$, as shown in Fig. 5-2b. Physically, this means that the top of the spring is suddenly raised by some constant amount X and then is held in that position. The equation to be solved for the response y is then

$$c \frac{dy}{dt} + ky = kX \qquad (5-1)$$

The complementary function y_c is obtained by solving the associated homogeneous equation

$$c \frac{dy}{dt} + ky = 0 = y(cp + k)$$

The characteristic equation and its root are

$$cp + k = 0 \qquad \text{and} \qquad p = -k/c$$

Therefore,

$$y_c = C_1 e^{pt} = C_1 e^{-kt/c}$$

The forcing function is constant and so the particular integral is assumed to be constant; $y_p = A$. The derivative is $dy_p/dt = 0$. Substituting in Eq. 5-1

$$c(0) + kA = kX$$

$$y_p = A = X$$

The complete solution is

$$y = y_c + y_p = C_1 e^{-kt/c} + X$$

From the initial condition $y = 0$ when $t = 0$,

$$0 = C_1 + X \qquad \text{and} \qquad C_1 = -X$$

Therefore,

$$y = X(1 - e^{-kt/c}) \qquad (5-2)$$

The response is illustrated in Fig. 5-2c; y starts at zero with $t = 0$ and approaches the value X for "large" values of time.

Could we have predicted the general character of the curve shown in Fig. 5-2c? The answer is yes.[2] One knows that the system will continue to respond until the spring is again relaxed; therefore y must ultimately equal X. Also, the velocity of the system (dy/dt) will be greatest when the spring force applied to the dashpot is greatest; this occurs when $t = 0$. The velocity will decrease as the system responds because the spring force applied to the dashpot decreases as y approaches X. In terms of the curve shown in Fig. 5-2c, the slope (dy/dt) is greatest at $t = 0$ and approaches zero as y approaches X. Thus one could have predicted the nature of the response curve.

[2]A common-sense analysis is presented because the authors believe that it is important to gain some intuitive feeling for system response as well as to be able to handle the analysis mathematically.

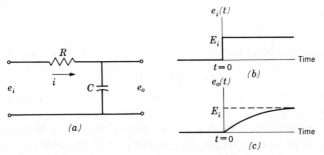

FIG. 5-3. Electrical system subjected to step input E_i.

Electrical System. The electrical network shown in Fig. 5-3a is equivalent to that in Fig. 2-13. The input voltage is e_i and the output voltage is e_o; the capacitor is initially discharged. An analysis of the circuit not only provides insight into first-order system response but also illustrates the versatility of the p-operator. The two equations required are (see Eqs. 2-20 and 2-21)

$$e_i = Ri + \frac{1}{C} \int_0^t i\, dt \quad \text{and} \quad e_o = \frac{1}{C} \int_0^t i\, dt$$

From Eq. 3-2,

$$e_i = Ri + \frac{i}{Cp} \quad \text{and} \quad i = \left(\frac{Cp}{RCp + 1}\right) e_i$$

$$e_o = \frac{i}{Cp} = \frac{1}{Cp}\left(\frac{Cp}{RCp + 1}\right) e_i = \frac{e_i}{RCp + 1}$$

By cross-multiplying and using Eq. 3-1,

$$(RCp + 1)e_o = e_i$$

$$RC\frac{de_o}{dt} + e_o = e_i(t)$$

The result is a differential equation relating the output voltage with the input voltage. Note that, in obtaining the equation, p initially operated on the current i and finally operated on the output voltage e_o, illustrating the versatility of the p-operator.

For a step input, $e_i(t) = E_i$ a constant (d-c) voltage; E_i is plotted in Fig. 5-3b. The equation to be solved is

$$RC\frac{de_o}{dt} + e_o = e_i(t) = E_i \tag{5-3}$$

The solution proceeds as before with the complementary function and the particular integral being

$$e_{oc} = C_1 e^{-t/RC} \quad \text{and} \quad e_{op} = E_i$$

Therefore,

$$e_o = C_1 e^{-t/RC} + E_i$$

From the initial condition, $e_o = 0$ at $t = 0$, the complete solution is

$$e_o = E_i(1 - e^{-t/RC}) \tag{5-4}$$

Figure 5-3c shows a plot of this expression which is the response of the system to a step input.

The mathematical response can be given a physical interpretation. The output voltage is that across the capacitor. Initially it is zero (capacitor discharged) and finally it becomes equal to the d-c input voltage (capacitor fully charged). The charging rate is dependent upon the current (dq/dt). At $t = 0$, the current i has its greatest value because the full voltage drop is across the resistor $(i = E_i/R)$. This condition corresponds to the greatest charging rate $(de_o/dt$ a maximum). As the capacitor accumulates charge, the voltage drop across it increases $[e = q(t)/C]$, resulting in a smaller drop across the resistor, a lower value of current, and a slower charging rate. Ultimately, $i = 0$ and $de_o/dt = 0$ when $e_o = E_i$.

In conclusion, note the response curves shown in Figs. 5-2c and 5-3c. Although the two systems are dissimilar, the responses are strikingly similar. The similarity is no accident; it will be established shortly that all first-order systems have the same response to a step input.

Transient and Steady-State Response. This section represents a slight digression from first-order system response but is necessary to introduce an important concept, that of terminology. The complementary functions for Eqs. 5-1 and 5-3 are

$$y_c = C_1 e^{-kt/c} \quad \text{and} \quad e_{oc} = C_1 e^{-t/RC}$$

Notice that in both cases, the expressions decrease in magnitude with increasing values of time and ultimately vanish. The expressions are transient (in the usual meaning of the word) in nature. Thus, control engineers usually call these expressions the *transient solutions*.

The particular integrals for Eqs. 5-1 and 5-3 are

$$y_p = X \quad \text{and} \quad e_{op} = E_i$$

Note that the expressions are not modified by the time, and in addition, yield the ultimate values $(t \longrightarrow \infty)$ of the complete solutions. These expressions are given the name *steady-state solutions* and represent the *steady-state* responses of the systems. In the future, we will speak of the *transient solution* instead of the *complementary function* and the *steady-state solution* rather than the *particular integral*. Of course, nothing has been changed other than terminology.

Time Constant. The idea of a *time constant* is useful in studying the response of first-order systems. Equations 5-1 and 5-3, by making the coefficient of the zeroth-derivative term unity, can be written as

$$\frac{c}{k}\frac{dy}{dt} + y = X \qquad \text{(dividing through by } k\text{)}$$

$$RC\frac{de_o}{dt} + e_o = E_i \qquad \text{(already in the right form)}$$

Let us investigate the units of the coefficients of the first-derivative terms.

$$\left[\frac{c}{k}\right] = \left[\frac{\text{lb-sec/in.}}{\text{lb/in.}}\right] = [\text{sec}]$$

$$[RC] = [\text{volt-sec/coulomb}][\text{coulomb/volt}] = [\text{sec}]$$

In both illustrations, the constant coefficients have the units of seconds; this will always be the case with first-order equations for physical systems. Because of the units (sec), the name *time constant* is given to the coefficient of the first-derivative term and is denoted by the symbol τ. Using this notation for c/k and RC, the characteristic equation for both of the previous equations becomes

$$\tau p + 1 = 0$$

from which the important relationships

$$p = -\frac{1}{\tau} \qquad \text{and} \qquad \tau = -\frac{1}{p}$$

result. The transient solutions of the previous examples are

$$y_t = C_1 e^{-t/\tau} \qquad \text{and} \qquad e_{ot} = C_1 e^{-t/\tau}$$

where the arbitrary constants and time constants pertain to the particular equation involved. Completing the solutions yields expressions equivalent to Eqs. 5-2 and 5-4:

$$y = X(1 - e^{-t/\tau}) \tag{5-5}$$

$$e_o = E_i(1 - e^{-t/\tau}) \tag{5-6}$$

Quite obviously the value of the time constant influences the response of the system. When $t = \tau$, the output will have reached

$$(1 - e^{-1}) = (1 - 0.368) = 0.632$$

of its final (steady-state) value. When $t = 4\tau$, the output will have reached

$$(1 - e^{-4}) = (1 - 0.018) = 0.982$$

of its final value. These properties of the time constant are quite useful in estimating the response time of a first-order system. When the value of time has reached that of the time constant, the response has proceeded 63.2% of the way to its steady-state value. Also, a reasonable estimate of the time required for a physical system to respond completely is four time constants (98.2% of the way mathematically).

It should be realized that a time constant is the negative reciprocal of the coefficient of t in the exponential term e^{pt} in the transient solution. Further, the response of a system would reach its final value with $t = \tau$ *if* it maintained its initial rate of response. The student should verify this point.

Dimensionless Response Equation. The time has come to show that the response of all first-order systems to a step input is the same. Equations 5-5 and 5-6 can be written as

$$\frac{y}{X} = 1 - e^{-t/\tau}$$

$$\frac{e_o}{E_i} = 1 - e^{-t/\tau}$$

Now the equations are in dimensionless form; the ratios y/X, e_o/E_i, and t/τ are all dimensionless. In general form, the response of all first-order systems to a step input can be expressed as

$$\frac{f}{K} = 1 - e^{-t/\tau} \tag{5-7}$$

where f is some function and K is its steady-state value. A plot of Eq. 5-7 is shown in Fig. 5-4. The curve provides the mathematical response for all first-order systems subjected to a step input. A similar analysis can be made for the response to a ramp input [14].

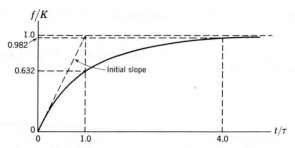

FIG. 5-4. Mathematical response of first-order systems with step inputs ($f/K = 1 - e^{-t/\tau}$).

5-3. Second-Order System Response

This section will deal with the response of second-order systems. The approach will be the same as in Sec. 5-2. A specific mechanical system will be analyzed for a step input, and the analysis will then be generalized to provide a solution for all second-order systems subjected to a step input. In addition, a second mechanical system with a ramp input will be considered.

Mechanical System with Step Input. The system shown in Fig. 5-5a will be analyzed. It is the same as that given in Fig. 2-6a in which x is the input and y the response. A step input X will be applied. Because the system has two energy storage elements (the spring and the mass), it *may* oscillate, the energy being alternately stored in the spring and in the mass. The type of response depends on the values of the physical parameters, that is, the values M, k, and c.

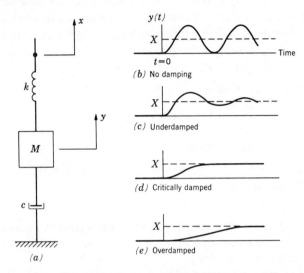

FIG. 5-5. Mechanical system subjected to step input X.

The possibility of various responses can be established from the mathematical viewpoint. The system equation is a second-order differential equation; thus, there will be two roots of the characteristic equation. Depending on the nature of these roots (real and unequal, real and equal, or complex conjugates), the form of the solution varies. Four possible responses will be investigated by permitting the damping coefficient c to change while keeping M and k constant.

No Damping. If the damping coefficient c is made zero, the system differential equation (see Eq. 2-6) is

$$M \frac{d^2 y}{dt^2} + ky = kx(t) = kX \tag{5-8}$$

The characteristic equation and its roots are

$$Mp^2 + k = 0 \quad \text{and} \quad p = \pm j \left(\frac{k}{M}\right)^{1/2} = \pm j\omega_n$$

where ω_n is the *natural frequency* (or *natural circular frequency*) expressed in radians per second. Therefore, the transient solution is

$$y_t = C_1 \sin \omega_n t + C_2 \cos \omega_n t$$

The steady-state solution y_{ss}, by the method previously described, is

$$y_{ss} = X$$

The complete solution is

$$y = C_1 \sin \omega_n t + C_2 \cos \omega_n t + X$$

With the system initially at rest, the initial conditions are now determined. At $t = 0$, $y = 0$ and $dy/dt = 0$ because the mass cannot be given an instantaneous velocity with a finite force (although there is an instantaneous acceleration). Substituting in the solution,

$$0 = C_1(0) + C_2(1) + X \qquad \text{and} \qquad C_2 = -X$$

The derivative of the solution is

$$\frac{dy}{dt} = C_1 \omega_n \cos \omega_n t - C_2 \omega_n \sin \omega_n t$$

Substituting the initial condition,

$$0 = C_1 \omega_n (1) - C_2 \omega_n (0) \qquad \text{and} \qquad C_1 = 0$$

Therefore, the solution is

$$y = X(1 - \cos \omega_n t)$$

The system response is plotted in Fig. 5-5*b*. The persistent sinusoidal oscillation at the system's natural frequency is what one would expect from an undamped spring-mass system.

Small Amount of Damping. The differential equation for the system, including damping, is

$$M \frac{d^2 y}{dt^2} + c \frac{dy}{dt} + ky = kX \tag{5-9}$$

The characteristic equation and its roots are

$$Mp^2 + cp + k = 0 \qquad \text{and} \qquad p = \frac{-c}{2M} \pm \frac{(c^2 - 4Mk)^{1/2}}{2M} \tag{5-10}$$

If the amount of damping is small, the quantity raised to the one-half power will be negative. For this case,

$$p = \frac{-c}{2M} \pm j \frac{(4Mk - c^2)^{1/2}}{2M} = -a \pm j\omega_d$$

where ω_d is the *damped natural frequency* (radians/sec) and $a = c/2M$. The steady-state solution is again $y_{ss} = X$. Therefore, the complete solution is

$$y = e^{-at}(C_1 \sin \omega_d t + C_2 \cos \omega_d t) + X$$

Without solving for the arbitrary constants, it can be seen that this solution leads to the response illustrated in Fig. 5-5*c*. The response is oscillatory, but the amplitude of oscillation decreases with time because of the exponential term, e^{-at}, provided by damping in the system. The mathematical response is entirely plausible from the physical viewpoint: the addition of damping dissipates energy from the system with the result that oscillation ultimately ceases. By definition, a system whose response is oscillatory is called an *underdamped system.*

An estimate, based on the idea of the time constant, can be made of the time required for the oscillations to cease. Recall that a time constant is the negative reciprocal of the coefficient of t in the exponential term of the transient solution. In this case,

$$\tau = -\left(-\frac{1}{a}\right) = \frac{1}{a} = \frac{2M}{c}$$

Thus the approximate time required for oscillations to decay to a negligible amount is equal to 4τ (amplitude has decayed to e^{-4} or 1.8% of its initial value) and

$$4\tau = \frac{8M}{c}$$

The example provides another illustration of the use of the time constant in predicting system response.

Critical Damping. An oscillatory response will result as long as the value of the damping coefficient c is small enough for $c^2 - 4kM$ in Eq. 5-10 to be negative. Now assume a value of c sufficient to cause this quantity to be zero. Here

$$p = \frac{-c}{2M} \pm 0$$

and the solution for the system differential equation is

$$y = C_1 e^{-ct/2M} + C_2 t e^{-ct/2M} + X$$

Note that the response is no longer oscillatory and appears as shown in Fig. 5-5d. The system is now *critically damped.* The approximate time required for the system to respond completely is

$$4\tau = \frac{8M}{c}$$

Although it would appear that the response time of the critically damped system is the same as that of the underdamped system, remember that c is being varied and is now of greater magnitude. A critically damped system will respond completely in less time than an underdamped system because the value of c is greater in the expression $4\tau = 8M/c$.

Overdamped Response. For damping greater than that required for critical damping, the roots of the characteristic equation are

$$p = \frac{-c}{2M} \pm \frac{(c^2 - 4Mk)^{1/2}}{2M} = -a \pm b$$

where b is now a real number with a magnitude greater than zero but always less than a. The solution for the system equation is

$$y = C_1 e^{(-a+b)t} + C_2 e^{(-a-b)t} + X$$

The response is shown in Fig. 5-5*e* and is quite slow. An estimate of the response time can be obtained from the *larger* of the two time constants; thus,

$$\tau = -\left(\frac{1}{-a+b}\right) \quad \text{and} \quad 4\tau = \frac{4}{a-b}$$

This concludes the discussion of the types of responses possible from the second-order mechanical system subjected to a step input. It may be correctly inferred that these responses are possible from all second-order systems with a step input.

Damping Ratio. A convenient concept for studying the response of second-order systems is that of the *damping ratio* which is defined as the ratio of the actual damping in the system, c_a, to that damping which would produce critical damping, c_c. It is given the symbol ζ.

$$\zeta \equiv \frac{c_a}{c_c} \tag{5-11}$$

For the mechanical system under consideration, $c_a = c$. The quantity raised to the one-half power in Eq. 5-10 must be zero for critical damping; thus,

$$c_c = 2(Mk)^{1/2}$$

Therefore,

$$\zeta = \frac{c}{2(Mk)^{1/2}} \tag{5-12}$$

From the definition, a system with $\zeta < 1$ will have an oscillatory response (underdamped), one with $\zeta = 1$ will be critically damped, while one with $\zeta > 1$ will be sluggish in response (overdamped).

General Response Equation. Equation 5-9 can be written as

$$\frac{d^2y}{dt^2} + \frac{c}{M}\frac{dy}{dt} + \frac{k}{M}y = \frac{k}{M}X$$

One notes that

$$\frac{k}{M} = \omega_n^2$$

and

$$\frac{c}{M} = \frac{2c}{2M} = \frac{2ck^{1/2}}{2M^{1/2}k^{1/2}M^{1/2}} = 2\zeta\omega_n$$

Therefore,

$$\frac{d^2y}{dt^2} + 2\zeta\omega_n\frac{dy}{dt} + \omega_n^2 y = \omega_n^2 X$$

All the second-order equations for systems subjected to a step input can be expressed in this form, the general equation being

$$\frac{d^2f}{dt^2} + 2\zeta\omega_n\frac{df}{dt} + \omega_n^2 f = \omega_n^2 K \tag{5-13}$$

where f is any function and K is its steady-state value. The characteristic equation is

$$p^2 + 2\zeta\omega_n p + \omega_n^2 = 0 \tag{5-14}$$

from which

$$p_1 = -[\zeta + (\zeta^2 - 1)^{1/2}]\omega_n$$
$$p_2 = -[\zeta - (\zeta^2 - 1)^{1/2}]\omega_n$$

If desired, Eq. 5-13 can be solved for $\zeta < 1$, $\zeta = 1$, and $\zeta > 1$ to yield response equations for the underdamped, critically damped, and overdamped cases, respectively. The nature of the responses for three values of ζ is shown in Fig. 5-6 with the assumption that the system is initially at rest.

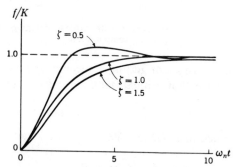

FIG. 5-6. Mathematical response of second-order systems with step inputs.

The important features of these responses are that a critically damped system responds completely without oscillation in the least amount of time, an underdamped system gets close to the final value more rapidly although taking somewhat longer to reach equilibrium, and an overdamped system is sluggish in responding. Note that the expressions f/K and $\omega_n t$ are dimensionless. It follows that the response of all second-order systems to a step input is the same.

Mechanical System with Ramp Input. Attention is now turned to the response of second-order systems to ramp inputs. The response will be illustrated by considering a specific mechanical system, that shown in Fig. 5-7a which is the same as that given in Fig. 2-7a. The differential equation (Eq. 2-8) is

$$I\frac{d^2\theta_o}{dt^2} + b\frac{d\theta_o}{dt} + G\theta_o = G\theta_i(t) \tag{5-15}$$

The input is to be a ramp input with $\theta_i(t) = Kt$, as shown in Fig. 5-7b. Thus the equation to be solved is

$$I\frac{d^2\theta_o}{dt^2} + b\frac{d\theta_o}{dt} + G\theta_o = GKt \tag{5-16}$$

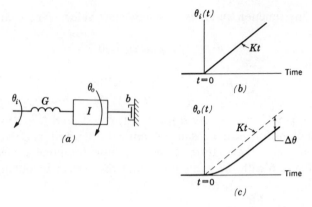

FIG 5-7. Mechanical system with critical damping subjected to a ramp input Kt.

Assume, for the purpose of getting a specific response, that the values of I, b, and G are such that critical damping exists. Then the transient solution is

$$\theta_{ot} = C_1 e^{-bt/2I} + C_2 t e^{-bt/2I}$$

The forcing function is GKt. Therefore, the steady-state solution is assumed to be

$$\theta_{oss} = At + B$$

$$\frac{d\theta_{oss}}{dt} = A \quad \text{and} \quad \frac{d^2\theta_{oss}}{dt^2} = 0$$

Substituting in Eq. 5-16,

$$I(0) + bA + GAt + GB = GKt$$

from which

$$A = K \quad \text{and} \quad B = -\frac{bK}{G}$$

The complete solution is

$$\theta_o = C_1 e^{-bt/2I} + C_2 t e^{-bt/2I} + Kt - \frac{bK}{G} \qquad (5\text{-}17)$$

If the arbitrary constants are determined from the initial conditions, assuming that the system is initially at rest, the response is that shown in Fig. 5-7c. One sees that, after the transients have disappeared, the response faithfully follows the input but lags by an angle $\Delta\theta$. Equation 5-17 may be evaluated as t approaches infinity to give this result:

$$\theta_o = Kt - \frac{bK}{G} = \theta_i - \frac{bK}{G}$$

or

$$\theta_i - \theta_o = \Delta\theta = \frac{bK}{G}$$

The angle $\Delta\theta$ can be given physical interpretation. A ramp positional input corresponds to a step velocity input; with $\theta_i = Kt$, $d\theta_i/dt = \omega_i = K$. Thus K is the angular velocity. A torque is required to drive the mass at a constant velocity because of the action of the rotary dashpot. The torque required is bK and must be supplied through the torsion spring, requiring a deflection $\Delta\theta$. The torque transmitted is $G(\Delta\theta)$. Equating the expressions for torque

$$G(\Delta\theta) = bK \qquad \text{or} \qquad \Delta\theta = \frac{bK}{G}$$

The response of second-order systems to ramp inputs can be generalized and plotted in dimensionless form in a manner similar to that used with step inputs [14].

5-4. The Concept of System Stability

A control system, if it is to be of any practical value, must be stable. In Chapter 1 this was stated to mean that, in response to some input, the system will not oscillate violently or drive itself to some limiting value of the controlled variable but rather will attain some useful response. A more specific definition of stability can now be introduced. A stable system is one in which transients decay, that is, the transient response disappears with increasing values of time. Such a condition requires that the coefficients of t in the exponential term(s) of the transient solution be negative real numbers or complex numbers with negative real parts; for example, e^{-2t} decays while e^{+2t} grows as time advances. These coefficients are, of course, the roots of the characteristic equation. Therefore, from the physical viewpoint, a stable system is one in which the transients die out and the system "settles down" to some useful response. Mathematically, a system is said to be stable if the roots of the characteristic equation are negative real numbers or complex numbers with negative real parts. The latter statement constitutes the basic concept of stability. Although there are many techniques for determining stability, all are based on the previous definition.

A review of the systems analyzed in this chapter will show that they are stable (with the exception of one which will be discussed shortly). Two differential equations for unstable systems will illustrate the nature of unstable response. Consider the equation

$$\frac{dy}{dt} - y = X$$

Here $p = 1$ and the transient solution is

$$y_t = C_1 e^t$$

It can be seen that the response does not vanish, but instead grows rapidly with increasing values of time ($y_t \rightarrow \infty$ as $t \rightarrow \infty$). The response of the actual system would not, of course, approach infinity but rather would proceed to some extreme limiting value determined by the physical nature of the system. Such a response is of limited practical value.

As a second example, consider the equation

$$\frac{d^2y}{dt^2} - 2\frac{dy}{dt} + 5y = Kt$$

Here $p = 1 \pm j2$ and the transient solution is

$$y_t = e^t(C_1 \sin 2t + C_2 \cos 2t)$$

An oscillatory response is indicated with an amplitude which increases with time rather than decreases. The physical system would tend to oscillate violently, perhaps destroying itself.

The two examples illustrate the response of systems when the roots of the characteristic equation are positive real numbers or complex numbers with positive real parts, that is, unstable systems. It should be noted that the system input has no effect on stability. A linear system which is stable to one input is stable to all inputs.

So far the possibility of having complex roots in which the real part is zero has been overlooked. This situation occurred in the solution for Eq. 5-8. The response turned out to be a persistent oscillation as shown in Fig. 5-5b. A system with this response is said to possess *limited stability*, the amplitude of oscillation neither decaying nor growing with time. From a practical viewpoint, however, the response may well be considered unstable; a position X is required which the system cannot attain. Nevertheless, the borderline case between absolute stability and instability is called *limited stability*.

One final example will emphasize the basic concept of stability. Envision a 100th degree characteristic equation; suppose that 99 of its roots are either negative real numbers or pairs of complex numbers with negative real parts but that a single root is a positive number. The system is unstable because there will be one exponential term in the transient solution which will grow with time and so the transient solution will approach infinity as time approaches infinity. *For absolute stability, all roots must be negative real numbers or complex numbers with negative real parts.*

Since solving a higher-degree characteristic equation for its roots is time consuming, the *Routh criterion* can be used to determine whether or not there are positive roots without actually solving for the roots. Routh's criterion for stability is presented in Appendix C.

5-5. Analysis of Two Control Systems

Two simple control systems will now be analyzed, a first-order and a second-order system. Of interest will be the stability and response of these systems. The response to two kinds of input will be considered to answer two basic questions. When the reference input (input) is changed, what does the controlled variable (output) do? If the reference input is not changed but some disturbance occurs, what does the controlled variable do? These important questions will be answered for a hydromechanical and an electromechanical system.

Hydromechanical System. The system to be analyzed is shown in Fig. 5-8a. It is called *hydromechanical* because it is composed of hydraulic and mechanical parts. The system was first introduced in Fig. 2-17 and the appropriate differential equation (Eq. 2-30) is

$$\frac{Ab}{aK_v} \frac{dy}{dt} + y = \frac{b}{a} x(t) \tag{5-18}$$

To investigate the response of the system to a step change of reference input, that is, $x(t) = X$, the valve spool is suddenly displaced. Assume that the system is initially at rest and consider $x = y = 0$ in this equilibrium position. Solving Eq. 5-18, with $x(t) = X$, we find $p = -aK_v/Ab$. Therefore, the system is stable, the root of the characteristic equation being a negative real number. Completion of the solution yields

$$y = \frac{bX}{a} (1 - e^{-t/\tau}) \qquad \tau = \frac{Ab}{aK_v}$$

The response is shown in Fig. 5-8b and is characteristic of that of all first-order systems. The approximate time required for the system to respond completely is equal to 4τ.

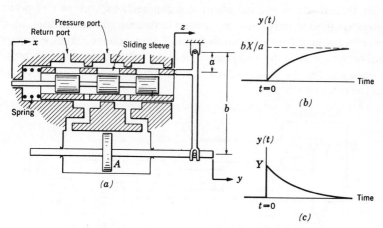

FIG. 5-8. Response of hydromechanical control system.

It is worth while to study the expression for τ to see the effect of the various system parameters on response time. If the piston area A were increased, while holding the other parameters constant, τ would be increased, resulting in a longer response time. This conclusion can be verified from physical considerations by noting that, for a given flow rate through the valve, the larger the piston area, the lower will be the velocity ($v = q/A$). Returning to the expression for τ, an increase in the valve constant K_v would result in a smaller time constant and in a faster responding system. The conclusion is reasonable because an increased flow rate into a given cylinder will produce an increased velocity. Similar arguments can be advanced for the effects of the lever dimensions a and b.

Attention is now turned to the possibility of a disturbance in the system. Assume that the system is at rest and that the reference input is to be unchanged. Further assume that the piston is suddenly displaced in some manner to a position Y (previously, $y = 0$). What will be the response? The question can be answered by solving Eq. 5-18 with $x(t) = 0$ (the reference input is not changed and is still zero) and with the initial condition $y = Y$ at $t = 0$. The solution is

$$y = Ye^{-t/\tau} \qquad \tau = \frac{Ab}{aK_v}$$

This response is shown in Fig. 5-8c.

Although Eq. 5-18 provides an oversimplified mathematical model for the hydromechanical system, its use does give a first approximation to the response of the system under the conditions specified.

Electromechanical System. An electromechanical positioning system is shown in Fig. 5-9. The desired angular position input is denoted as θ_i. The controlled output angular position is denoted as θ_o. A pair of potentiometers are used to make the comparison between input and output. A difference between these positions will result in a voltage signal e (proportional to the error) to the amplifier. The signal is then amplified to provide a voltage v to the d-c motor which in turn supplies a torque T which positions the rotating mass.

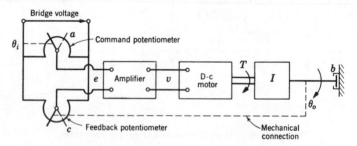

Fig. 5-9. Electromechanical positioning system.

A schematic drawing of a potentiometer is given in Fig. 5-10. The potentiometer illustrated is a wire-wound resistor with a movable slider, or wiper, which makes electrical contact with the resistance winding. The

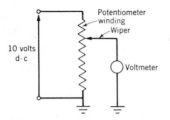

FIG. 5-10. Potentiometer schematic drawing.

purpose of the device is to provide a voltage output proportional to the
displacement of the wiper. Assume that 10 volts d-c is applied across the
potentiometer and that a voltmeter is attached to the wiper as shown. If
the wiper is positioned at the top of the winding, the voltage reading will be
10 volts; if at the bottom, 0 volts; if midway between top and bottom,
5 volts, and so forth. Thus the potentiometer provides an output voltage
proportional to the displacement of the wiper. This assumes that the same
current flow passes through both the upper and lower portions of the resistance winding; if such is not the case, the proportionality, based on the
relative resistance values, does not exist. Therefore, there must be a negligible current flow through the voltmeter, that is, the voltmeter must provide
a high impedance so that it does not *load* the potentiometer. Potentiometers are commercially available in which the displacement can be either
translation or rotation.

A pair of potentiometers can be utilized to compare two displacements
as shown in Fig. 5-9. Such an arrangement is commonly called a *bridge
circuit* because it behaves as a Wheatstone bridge. If the two wipers are in
corresponding positions, there is no potential difference between them and
the error voltage e is zero. Now assume that the wiper of the command
(input) potentiometer is moved manually to position a. An error voltage e,
of a polarity depending on the sign (plus or minus) of the error and of a
magnitude proportional to the magnitude of the error, will be introduced to
the amplifier. The result will be a movement of the rotating mass. The
position of the mass is detected by the feedback potentiometer, and the
system will continue to respond until the wiper of the feedback potentiometer comes to rest at position c.

The system error is defined as the difference between the input and the
controlled variable. In this case, e is a voltage signal proportional to error
and is called the *actuating signal.* It can be expressed as

$$e = K_1(\theta_i - \theta_o) \qquad (5\text{-}19)$$

where K_1 is the proportionality factor.

The amplifier will be considered as a device which receives a small
voltage signal (usually in the millivolt range) and provides a voltage output

at a higher power level. As with the voltmeter used in conjunction with the potentiometer shown in Fig. 5-10, it is important that the amplifier have a high input impedance. The amplifier may be either d-c or a-c depending on the choice of bridge voltage. The operation of the amplifier can be expressed as

$$v = K_2 e \tag{5-20}$$

where K_2 is the proportionality factor associated with the amplifier. The factor K_2 is commonly called the *gain* of the amplifier and is usually adjustable so that its value can be changed to provide optimum system response.

The d-c motor was discussed in some detail in Chapter 2. The output torque is proportional to the armature current $(T = K_t i_a)$ and, if we neglect armature inductance and the effects of counter emf $(e_c = K_e \omega)$, is

$$T = K_3 v \tag{5-21}$$

where $K_3 = K_t/R$, R being the armature resistance and v the armature voltage.

Considering the rotating mass and dashpot,

$$T = I\frac{d^2\theta_o}{dt^2} + b\frac{d\theta_o}{dt} \tag{5-22}$$

Combining Eqs. 5-19, 5-20, 5-21, and 5-22 yields the system equation

$$I\frac{d^2\theta_o}{dt^2} + b\frac{d\theta_o}{dt} + K\theta_o = K\theta_i(t) \tag{5-23}$$

where $K = K_1 K_2 K_3$. Equation 5-23 is identical in form to Eqs. 5-9 and 5-15 and so the response of the electromechanical positioning system to a step or ramp reference input is similar to those previously described for second-order systems.

It is important to investigate the effect of a torque disturbance occurring at the mass on the output position θ_o. Assume a positive step disturbance as shown in Fig. 5-11a. The addition of this torque requires that Eq. 5-22 become

$$T = I\frac{d^2\theta_o}{dt^2} + b\frac{d\theta_o}{dt} - T_D$$

and that Eq. 5-23 be

$$I\frac{d^2\theta_o}{dt^2} + b\frac{d\theta_o}{dt} + K\theta_o = K\theta_i(t) + T_D \tag{5-24}$$

If the system were initially at rest, the error being zero and considering $\theta_i = \theta_o = 0$ in this equilibrium position, Eq. 5-24 becomes

$$I\frac{d^2\theta_o}{dt^2} + b\frac{d\theta_o}{dt} + K\theta_o = 0 + T_D \tag{5-25}$$

The equation can be solved for $\theta_o(t)$; the response is shown in Fig. 5-11*b* for an underdamped system. Note that the steady-state position $\theta_{oss} = T_D/K$. The system error $\epsilon(t)$ is shown in Fig. 5-11*c*. The curve is simply the re-

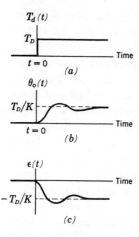

FIG. 5-11. System error resulting from torque disturbance T_D.

flection of $\theta_o(t)$ about the time axis. The reason is that the error is defined as the difference between the input and the controlled variable. The reference input is zero, and

$$\epsilon(t) = \theta_i(t) - \theta_o(t) = 0 - \theta_o(t)$$

Also,

$$\epsilon_{ss} = 0 - \theta_{oss} = -\frac{T_D}{K}$$

The steady-state error can be verified from physical considerations. At equilibrium, the d-c motor must provide an output torque T_{ss} equal in magnitude and opposite in sign to T_D. Therefore,

$$T_{ss} = K(\theta_{iss} - \theta_{oss}) = -T_D$$

Since $\theta_{iss} - \theta_{oss} = \epsilon_{ss}$,

$$\epsilon_{ss} = -\frac{T_D}{K}$$

Usually, the system error is to be minimized. The steady-state error resulting from the torque disturbance can be decreased by increasing the system gain K, accomplished by increasing the amplifier gain K_2. It would seem that ϵ_{ss} could easily be reduced to any desirable value thus improving the accuracy of the system. Recall, however, that in Chapter 1 the statement was made that accuracy and stability tend to be incompatible requirements. This can now be shown from mathematical considerations. Based on Eq. 5-11 the damping ratio for the system is

$$\zeta = \frac{b}{2(IK)^{1/2}}$$

It can be seen that increasing the value of K will decrease the damping ratio which indicates that the response of the system will become more oscillatory. Although the usual second-order system cannot become unstable (at worst it reaches limited stability), the *degree of stability* as $\zeta \to 0$ becomes intolerable because of persistent oscillations and the extended period of time required for the system to respond completely. Some compromise between accuracy, degree of stability, and response time is necessary.

It is worth remarking that we have gained some insight into the response of the electromechanical system by treating it as a second-order system. However, one should not be surprised by cases where the system becomes unstable with sufficient gain. Remember, it was a second-order system because of our assumptions, and these assumptions are completely ignored by the system itself. This is not to say that one should attempt to include all factors in the system equation; such a procedure is usually impractical if not downright impossible. The point is that the results obtained from the mathematical model are valid only to the extent that the model approximates the physical system.

5-6. Closure

The most basic fundamentals of linear system analysis have now been presented. Many control systems can be handled (although not in the best possible manner) with the tools already provided. These tools include the ability to write and solve differential equations, and an understanding of system response and the concept of stability. Although system response was illustrated with first-order and second-order systems, higher-order systems can be treated in a similar manner. More work is involved but that is a matter of detail and not of concept.

Throughout the chapter, examples of a common-sense approach for determining certain aspects of system response have been presented. The reader is encouraged to use this method whenever possible because it serves as a check on a purely mathematical approach and also leads to an intuitive understanding of system behavior. Both results of the common-sense approach are beneficial to the control engineer.

The remainder of the book will introduce additional techniques and concepts which will facilitate the analysis and design of higher-order control systems. To give an idea of what lies ahead, the student will learn such things as: another method for solving differential equations, how to represent systems in block-diagram form, the ways in which an error signal can be utilized to obtain the desired system performance, how to get the response to a sinusoidal input without solving the differential equation, how to determine system stability without finding the roots of the characteristic equation, how to set the system gain for best response, how to modify a system to obtain better performance, and how to deal with certain nonlinearities when their effect cannot be neglected.

Problems

5-1. Determine the expression for the response of each of the following systems to the input stated. Sketch input vs. time and response vs. time curves. What is the time constant for each system?

a) In part *a* of the accompanying figure the rotating speed ω is the response and a step input torque T is applied. Assume system initially at rest.

b) In part *b* of the figure the head h is the response and a step input flow rate Q_i is the input. Assume the tank initially empty and the valve resistance linear.

c) In part *c* of the figure the current i is the response and a step input voltage E is applied. Assume capacitor initially discharged. Suggestion: Use the relationship $i = dq/dt$.

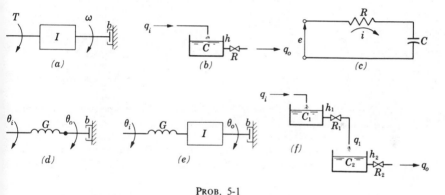

PROB. 5-1

5-2. Determine the expression for the response of each of the following systems to the input stated. Sketch input vs. time and response vs. time curves. What is the time constant for each system?

a) In part *d* of the figure for Prob. 5-1 the angular position θ_o is the response and a ramp input $\theta_i = Kt$ is applied. Assume system initially at rest.

b) In part *b* of the figure the output flow rate q_o is the response and a ramp input flow rate $q_i = Kt$ is applied. Assume the tank initially empty and the valve resistance linear.

5-3. It was stated that the response of a first-order system to a step input would reach the final value with $t = \tau$ if the response continued at its initial rate. Verify this statement by showing that the line marked "initial slope" in Fig. 5-4 passes through the point 1,1.

5-4. Obtain a dimensionless response equation for a first-order system subjected to a ramp input. Make a sketch similar to Fig. 5-4 showing input and response. HINT: Solve the equation $\tau df/dt + f = Kt$.

5-5. Determine the expression for the response of each of the following systems to the input stated. Sketch input vs. time and response vs. time curves. What is the damping ratio for each system?

a) In part e of the figure for Prob. 5-1 the angular position θ_o is the response and a step input position Θ_i is applied. Assume an underdamped system initially at rest.

b) In part f of the figure the head h_2 is the response and a step input flow rate Q_i is applied. Assume tanks initially empty, valve resistances linear, and that $R_1C_1 = R_2C_2 = RC$.

c) In part e of the figure a ramp input $\theta_i = Kt$ is applied. Assume an underdamped system initially at rest.

5-6. Determine the arbitrary constants C_1 and C_2 in Eq. 5-17. Assume the system initially at rest.

5-7. Suppose that the differential equations given describe certain control systems. Determine whether or not these systems are stable. Sketch the responses to a step input.

a) $\dfrac{d^2y}{dt^2} + Ky = Kx(t)$
                                         b) $\dfrac{d^2y}{dt^2} + c\dfrac{dy}{dt} + Ky = Kx(t)$

c) $\dfrac{dy}{dt} - Ky = Cx(t)$
                                          d) $-\dfrac{dy}{dt} - Ky = Cx(t)$

5-8. Write Eq. 5-23 in terms of $\theta_i(t)$ and $\epsilon(t)$. Determine the steady-state error ϵ_{ss} for a) a step input Θ_i and b) for a ramp input Ct. HINT: $\theta_o = \theta_i - \epsilon$.

5-9. Write Eq. 5-24 in terms of $\theta_i(t)$ and $\epsilon(t)$. Determine the steady-state error ϵ_{ss} for a step input disturbance T_D. Compare with Fig. 5-11. Assume $\theta_i(t) = 0$.

5-10. Obtain the complete solution for Eq. 5-13 with a) $\zeta = 1$ and b) $\zeta < 1$. Assume the system initially at rest.

5-11. For the tank system shown:

$C = 350 \text{ in.}^2$ $h_1(0) = 12 \text{ in.}$
$R = 0.075 \text{ sec/in.}^2$ $h_2(0) = 0 \text{ in.}$
$R' = 0.015 \text{ sec/in.}^2$

Determine: a) $h_2(t)$, b) $h_2(\text{max})$, and c) T_e = approximate time required for emptying system.

NOTE: The system given is actually a simplified representation of laundry tubs. The problem was prompted by the common household experience of removing the plug from the full side of a pair of laundry tubs and finding that water enters the other (previously empty) side.

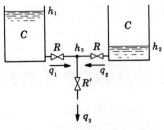

5-12. From the following characteristic equations, determine whether the corresponding control systems are stable or unstable. Consider that limited stability is instability.

a) $p^4 + 4p^3 + 8p^2 + 8p + 3 = 0$

b) $p^5 + 2p^4 + 3p^3 + 8p^2 + p + 4 = 0$

c) $p^4 + 7p^3 - 3p^2 + 2p + 5 = 0$

d) $p^5 + 9p^4 + 7p^2 + 2p + 1 = 0$

e) $p^4 + 2p^3 + 2p^2 + 4p + 3 = 0$

f) $p^5 + 2p^4 + 2p^3 + 4p^2 + 3p + 6 = 0$

Chapter 6

Laplace Transformation

6-1. Introduction

The Laplace transform is a mathematical tool that is extremely useful in the analysis of automatic control systems. Let us look first at what the Laplace transform is, then at why it is so useful, and finally at how it can be used in the study of automatic control systems.

The Laplace transform is one area of operational calculus wherein a function is transformed from dependency on one variable to dependency on a new variable. In the case of automatic control systems, one is concerned with transforming some functional relationship in the *time domain* (such as the differential equation defining the behavior of a control system) to a new "*s*" domain in which *s* rather than *t* is the independent variable. The transformation is accomplished through the use of a basic definition of the transform which defines as the independent variable, the complex variable *s*. It is then said that the function is transformed from the *t domain* to the *s domain*. Perhaps the concept of transformation is better understood when related to a more common mathematical operation. The analogy has frequently been made between taking the Laplace transform of a function and obtaining the logarithm of a number. To obtain the logarithm of a number, one finds the power to which 10 must be raised to get the specific number. It is then possible to return to the original number by the antilogarithm. Similarly, in operational calculus, one returns to the original function by taking the inverse transform.

The concept of transformation is employed in the solution of problems because it places the problem in a new domain in which, it is hoped, the method of computation is simpler. When one works with logarithms of numbers, the operation of multiplication becomes the simpler process of addition, division becomes subtraction, and raising to a power becomes multiplication by that power. When all of the simpler operations with logarithms are complete, the answer is obtained by finding the antilogarithm. The Laplace transform provides a similar type of simplifying media for the solution of linear differential equations. Namely, an equation expressed in terms of derivatives with respect to an independent variable becomes an algebraic expression in *s* when its Laplace transform is taken.

Thus, one can manipulate this new expression with the axioms of algebra rather than with the more complicated axioms of differential equations. A final result in the original domain is obtained by taking the inverse transform.

Manipulation with logarithms is facilitated by the use of tables of logarithms of numbers; similarly, the use of the Laplace transform is facilitated by tables of transform pairs.

The Laplace transform has become a widely used technique in dealing with automatic control systems for still another reason. Very often, in the design and analysis of control systems, it is desirable to predict system performance with regard to stability, accuracy, and speed of response without actually solving the system equations. The Laplace transform method allows the use of many analytical and graphical techniques for this purpose.

This chapter will present the Laplace transform with the attitude that it is a mathematical tool useful to the control engineer in studying and designing control systems. Emphasis is therefore placed on the application of the Laplace transform and on techniques, such as the use of tables of transform pairs, which facilitate rapid solution.

6-2. Definition of the Laplace Transform

A function $f(t)$, multiplied by e^{-st} and then integrated from $t = 0$ to $t = \infty$, may form a new function of the variable s. Thus, by definition

$$F(s) = \int_0^\infty f(t)e^{-st}\,dt \qquad (6\text{-}1)$$

where s is a complex variable defined by $s = \sigma + j\omega$. This process is restricted to functions $f(t)$ which are transformable (that is, the new function $F(s)$ exists). Frequently, Eq. 6-1 is expressed by the shorthand notation

$$\mathcal{L}[f(t)] = F(s) \qquad (6\text{-}2)$$

It is possible that the integral of Eq. 6-1 does not exist. This condition may occur when the function $f(t)$ does not behave properly near $t = 0$, or for large values of t, or because there are an infinite number of discontinuities in $f(t)$ for positive t. Although the existence of this integral is a matter of concern to the mathematician, it need not be a problem in an introductory study of linear control systems. In most cases, if $f(t)$ is a linear differential equation with a finite number of terms, transformation by Eq. 6-1 is unrestricted. As shown in previous chapters, the study of linear control systems is the study of linear differential equations, and in a physical system the number of terms in these equations is finite.

6-3. Complex Numbers and s as a Complex Variable

There are real numbers (for example, 5, 43, 146) and complex (not complicated) numbers. Unlike real numbers, a complex number is composed of two parts, a real part and an imaginary part in the form:

$$\text{Real part} + j \text{ Imaginary part} = \text{Re} + j\text{Im}$$

where $j = (-1)^{1/2}$. As an example consider the complex number $3 + j4$; 3 is the real part and 4 is the imaginary part. Any complex number may be represented as a point on the complex plane shown in Fig. 6-1.

A complex number may be represented in three forms; *rectangular*, *polar*, and *exponential*. The rectangular form expresses the complex number in its real and imaginary parts $(\text{Re} + j\text{Im})$ as already discussed.

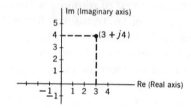

FIG. 6-1. The complex plane.

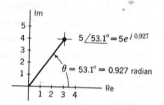

FIG. 6-2. Polar and exponential form on the complex plane.

The polar form expresses the complex number in terms of a magnitude and an angle on the complex plane as $A \underline{/\theta°}$. The complex number of Fig. 6-1 is illustrated in polar form in Fig. 6-2. The equivalence of the two notations is shown by the geometry of the figure to be

$$3 + j4 = (3^2 + 4^2)^{1/2} \underline{/\tan^{-1} \tfrac{4}{3}} = 5\underline{/53.1°}$$

The exponential representation of a complex number is made by taking the product of a magnitude A and a unit vector $e^{j\theta}$ (Fig. 6-3). Thus, the exponential form of a complex number is $Ae^{j\theta}$, where A is the radial distance from the origin to the point on the complex plane, e is the base of the natural logarithm, and θ is the angle measured from the positive real axis expressed in radians. This form may also be plotted on the complex plane in a fashion similar to the polar form, Fig. 6-2. For example, the complex

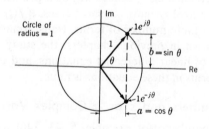

FIG. 6-3. The unit vector.

number $3 + j4$ becomes

$$5e^{j0.927}$$

The equivalence of the rectangular and exponential forms can be further grasped by noting that the unit vector $e^{j\theta} = \cos\theta + j\sin\theta$ on the complex plane, Fig. 6-3. Thus, for the above example,

$$5e^{j0.927} = 5(\cos 0.927 + j\sin 0.927) = 3 + j4$$

Operations with Complex Numbers. Complex numbers may be added, subtracted, multiplied, and divided in all three forms. However, various forms are best suited for various operations.

Addition and Subtraction. Addition and subtraction of complex numbers are most easily accomplished in the rectangular form. The algebraic sum of two complex numbers is the algebraic sum of its real and imaginary parts. Thus,

$$(1 + j2) + (4 + j1) = 5 + j3 \quad \text{and} \quad (1 - j2) - (3 - j1) = -2 - j1$$

In general then,

$$(a + jb) + (c + jd) = (a + c) + j(b + d) \qquad (6\text{-}3)$$

Multiplication and Division. Multiplication and division, on the other hand, are most easily carried out in exponential or polar form. The multiplication process is illustrated by the following:

$$(5e^{j1})(2e^{j2}) = 10e^{j(2+1)} = 10e^{j3}$$

and

$$(5\underline{/57.3^\circ})(2\underline{/114.6^\circ}) = 10\underline{/171.9^\circ}$$

In general then,

$$(Ae^{j\theta})(Be^{j\phi}) = ABe^{j(\theta+\phi)} \qquad (6\text{-}4)$$

Division is accomplished by recognizing that

$$\frac{1}{Ae^{j\theta}} = \frac{1}{A}e^{-j\theta} \qquad (6\text{-}5)$$

Thus, as an example,

$$\frac{5e^{j1}}{2e^{j2}} = 2.5e^{-j1}$$

In polar form this is illustrated by the following example:

$$\frac{5\underline{/57.3^\circ}}{2\underline{/114.6^\circ}} = 2.5\underline{/-57.3^\circ}$$

Multiplication may be accomplished in rectangular form by multiplying the complex quantities algebraically. For example,

$$(1 + j)(0 + j2) = 0 + j2 + j0 + j^2 2$$

Since $j^2 = -1$, this expression becomes

$$-2 + j2 = 2.828\underline{/135^\circ}$$

Conjugate of a Complex Number. The conjugate of a complex number is another complex number with the same real part as the first, but with the negative of the imaginary part. Thus, the conjugate of $3 - j2$ is $3 + j2$. In general,

$$\text{Conjugate of } (a + jb) = (a - jb) \tag{6-6}$$

Useful Relationships for Complex Numbers. From Fig. 6-3 it can be seen that the unit vector $e^{j\theta}$ may be represented by $a + jb$ or

$$e^{j\theta} = \cos \theta + j \sin \theta \tag{6-7}$$

The conjugate is, as illustrated in Fig. 6-3,

$$e^{-j\theta} = \cos \theta - j \sin \theta \tag{6-8}$$

If Eq. 6-8 is subtracted from Eq. 6-7, then

$$\frac{e^{j\theta} - e^{-j\theta}}{2j} = \sin \theta \tag{6-9}$$

If Eqs. 6-7 and 6-8 are added,

$$\frac{e^{j\theta} + e^{-j\theta}}{2} = \cos \theta \tag{6-10}$$

Equations 6-7 through 6-10 will prove extremely useful in the work to come.

The Complex Variable s. It was mentioned in Secs. 6-1 and 6-2 that s is a complex variable of the form

$$s = \sigma + j\omega \tag{6-11}$$

Then s may, as can any complex number, be represented as a point on the complex plane. In this case, the complex plane may be called the s *plane* with real and imaginary axes. As s is a *complex variable* it may take on many different values as complex numbers. Two specific values of s have been plotted on the complex plane in Fig. 6-4, namely,

$$s_1 = \sigma_1 + j\omega_1 \qquad \text{and} \qquad s_2 = \sigma_2 + j\omega_2$$

s_1' and s_2' are the complex conjugates of s_1 and s_2 respectively. The s *plane* is a field on which many games of analysis and synthesis are played by the control analyst. A deeper understanding of its meaning and a fuller understanding of its use will be gained in later chapters.

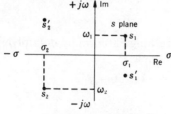

FIG. 6-4. The s plane.

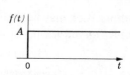

FIG. 6-5. The constant A defined for positive t.

6-4. Transforms and Theorems

The use of the basic definition of the Laplace transform given by Eq. 6-1 may be illustrated by actually transforming several common functions. Consider first, the Laplace transform of a constant A defined for positive values of time t, as shown in Fig. 6-5. Then,

$$f(t) = A \quad \text{for } 0 \leq t \tag{6-12}$$

By Eqs. 6-1 and 6-2,

$$\mathcal{L}(A) = F(s) = \int_0^\infty A e^{-st}\, dt = A \int_0^\infty e^{-st}\, dt = \frac{-A e^{-st}}{s} \Big|_0^\infty$$

which becomes

$$\mathcal{L}(A) = \frac{A}{s} \tag{6-13}$$

The Laplace transform may also be determined for the case where the function is defined by

$$f(t) = A \quad \text{for } 0 \leq t \leq t_1$$
$$\text{and} \qquad f(t) = 0 \quad \text{elsewhere} \tag{6-14}$$

This function, known as a *pulse*, is plotted in Fig. 6-6. The Laplace transform of this function is

$$F(s) = \int_0^{t_1} A e^{-st}\, dt = -\frac{A e^{-st}}{s} \Big|_0^{t_1}$$

or

$$F(s) = A \frac{(1 - e^{-st_1})}{s} \tag{6-15}$$

An interesting variation of the function of Eq. 6-14 is obtained by maintaining the area a under the pulse constant and letting t_1 approach 0. In the limit when $t_1 = 0$, this function is called an *impulse*. If the duration of the impulse is t_1 and the area is a, then the height is a/t_1, Fig. 6-7. The transform of this function is given by

$$F(s) = \lim_{t_1 \to 0} \left[\int_0^{t_1} \frac{a}{t_1} e^{-st}\, dt \right] = \lim_{t_1 \to 0} \left[\frac{-a}{st_1} e^{-st} \Big|_0^{t_1} \right]$$

or

$$F(s) = \lim_{t_1 \to 0} \left[\frac{a(1 - e^{-st_1})}{st_1} \right]$$

which is an indeterminate limit in its present form. By L'Hopital's rule, the numerator and denominator may be differentiated with respect to t_1 to yield

$$F(s) = \lim_{t_1 \to 0} \left(\frac{ase^{-st_1}}{s} \right) = a \tag{6-16}$$

In other words, the Laplace transform of an impulse is the area under the impulse curve. If that area is equal to unity, then it is known as a *unit impulse*.

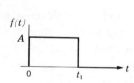

FIG. 6-6. Plot of a pulse function.

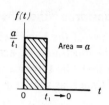

FIG. 6-7. The impulse function.

It may seem somewhat superfluous to be considering a time function of zero duration, infinite height, and area of unity (the unit impulse) as it would appear somewhat difficult to obtain the physical event. The unit impulse is, however, a powerful mathematical tool for studying the transient behavior of a real physical system.

Consider next, the Laplace transform of the function

$$f(t) = At \tag{6-17}$$

Then,

$$\mathcal{L}[f(t)] = \int_0^\infty At e^{-st}\, dt$$

This integral may be evaluated by integrating by parts. Let $u = At$ and $dv = e^{-st}\, dt$. The rule for integrating by parts is

$$\int_a^b u\, dv = uv \Big|_a^b - \int_a^b v\, du$$

Thus, applying the above rule,

$$\int_0^\infty At e^{-st}\, dt = \frac{-At e^{-st}}{s}\Big|_0^\infty - A \int_0^\infty \left(-\frac{e^{-st}}{s}\right) dt$$

$$= \frac{-At e^{-st}}{s}\Big|_0^\infty - \frac{A e^{-st}}{s^2}\Big|_0^\infty$$

Evaluating the limits,

$$\mathcal{L}(At) = \frac{A}{s^2} \tag{6-18}$$

In general, it may be shown that

$$\mathcal{L}(t^n) = \frac{n!}{s^{n+1}} \quad \text{for } n > -1 \tag{6-19}$$

A function which commonly occurs in the study of control systems is $f(t) = \sin \omega t$. The transform is again obtained by Eq. 6-1. Thus,

$$\mathcal{L}(\sin \omega t) = \int_0^\infty (\sin \omega t) e^{-st}\, dt$$

Integration may be accomplished easily and quickly by a table of integrals [9], the use of which is encouraged in this work. Thus,

$$\int_0^\infty (\sin \omega t) e^{-st} \, dt = \frac{\omega}{s^2 + \omega^2}$$

and therefore,

$$\mathcal{L}(\sin \omega t) = \frac{\omega}{s^2 + \omega^2} \tag{6-20}$$

Since the study of automatic control systems involves differential equations, it is necessary to transform the derivative of a function. Consider $f(t) = dx(t)/dt$ when $x(t)$ has the value of $x(0)$ at $t = 0$. Then,

$$\mathcal{L}\left[\frac{dx(t)}{dt}\right] = \int_0^\infty \frac{dx(t)}{dt} e^{-st} \, dt$$

The right-hand side may be integrated by parts, choosing $u = e^{-st}$ and $dv = [dx(t)/dt]\,dt$. Then,

$$\mathcal{L}\left[\frac{dx(t)}{dt}\right] = e^{-st} x(t) \Big|_0^\infty + s \int_0^\infty x(t) e^{-st} \, dt$$

Which, if $x(t)$ is a transformable function, becomes

$$\mathcal{L}\left[\frac{dx(t)}{dt}\right] = -x(0) + sX(s)$$

or

$$\mathcal{L}\left[\frac{dx(t)}{dt}\right] = sX(s) - x(0) \tag{6-21}$$

The Laplace transform of the second derivative may be obtained by repeating the above process to yield

$$\mathcal{L}\left[\frac{d^2 x(t)}{dt^2}\right] = \mathcal{L}\left\{\frac{d[dx(t)/dt]}{dt}\right\} = s^2 X(s) - sx(0) - \frac{dx(0)}{dt} \tag{6-22}$$

where $dx(0)/dt$ is the value of the first derivative at time zero. In the special case where all the initial conditions of a variable and its derivatives are zero, Eq. 6-22 reduces to

$$\frac{d^2 x(t)}{dt^2} = s^2 X(s)$$

In general, then, with all initial conditions zero, the Laplace transform of a derivative of a function is

$$\mathcal{L}\left[\frac{d^n x(t)}{dt^n}\right] = s^n X(s) \tag{6-23}$$

The Laplace transform of a translated function is important in control work. The translated function is defined as

$$0 \text{ for } 0 \leq t < D$$

$$\text{(6-24)}$$

and $f(t - D)$ for $t \geq D$

The function $f(t)$ and the translated function $f(t - D)$ are illustrated in Fig. 6-8.

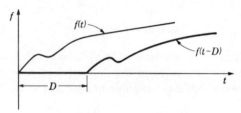

FIG. 6-8. The translated function.

The Laplace transform is obtained by noting that

$$\int_0^\infty f(t - D)e^{-st}\, dt = \int_D^\infty f(t - D)e^{-st}\, dt$$

If a change of variable is made by replacing t by $(t + D)$ and altering the limits accordingly

$$\mathcal{L}\left\{ \begin{matrix} f(t - D), \\ 0 \end{matrix} \right\} = \int_0^\infty f(t)\, e^{-s(t+D)}\, dt$$

$$= e^{-sD} \int_0^\infty f(t)\, e^{-st}\, dt$$

$$\mathcal{L}\left\{ \begin{matrix} f(t - D), \\ 0 \end{matrix} \right\} = e^{-sD} F(s) \qquad\qquad \text{(6-25)}$$

In automatic control systems this phenomenon is referred to as *dead time*. Another terminology frequently used in the process industry is transport delay.

Linearity of the Laplace Transform. If a function is multiplied by a constant, the Laplace transform of the product is the constant times the Laplace transform of the function. Thus,

$$\mathcal{L}[Kf(t)] = KF(s) \qquad\qquad \text{(6-26)}$$

where K is a constant.

Another useful theorem states that the Laplace transform of the sum of two functions is the sum of the Laplace transforms of the functions. Therefore,

$$\mathcal{L}[x(t) + y(t)] = X(s) + Y(s) \qquad\qquad \text{(6-27)}$$

Final Value and Initial Value Theorems. Two valuable theorems in control analysis relate the final ($t \rightarrow \infty$) and initial ($t \rightarrow 0$) values of a function of time to the value of the Laplace transform of the function as the complex variable s approaches zero and infinity, respectively.

The final value of a function of time, if the $\lim_{t \rightarrow \infty} f(t)$ exists, is the steady-state value of the function. As illustrated in Chapter 5, the final value, or steady-state value, is a quantity of considerable importance in the analysis of system behavior. The steady-state value of $f(t)$ may be obtained from the Laplace transform $F(s)$ by,

$$\lim_{t \rightarrow \infty} f(t) = \lim_{s \rightarrow 0} sF(s) \qquad (6\text{-}28)$$

With the initial value theorem, the value of $f(t)$ at $t = 0$ may be obtained from the transform $F(s)$ by

$$\lim_{t \rightarrow 0} f(t) = \lim_{s \rightarrow \infty} sF(s) \qquad (6\text{-}29)$$

The left-hand side of this equation is sometimes written as $\lim_{t=0+} f(t)$ where the 0+ indicates an infinitely small interval following the instant zero. Another way of properly restricting Eq. 6-29 is to specify that t approaches zero from positive values of time.[1]

Equations 6-28 and 6-29 constitute the first tools for the prediction of system behavior in the time domain without actually transforming functions in s back to the time domain. An example of this is given in Sec. 6-7.

6-5. Table of Laplace Transform Pairs

The use of the Laplace transform would be a tedious process indeed, if it were necessary to evaluate the integral of Eq. 6-1 each time the transform of a function was taken. As the transform of a function $f(t)$ depends only on $f(t)$, a table of transforms can be prepared for rapid reference. Such a table is called a table of *transform pairs* and is given in Table 6-1. As an example of the use of a table of pairs, consider the Laplace transform of the sinusoidal input function

$$f(t) = K \sin \omega t$$

The transform is obtained with the aid of item seventeen in Table 6-1. Thus,

$$F(s) = \frac{K\omega}{s^2 + \omega^2}$$

Further use of Table 6-1 will be explored in the following sections. As will be seen, it is an invaluable tool in handling complex problems. A more complete table of transforms, including many more sophisticated transformations, is cited in reference [21].

---

[1]Consider, for example, a unit step function $u(t)$ which has the values zero and unity at the instant of time zero. If zero is approached from positive values of time then $\lim_{t \rightarrow 0} u(t) = 1$. Thus $u(t = 0+) = 1$.

TABLE 6-1
Laplace Transform Pairs

Functions of Time: $f(t)$ for $0 \leq t$	Laplace Transforms: $\mathcal{L}[f(t)]$
1. $f(t)$	$\int_0^\infty f(t)e^{-st}\,dt = F(s)$
2. $x(t) + y(t)$	$X(s) + Y(s)$
3. $Kf(t)$	$KF(s)$
4. $\dfrac{df(t)}{dt}$	$sF(s) - f(0)$
5. $\dfrac{d^2f(t)}{dt^2}$	$s^2F(s) - sf(0) - \dfrac{df(0)}{dt}$
6. $\dfrac{d^nf(t)}{dt^n}$	$s^nF(s) - \displaystyle\sum_{i=1}^{n} s^{n-i}\dfrac{d^{i-1}f(0)}{dt^{i-1}}$
7. $\displaystyle\int_0^t f(t)\,dt$	$\dfrac{1}{s}F(s)$
8. 1 or $u(t)$	$\dfrac{1}{s}$
9. t	$\dfrac{1}{s^2}$
10. t^n for $n > -1$	$\dfrac{n!}{s^{n+1}}$
11. e^{-at}	$\dfrac{1}{s+a}$
12. te^{-at}	$\dfrac{1}{(s+a)^2}$
13. t^ne^{-at}	$\dfrac{n!}{(s+a)^{n+1}}$
14. $1 - e^{-at}$	$\dfrac{a}{s(s+a)}$
15. $e^{-at} - e^{-bt}$	$\dfrac{b-a}{(s+a)(s+b)}$
16. $ae^{-at} - be^{-bt}$	$\dfrac{(a-b)s}{(s+a)(s+b)}$
17. $\sin \alpha t$	$\dfrac{\alpha}{s^2+\alpha^2}$
18. $\cos \alpha t$	$\dfrac{s}{s^2+\alpha^2}$

TABLE 6-1 (*Continued*)

Functions of Time: $f(t)$ for $0 \leq t$	Laplace Transforms: $\mathcal{L}\,[f(t)]$
19. $t \sin \alpha t$	$\dfrac{2\alpha s}{(s^2 + \alpha^2)^2}$
20. $t \cos \alpha t$	$\dfrac{s^2 - \alpha^2}{(s^2 + \alpha^2)^2}$
21. $e^{-at} \sin \alpha t$	$\dfrac{\alpha}{(s + a)^2 + \alpha^2}$
22. $e^{-at} \cos \alpha t$	$\dfrac{s + a}{(s + a)^2 + \alpha^2}$
23. $e^{-\zeta \omega_n t} \sin \omega_n (1 - \zeta^2)^{\frac{1}{2}} t$ for $\zeta < 1$	$\dfrac{\omega_n (1 - \zeta^2)^{\frac{1}{2}}}{s^2 + 2\zeta\omega_n s + \omega_n^2}$
24. $e^{-\zeta \omega_n t} \sinh \omega_n (\zeta^2 - 1)^{\frac{1}{2}} t$ for $\zeta > 1$	$\dfrac{\omega_n (\zeta^2 - 1)^{\frac{1}{2}}}{s^2 + 2\zeta\omega_n s + \omega_n^2}$
25. $1 - \dfrac{e^{-\zeta \omega_n t}}{\sqrt{1 - \zeta^2}} \sin (\omega_n \sqrt{1 - \zeta^2}\, t + \phi)$ $\phi = \tan^{-1} \dfrac{\sqrt{1 - \zeta^2}}{\zeta}$ for $\zeta < 1$	$\dfrac{\omega_n{}^2}{s(s^2 + 2\zeta\omega_n s + \omega_n{}^2)}$
26. $\begin{Bmatrix} f(t - a) & \text{where } t > a \\ 0 & \text{where } t < a \end{Bmatrix}$	$e^{-as} F(s)$

6-6. The Inverse Transformation

Following the manipulation in the s domain, it may be desirable to return to the original variable t in the time domain. The process of converting a function back to dependence on t is called *inverse transformation*. The inverse transform is denoted by

$$\mathcal{L}^{-1}[F(s)] = f(t) \tag{6-30}$$

In general, if a function possesses a given transform, then it is the only possible transform for that function, and further, the inverse transform is usually a unique function in time from which the transform was derived. The latter fact makes it possible to proceed in either direction in Table 6-1. In taking the inverse transform, one finds the $F(s)$ in the right-hand column and then moves to the left to obtain the function of time from which that transform was derived. It should be noted that in no case in Table 6-1 does a transform of a function of time appear more than once. If, for example, one has the transform

$$F(s) = \frac{1}{(s + \gamma)(s + \beta)}$$

then item fifteen of Table 6-1 provides the function of time

$$f(t) = \frac{1}{(\beta - \gamma)} (e^{-\gamma t} - e^{-\beta t})$$

Partial Fraction Expansion. The function $F(s)$ for which an inverse transform is desired often appears as the ratio of two polynomials. Consider,

$$F(s) = \frac{N(s)}{D(s)} = \frac{a_i s^i + a_{i-1} s^{i-1} + \cdots + a_1 s + a_0}{b_n s^n + b_{n-1} s^{n-1} + \cdots + b_1 s + b_0} \qquad (6\text{-}31)$$

The indices i and n are real positive integers, the a's and b's are real constants and b_n is made unity. To obtain the inverse transform of this complicated function would be a difficult task. It is therefore desirable to manipulate Eq. 6-31 to a form that may be more readily handled. Specifically, it would be convenient to rearrange this ratio of two polynomials in groups of partial fractions each of which could be easily transformed by inspection.

It is important that the highest power of s in the denominator of Eq. 6-31 be greater than the highest power of s in the numerator. Specifically,

$$n > i \qquad (6\text{-}32)$$

If this is not the case, the numerator may be divided by the denominator as many times as necessary to produce a series of terms in s plus a ratio of polynomials as a remainder which satisfies the condition of Eq. 6-32. In general, in the study of automatic controls the nature of the system is such that Eq. 6-32 is satisfied and the division process is not necessary.

The denominator of Eq. 6-31 can be factored to yield the form

$$F(s) = \frac{N(s)}{(s + r_1)(s + r_2)(s + r_3) \cdots (s + r_n)} \qquad (6\text{-}33)$$

where $-r_1, -r_2, -r_3, \ldots, -r_n$ are roots of $D(s) = 0$ (or zeros of $D(s)$, that is, values of s for which $D(s) = 0$) and may be either real or complex. First, consider the case where all roots of $D(s) = 0$ are different (there are no repeated roots). Then the partial fraction expansion is given by

$$\frac{N(s)}{D(s)} = \frac{C_1}{s + r_1} + \frac{C_2}{s + r_2} + \cdots + \frac{C_n}{s + r_n} \qquad (6\text{-}34)$$

The nth constant C_n can be determined by multiplying both sides of the equation by the denominator of the nth fraction, namely $(s + r_n)$. Hence the nth factor may be canceled on both sides of Eq. 6-34. Next, replace s by $-r_n$ and solve for C_n. This process is expressed mathematically by the relationship

$$C_n = \lim_{s \to -r_n} \left[\frac{N(s)}{D(s)} (s + r_n) \right] \tag{6-35}$$

In particular, then,

$$C_1 = \frac{N(s)}{D(s)} (s + r_1) \Big|_{s = -r_1} \tag{6-36}$$

$$C_2 = \frac{N(s)}{D(s)} (s + r_2) \Big|_{s = -r_2} \tag{6-37}$$

EXAMPLE 6-1. Determine the partial fraction expansion and the inverse transform of the ratio of polynomials

$$F(s) = \frac{s + 2}{s^3 + 8s^2 + 19s + 12} = \frac{s + 2}{(s + 3)(s + 4)(s + 1)}$$

Solution: $F(s)$ is expanded in partial fractions as

$$F(s) = \frac{C_1}{s + 3} + \frac{C_2}{s + 4} + \frac{C_3}{s + 1}$$

Using Eq. 6-35,

$$C_1 = \frac{-3 + 2}{(-3 + 4)(-3 + 1)} = \frac{-1}{-2} = \frac{1}{2}$$

$$C_2 = \frac{-4 + 2}{(-4 + 3)(-4 + 1)} = \frac{-2}{-1(-3)} = -\frac{2}{3}$$

and

$$C_3 = \frac{-1 + 2}{(-1 + 3)(-1 + 4)} = \frac{1}{2(3)} = \frac{1}{6}$$

Therefore,

$$F(s) = \frac{1/2}{s + 3} - \frac{2/3}{s + 4} + \frac{1/6}{s + 1}$$

Using Table 6-1, item eleven, the inverse transform of $F(s)$ is obtained by inspection

$$f(t) = \mathcal{L}^{-1}[F(s)] = \tfrac{1}{2}e^{-3t} - \tfrac{2}{3}e^{-4t} + \tfrac{1}{6}e^{-t} \qquad Ans.$$

If $F(s)$ contains repeated roots in the denominator, the procedure for finding the C's for the partial fraction expansion must be modified. Consider,

$$\frac{N(s)}{D(s)} = \frac{N(s)}{(s + r_1)^m (s + r_{m+1})(s + r_{m+2}) \cdots (s + r_{m+n})} \tag{6-38}$$

The root $-r_1$ is repeated here m times, and there are n nonrepeated roots making a total of $(m + n)$ factors of $D(s)$. Expanding in partial fraction form, Eq. 6-38 becomes

$$\frac{N(s)}{D(s)} = \frac{C_1}{(s + r_1)^m} + \frac{C_2}{(s + r_1)^{m-1}} + \cdots + \frac{C_m}{s + r_1}$$

$$+ \frac{C_{m+1}}{s + r_{m+1}} + \cdots + \frac{C_{m+n}}{s + r_{m+n}} \tag{6-39}$$

The constants C_1, C_2, and C_m for the repeated roots are determined from the following relationships:

$$C_1 = \frac{N(s)}{D(s)} (s + r_1)^m \bigg|_{s = -r_1} \tag{6-40}$$

$$C_2 = \frac{d}{ds} \left[\frac{N(s)}{D(s)} (s + r_1)^m\right]\bigg|_{s = -r_1} \tag{6-41}$$

$$C_m = \frac{1}{(m - 1)!} \frac{d^{m-1}}{ds^{m-1}} \left[\frac{N(s)}{D(s)} (s + r_1)^m\right]\bigg|_{s = -r_1} \tag{6-42}$$

EXAMPLE 6-2. Find the inverse transform of the function

$$F(s) = \frac{s + 1}{(s + 2)^2(s + 3)}$$

Solution: $F(s)$ is expanded in partial fractions as

$$\frac{N(s)}{D(s)} = \frac{C_1}{(s + 2)^2} + \frac{C_2}{s + 2} + \frac{C_3}{s + 3}$$

From Eqs. 6-35, 6-40, and 6-41 the constants become

$$C_1 = \frac{-2 + 1}{-2 + 3} = -1$$

$$C_2 = \frac{d}{ds}\left[\frac{s + 1}{s + 3}\right]\bigg|_{s = -2} = \frac{(s + 3) - (s + 1)}{(s + 3)^2}\bigg|_{s = -2}$$

$$= \frac{(-2 + 3) - (-2 + 1)}{(-2 + 3)^2} = 2$$

$$C_3 = \frac{-3 + 1}{(-3 + 2)^2} = -2$$

The inverse transform is then obtained by inspection from items eleven and twelve of Table 6-1 as

$$f(t) = -te^{-2t} + 2e^{-2t} - 2e^{-3t} \qquad\qquad Ans.$$

6-7. Solution of Differential Equations by the Laplace Transform Method

Sections 6-1 through 6-6 present the tools necessary to solve a linear differential equation with constant coefficients by the Laplace transform method. It is now possible to solve a few of the fundamental system equations to illustrate the procedure.

Equation of a First-Order System. Section 5-2 discussed the class of physical systems defined by a first-order linear differential equation with constant coefficients. Typical of this class is a mechanical spring-damper system where one end of the spring is displaced at time zero, Fig. 6-9. The differential equation of motion defining the other end of the spring is given by Eq. 5-1, including the $u(t)$ notation indicating a step input, as

$$c \frac{dy}{dt} + ky = kXu(t) \tag{6-43}$$

As a result of the linearity properties, Eqs. 6-26 and 6-27, the Laplace transform of the left side of the equation is the sum of the Laplace transforms of the individual terms. Also, the transform of the left side equals the transform of the right. Thus, using Table 6-1 to transform term by term on both sides of the equation,

$$c[sY(s) - y(0)] + kY(s) = \frac{kX}{s}$$

Once the transform to the s domain is complete the equation in s may be manipulated algebraically. If the initial value of $y(t)$ is given as zero, then[2]

$$(cs + k)Y(s) = \frac{kX}{s}$$

or

$$Y(s) = \frac{kX}{s(cs + k)} = \frac{(k/c)X}{s(s + k/c)}$$

The latter expression may be expanded into partial fractions to give

$$Y(s) = \frac{k}{c} X \left(\frac{c/k}{s} + \frac{-c/k}{s + k/c} \right)$$

Thus,

$$Y(s) = X \left(\frac{1}{s} - \frac{1}{s + k/c} \right)$$

Turning to Table 6-1, the time function is determined by inspection to be

$$y(t) = X[1 - e^{-(k/c)t}] \tag{6-44}$$

This is seen to agree with Eq. 5-2 obtained by use of the p-operator.

Equation of a Second-Order System. Consider a mechanical spring-mass-damper system with a step input displacement at one end of the spring (Fig. 6-10). The equation of motion, given by Eq. 5-9 but using the $u(t)$ notation, is

$$M \frac{d^2y}{dt^2} + c \frac{dy}{dt} + ky = kXu(t) \tag{6-45}$$

---

[2]It is helpful to define, at this time, the characteristic equation in s analogous to the characteristic equation in p obtained in Sec. 3-2. If the homogeneous form of Eq. 6-43 is transformed with all initial conditions equal to zero, the result is

$$csY(s) + kY(s) = 0$$

Again, because y is a required nontrivial solution

$$cs + k = 0$$

This is, by definition, the characteristic equation in s.

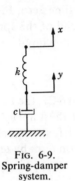

FIG. 6-9.
Spring-damper
system.

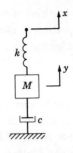

FIG. 6-10.
Spring-mass-
damper system.

Transforming with the aid of Table 6-1 yields

$$M\left[s^2 Y(s) - sy(0) - \frac{dy(0)}{dt}\right] + c[sY(s) - y(0)] + kY(s) = \frac{kX}{s} \qquad (6\text{-}46)$$

Again, if all initial conditions are specified as zero, then

$$(Ms^2 + cs + k)\,Y(s) = \frac{kX}{s}$$

or

$$Y(s) = \frac{kX}{s(Ms^2 + cs + k)} = \frac{(k/M)X}{s[s^2 + (c/M)s + k/M]} \qquad (6\text{-}47)$$

Substituting the definitions developed in Sec. 5-3 that

$$\omega_n = \left(\frac{k}{M}\right)^{\!1/2} \qquad (6\text{-}48)$$

and

$$\frac{c}{M} = 2\zeta\omega_n \qquad (6\text{-}49)$$

gives

$$Y(s) = \frac{\omega_n^2 X}{s(s^2 + 2\zeta\omega_n s + \omega_n^2)} \qquad (6\text{-}50)$$

The next step is to perform the inverse transformation. Consider, for example, the case where $\zeta = 1$ (as discussed in Chapter 5, the nature of the solution depends on ζ). Equation 6-50 becomes

$$Y(s) = \frac{\omega_n^2 X}{s(s^2 + 2\omega_n s + \omega_n^2)} = \frac{\omega_n^2 X}{s(s + \omega_n)^2} \qquad (6\text{-}51)$$

Expanding in partial fraction form gives

$$Y(s) = \omega_n^2 X\left[\frac{C_1}{s} + \frac{C_2}{(s + \omega_n)^2} + \frac{C_3}{s + \omega_n}\right]$$

and using Eqs. 6-35, 6-40, and 6-41,

$$Y(s) = \omega_n^2 X\left[\frac{1/\omega_n^2}{s} + \frac{-1/\omega_n}{(s + \omega_n)^2} + \frac{-1/\omega_n^2}{s + \omega_n}\right]$$

Thus, $$Y(s) = X\left[\frac{1}{s} - \frac{\omega_n}{(s + \omega_n)^2} - \frac{1}{s + \omega_n}\right] \qquad (6\text{-}52)$$

The inverse transform may now be completed by inspection using items eight, eleven, and twelve of Table 6-1. Thus,

$$y(t) = X[1 - e^{-\omega_n t} - \omega_n t\, e^{-\omega_n t}] \qquad (6\text{-}53)$$

It is interesting to make a test of the initial and final value theorems for the transformed form of the solution (Eq. 6-51). In the time domain it can be seen from Eq. 6-53 that at $t = 0$, $y(t) = 0$; and as t approaches infinity, $y(t) = X$. Turn next to Eq. 6-51. Applying the initial value theorem, Eq. 6-29, gives

$$\lim_{t\to 0} f(t) = \lim_{s\to\infty}\left[\frac{s\omega_n^2 X}{s(s^2 + 2\omega_n s + \omega_n^2)}\right] = 0$$

Thus,
$$\lim_{t\to 0} f(t) = 0$$

From the final value theorem, Eq. 6-28,

$$\lim_{t\to\infty} f(t) = \lim_{s\to 0}\left[\frac{s\omega_n^2 X}{s(s^2 + 2\omega_n s + \omega_n^2)}\right] = X$$

or
$$\lim_{t\to\infty} f(t) = X$$

which substantiates the results obtained from Eq. 6-53.

It is perhaps advantageous at this point to consider a detailed numerical example to further illustrate the use of the transform method.

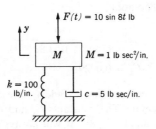

Fig. 6-11. Spring-mass-damper system with harmonic excitation.

EXAMPLE 6-3. The system to be studied is a spring-mass-damper system as shown in Fig. 6-11. The problem is to find the displacement $y(t)$ of the mass when it is subjected to a simple harmonic force. The initial value of y and its derivatives is zero. The differential equation of motion is

$$\frac{d^2 y}{dt^2} + 5\frac{dy}{d\iota} + 100 y = 10 \sin 8t$$

Solution: Transforming the equation with the aid of Table 6-1 yields

$$Y(s)[s^2 + 5s + 100] = \frac{80}{s^2 + 64}$$

or
$$Y(s) = \frac{80}{(s^2 + 64)(s^2 + 5s + 100)}$$

Factoring the quadratic terms,

$$Y(s) = \frac{80}{(s + j8)(s - j8)(s + 2.5 + j9.68)(s + 2.5 - j9.68)}$$

Then expressing in partial fraction form,

$$Y(s) = \frac{C_1}{s + j8} + \frac{C_2}{s - j8} + \frac{C_3}{s + 2.5 + j9.68} + \frac{C_4}{s + 2.5 - j9.68}$$

The constants C_1, C_2, C_3, and C_4 may be evaluated by applying Eq. 6-35.

$$C_1 = \frac{80}{(s - j8)(s^2 + 5s + 100)}\bigg|_{s = -j8} = \frac{80}{(-j16)(36 - j40)}$$

Converting to exponential notation, C_1 becomes

$$C_1 = \frac{80}{(16e^{-j1.57})(53.8e^{-j0.84})}$$

or
$$C_1 = 0.0929e^{j2.41}$$

Evaluating C_2 in the same manner,

$$C_2 = \frac{80}{(s + j8)(s^2 + 5s + 100)}\bigg|_{s = j8} = \frac{80}{(j16)(36 + j40)}$$

Converting to exponential notation,

$$C_2 = \frac{80}{(16e^{j1.57})(53.8e^{j0.84})}$$

or
$$C_2 = 0.0929e^{-j2.41}$$

The student should recognize that C_1 and C_2 are complex conjugates. This occurrence was not accidental. The constants C_1 and C_2 are associated with the conjugate factors $(s + j8)$ and $(s - j8)$. It will always be true that the two constants so associated with complex conjugate factors, will in themselves be complex conjugates.

Evaluating C_3 and C_4 in a like manner,

$$C_3 = 0.0768e^{-j0.45}$$

$$C_4 = 0.0768e^{j0.45}$$

The inverse transform of $Y(s)$ may be evaluated by first obtaining the inverse transforms of the individual terms from Table 6-1. Thus,

$$\mathcal{L}^{-1}\left[\frac{1}{s + j8}\right] = e^{-j8t}$$

$$\mathcal{L}^{-1}\left[\frac{1}{s - j8}\right] = e^{j8t}$$

$$\mathcal{L}^{-1}\left[\frac{1}{s + 2.5 + j9.68}\right] = e^{-(2.5+j9.68)t}$$

$$\mathcal{L}^{-1}\left[\frac{1}{s + 2.5 - j9.68}\right] = e^{-(2.5-j9.68)t}$$

The inverse transform $y(t)$ is then

$$y(t) = 0.0929(e^{j2.41}e^{-j8t} + e^{-j2.41}e^{j8t})$$
$$+ 0.0768[e^{-j0.45}e^{-(2.5+j9.68)t} + e^{j0.45}e^{-(2.5-j9.68)t}]$$

Rearranging,

$$y(t) = 0.0929[e^{-j(8t-2.41)} + e^{j(8t-2.41)}] + 0.0768e^{-2.5t}[e^{-j(9.68t+0.45)} + e^{j(9.68t+0.45)}]$$

Although this is a complete expression for $y(t)$, it is not presented in the most convenient or meaningful form. The equation for $y(t)$ may be expressed in trigonometric terms by using Eq. 6-10.

$$\frac{e^{j\theta} + e^{-j\theta}}{2} = \cos\theta$$

Thus,

$$y(t) = 0.186\cos(8t - 2.41) + 0.154e^{-2.5t}\cos(9.68t + 0.45)$$

or if the phase angle is expressed in degrees,

$$y(t) = 0.186\cos(8t - 138.1°) + 0.154e^{-2.5t}\cos(9.68t + 25.8°)$$

Noting also that $\cos\theta = \sin(\theta + 90°)$, the final expression for $y(t)$ becomes

$$y(t) = 0.186\sin(8t - 48.1°) + 0.154e^{-2.5t}\sin(9.68t + 115.8°) \qquad Ans.$$

This solution may be verified by the methods of Chapter 5. The term of the solution including the exponential $e^{-2.5t}$ is the transient solution which will, for all practical purposes, decay during a period of four time constants or when $t = 4/2.5$ seconds. The term $0.186\sin(8t - 48.1°)$ is the steady-state solution and will persist as long as the forcing function is present.

6-8. The Convolution Integral

If, in the process of inversion of the Laplace transform $F(s)$ to the time domain, the inverse transform of $F(s)$ is not known, the convolution integral often provides a direct solution. In particular, when $F(s)$ is composed of the product of two transforms $X(s)$ and $Y(s)$ whose inverse transforms are known to be $x(t)$ and $y(t)$ respectively, then,

$$F(s) = X(s) Y(s) \tag{6-54}$$

and $$\mathcal{L}^{-1}\{F(s)\} = \mathcal{L}^{-1}\{X(s) Y(s)\}$$

The convolution theorem states that

$$\mathcal{L}^{-1}\{X(s) Y(s)\} = \int_0^t x(t - \tau) y(\tau) d\tau \tag{6-55}$$

where t and τ are functions of time.

Hence $$\mathcal{L}^{-1}\{F(s)\} = \int_0^t x(t - \tau) y(\tau) d\tau \tag{6-56}$$

EXAMPLE 6-4. Find the inverse transform of $F(s) = \dfrac{10}{s(s + 2)}$ with the aid of the convolution integral. It is known that

$$\mathcal{L}^{-1}\left\{\frac{10}{s}\right\} = 10 \quad \text{and} \quad \mathcal{L}^{-1}\left\{\frac{1}{s + 2}\right\} = e^{-2t}$$

Solution:

$$\mathcal{L}^{-1}\left\{\frac{10}{s(s + 2)}\right\} = \int_0^t 10 e^{-2\tau} d\tau$$

$$= \frac{10 e^{-2\tau}}{(-2)}\bigg|_0^t = -5 e^{-2t} + 5$$

$$= 5(1 - e^{-2t}) \hspace{4cm} Ans.$$

It should be pointed out that the independent variables $t - \tau$ and τ may be interchanged in the functions in the integral on the right of Eq. 6-56. That is,

$$\mathcal{L}^{-1}\{F(s)\} = \int_0^t y(t - \tau) x(\tau) d\tau$$

Therefore, Example 6-4 could have been solved in the following manner.

$$\mathcal{L}^{-1}\left\{\frac{10}{s(s + 2)}\right\} = \int_0^t 10 e^{-2(t - \tau)} d\tau$$

$$= \int_0^t 10 e^{-2t} e^{2\tau} d\tau$$

$$= 10 e^{-2t} \int_0^t e^{2\tau} d\tau = 10 e^{-2t} \frac{e^{2\tau}}{2}\bigg|_0^t$$

$$= 5 e^{-2t} [e^{2t} - 1] = 5 [1 - e^{-2t}]$$

It should be readily seen that a diligent choice of the variables could save manipulation. Also, integration by parts would generally be required except in the simplest of cases as cited.

The convolution integral has particular significance in regard to expressing the general response of a linear system.

For example, consider a linear spring-mass-dashpot system in terms of the transformed differential equation

$$(s^2 + 2\zeta \omega_n s + \omega_n^2) Y(s) = \omega_n^2 X(s)$$

where $Y(s)$ is the Laplace transform of the response, $X(s)$ the Laplace transform of the input with the system initially at rest in equilibrium. Forming the ratio of the Laplace transforms of the output to the input yields the transfer function for the system $G(s)$.

$$\frac{Y(s)}{X(s)} = \frac{\omega_n^2}{s^2 + 2\zeta \omega_n s + \omega_n^2} = G(s) \tag{6-57}$$

(System transfer functions will be treated in depth in Chapter 7.) Thus, the system response is given by

$$Y(s) = G(s) X(s)$$

If the inverse transform of $G(s)$ from Eq. 6-57 is known and the inverse transform of $X(s)$ is known, then $y(t)$ may be obtained from the convolution

$$y(t) = \int_0^t x(t - \tau) g(\tau) \, d\tau \tag{6-58}$$

The inverse transform of the transfer function is

$$\mathcal{L}^{-1}\{G(s)\} = g(t) \tag{6-59}$$

where $g(t)$ is the *impulse response* and is determined solely by the parameters of the system (i.e., as noted in Eq. 6-57, $g(t)$ would be a function of ζ and ω_n).

The term impulse response is derived from the fact that when $x(t)$ is a unit impulse

$$X(s) = 1$$

Thus,

$$Y(s) = G(s)$$

and

$$\mathcal{L}^{-1}\{Y(s)\} = y(t) = g(t)$$

The impulse response is a powerful analysis tool as it completely identifies the dynamic behavior of a linear system.

6-9. Closure

The concepts and techniques of the Laplace transform method have been presented. The use of the transformation for the solution of the types of differential equations encountered in simple control systems was illustrated in Sec. 6-7. The procedure for the solution of these equations is simple; first transform the equation from the t domain to the s domain. Second, with the axioms of algebra, manipulate this expression in s into a form which may be readily inversely transformed to the t domain by inspection with the aid of a table of transform pairs. The expression obtained is the complete solution of the differential equation.

It was pointed out in the introductory paragraphs that the Laplace transform is useful in many ways other than to obtain the complete solution of an equation in the time domain. This point was illustrated in Sec. 6-4 by the presentation of the initial and final value theorems and the application of these theorems to the analysis of the transform of Eq. 6-45. In the following chapters considerable use will be made of the transformation in specialized techniques which have been developed to aid in the ultimate goal of all of this work—the determination and prediction of control system dynamic behavior.

Problems

6-1. Represent the following complex numbers in both polar and exponential form (plot on the complex plane):

   a) $3 + j7$ b) $2 - j4$ c) $-5 + j11$ d) $-6 - j3$

6-2. Represent the following complex numbers in rectangular form (plot on the complex plane):

   a) $105\underline{/128°}$ b) $17e^{-j3.74}$ c) $10e^{j8.95}$ d) $5e^{j0.1}$

6-3. a) Perform the indicated multiplication using the rectangular form. Check the result by using the polar form.

$$(3 + j7)(2 - j4)$$

b) Perform the indicated division using the rectangular form. (HINT: Multiply both numerator and denominator by the conjugate of the denominator.) Check the result by using the polar form.

$$\frac{3 + j7}{2 - j4}$$

6-4. Transform the following functions of t by Eq. 6-1:

a) $f(t) = \cos t$
b) $f(t) = \sin t$

c) $f(t) = 5\dfrac{d^2x}{dt^2} + 10\dfrac{dx}{dt} + x - 3t$

  with $\dfrac{dx}{dt}\bigg|_{t=0} = 10$ and $x\big|_{t=0} = 2$

d) $f(t) = 800$ for $0 \le t \le 5$

6-5. The Laplace transform of the translated function is a useful technique for obtaining the transform for simple discontinuous functions. For example, the function in the figure for Prob. 6-5a is discontinuous at $t = D$. It may be thought of as being composed of two separate functions f_1 and f_2 as shown in Prob. 6-5b.

a) Using the Laplace transform of a translated function, find the Laplace transform of $f(t) = f_1(t) + f_2(t)$ in Prob. 6-5a and 6-5b.   b) Using the Laplace transform of translated functions, transform the function of the figure for Prob. 6-5c.

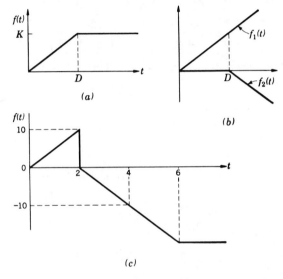

(a)

(b)

(c)

PROB. 6-5

6-6. Apply the initial and final value theorems to the following functions of s:

a) $F(s) = \dfrac{1}{s(s + 1)}$

b) $F(s) = \dfrac{10}{s(s^2 + 2s + 5)}$

c) $F(s) = \dfrac{10(s^2 + 2s + 1)}{s(0.1s^2 + s + 1)}$

d) $F(s) = \dfrac{100}{(s + 5)^4}$

e) $F(s) = \dfrac{s + 5}{s(s + 1)^2(s^2 + 2s + 1)}$

6-7. Expand the following functions of s into partial fractions:

a) $F(s) = \dfrac{s + a}{(s + b)(s + c)}$

b) $F(s) = \dfrac{s^2 + 2s + 1}{s(s + 3)(s + 1)}$

c) $F(s) = \dfrac{s + 1}{s(s^2 + 10s + 49)}$

d) $F(s) = \dfrac{(s + 2)(s + 3)}{s^3 + 6s^2 + 11s + 6}$

e) $F(s) = \dfrac{1}{s^2(s^2 + 10s + 205)}$

f) $F(s) = \dfrac{(s + 4)^2}{s(s + 2)^2(s + 3)^3}$

6-8. Find the inverse transform of the following:

a) $F(s) = \dfrac{1}{s + 1} - \dfrac{3}{s + 10} + \dfrac{5}{s^2 + 2s + 8}$

b) $F(s) = \dfrac{1070}{s^2(s + 3)^3}$

c) $F(s) = \dfrac{s + 2}{(s + j6)(s - j6)}$

d) $F(s) = \dfrac{1}{s} + \dfrac{5}{s + 6} + \dfrac{2e^{-j1}}{s + 3 + j4} + \dfrac{2e^{j1}}{s + 3 - j4}$

6-9. Find the inverse transform of the following functions using the convolution integral.

a) $F(s) = \dfrac{8}{(s + 4)(s^2 + 16)}$

b) $F(s) = \dfrac{10}{s(s + 2)(s^2 + 0.659)}$

6-10. Find the impulse response for systems with the following transfer function.

a) $G(s) = \dfrac{100}{(s + 2)(s + 5)(s + 5)}$

b) $G(s) = \dfrac{1}{.04s^2 + .08s + 1}$

6-11. Using the Laplace transform method solve the following differential equations:

a) $\dfrac{dy}{dt} + 10y = 2u(t)$ with $y|_{t=0} = 0$

b) $\dfrac{d^2x}{dt^2} + 3\dfrac{dx}{dt} + x = 5u(t)$ with $x|_{t=0} = 0$ and $\dfrac{dx}{dt}\bigg|_{t=0} = 0$

c) $\dfrac{d^2z}{dt^2} + 2\dfrac{dz}{dt} + z = 800 \sin 2t$ with $\dfrac{dz}{dt}\bigg|_{t=0} = 0$ and $z|_{t=0} = 0$

d) $\dfrac{dx}{dt} + 100x = 300u(t)$ with $\dfrac{dx}{dt}\bigg|_{t=0} = 20$

Chapter 7

Block Diagrams and Transfer Functions

7-1. Introduction

Block diagrams provide a convenient means for visualizing and analyzing control systems. These diagrams are obtained by first writing the equations which describe the behavior of each of the elements which make up the system. Then, the information contained in each of the equations is put in the form of a ratio of some output quantity to some input quantity. The relationship obtained is called a *transfer function* and is the mathematical representation of the particular element which is placed in the block. When all the elements in a system are represented in suitably related blocks, the over-all system equation can be obtained by a manipulation of the block diagram rather than by the simultaneous solution of the system equations by the usual mathematical methods. There are also other uses for the information contained in block diagrams, but these uses will be presented in later chapters.

This chapter first develops the concept of a transfer function and its relationship to individual blocks. Blocks representing typical elements found in linear systems are explored. Then, the over-all block diagram is explained and its use illustrated with an example. Finally, some transfer functions are developed for use in later portions of the book.

7-2. The Transfer Function

The transfer function for a linear element, component, or system can be defined as the ratio of the transform of the output to the transform of the input with the assumption that all initial conditions are zero. The transfer function for a particular element can be obtained by using the following three steps.

1. Write the appropriate equation which defines the behavior of the element.

2. Transform this equation assuming all initial conditions to be zero.

3. Form the ratio of output $O(s)$ to input $I(s)$ as

$$\frac{O(s)}{I(s)} = F(s) \qquad (7\text{-}1)$$

The transfer function is $F(s)$.

The element under consideration can be represented in block form as shown in Fig. 7-1. This block is interpreted to mean that the output $O(s)$ is obtained by multiplying the transfer function $F(s)$ by the input $I(s)$, that is,

$$O(s) = F(s)I(s) \quad \text{or} \quad \frac{O(s)}{I(s)} = F(s)$$

The latter expression is the definition of a transfer function. Thus one can put a transformed equation into block form and can also recover the equation from the block.

7-3. Blocks Representing Typical Elements

There are four basic blocks which occur in block diagrams representing linear control systems. Each block will be identified and illustrated by an example.

Proportional Block. The transfer function for many elements is simply a constant which indicates a proportionality between the output and the input. The block which represents such an element will be called a *proportional block*.

Consider the potentiometer shown in Fig. 7-2a. A constant voltage V is applied across the resistance winding which is L units long. The displacement of the wiper, the input, is denoted by x. The output voltage e_o is measured relative to the reference (ground) indicated. Thus the potentiometer is being used as a *displacement transducer*, that is, a device which provides an output voltage proportional to an input displacement.

The equation defining the behavior of the device is obtained as follows:

$$\frac{e_o}{V} = \frac{x}{L} \quad \text{or} \quad e_o = \frac{V}{L} x$$

Transforming and forming the ratio of output to input to get the transfer function yields

$$E_o(s) = \frac{V}{L} X(s) \quad \text{and} \quad \frac{E_o(s)}{X(s)} = \frac{V}{L} \tag{7-2}$$

The block representing the transducer is shown in Fig. 7-2b. The gain constant V/L is frequently designated as K; thus, the block merely relates the

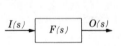

FIG. 7-1. An element represented in block form; $F(s)$ is the transfer function.

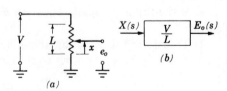

(a)

(b)

FIG. 7-2. A displacement transducer and its block representation.

output to the input by a constant. Proportional blocks may relate information from one form of signal to another (mechanical to electrical in the example used), indicate a gear or linkage ratio, or represent a source of energy.

Time Constant Block. The equation which defines the response of some elements is a first-order differential equation with all terms present. In this case, the transfer function contains a time constant, and so a block which represents a first-order element or system will be called a *time constant block.*

The rotating system shown in Fig. 7-3a will be used as an example. The applied torque T is the input and the rotating speed ω is the output. The appropriate system equation is

$$I \frac{d\omega}{dt} + b\omega = T$$

which can be written as

$$\tau \frac{d\omega}{dt} + \omega = \frac{T}{b}$$

where the time constant $\tau = I/b$. Transforming this equation, remembering that the initial conditions are assumed to be zero, yields

$$\tau s \Omega(s) + \Omega(s) = \Omega(s)[\tau s + 1] = \frac{1}{b} T(s)$$

The transfer function is

$$\frac{\Omega(s)}{T(s)} = \frac{1/b}{\tau s + 1} \tag{7-3}$$

The block representing the system is shown in Fig. 7-3b and is typical of time constant blocks. Further examples of time constant blocks include those representing a spring-damper system, an electrical RC circuit, and a simple liquid-level system.

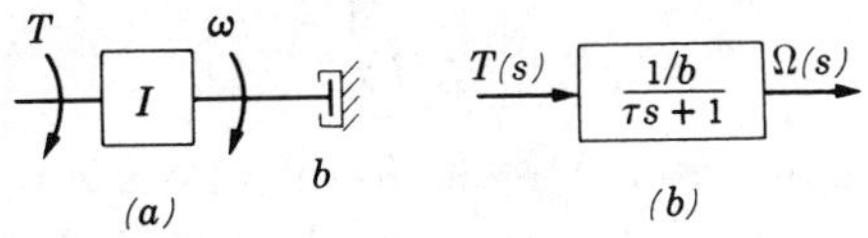

FIG. 7-3. A rotating system and its block representation.

Integrating Block. A third type of block represents an element for which a constant input produces a constant time rate of change of the output. Such a device is said to be an *integrating* element for a reason which will become apparent by considering an example.

A hydraulic cylinder is shown in Fig. 7-4a; the rate of fluid flow q to

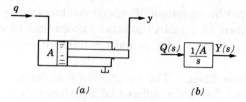

FIG. 7-4. A hydraulic cylinder and its
block representation.

the cylinder is the input, and the piston rod displacement y is the output.
The piston area is given the symbol A. The equation defining the response
of the cylinder is

$$q = A \frac{dy}{dt} \tag{7-4}$$

By transforming, assuming initial conditions zero, one obtains

$$Q(s) = As\,Y(s)$$

and the transfer function is

$$\frac{Y(s)}{Q(s)} = \frac{1/A}{s} \tag{7-5}$$

The block representing the hydraulic cylinder is shown in Fig. 7-4b and is
typical of integrating blocks.

But, one may question, where is the integration? This is found by
working from the information contained in the block to an equation which
describes the behavior of the element. Recall that the block is interpreted
to mean that the output is obtained by multiplying the transfer function
by the input; therefore

$$Y(s) = \frac{1/A}{s} Q(s)$$

Performing the inverse transformation yields

$$y = \frac{1}{A} \int_0^t q\, dt \tag{7-6}$$

Of course, this equation is but another form of Eq. 7-4 which was the starting point. However, Eq. 7-6 brings out the fact that an integration does
occur, that is, the input is integrated to obtain the output.

When looking at some physical device, an integrating element is probably most easily recognized by remembering that for a constant input, a
constant time rate of change of output will occur. For example, the hydraulic cylinder being considered will produce a constant output velocity
(that is, dy/dt) for a constant input flow rate.[1]

[1]This statement is no longer true should the piston reach the end of the cylinder. This
points out the fact that physical elements are linear only over some finite range.

Quadratic Block. The equation which defines the response of some elements is a second-order differential equation for which the characteristic equation is a quadratic. The response may well be oscillatory (but not necessarily so). A block representing such an element will be called a *quadratic block.*

Consider the familiar spring-mass-damper system shown in Fig. 7-5*a*. The input displacement is x and the output displacement is y. The appropriate differential equation is

$$M \frac{d^2y}{dt^2} + c \frac{dy}{dt} + ky = kx$$

This equation can be written as

$$\frac{M}{k} \frac{d^2y}{dt^2} + \frac{c}{k} \frac{dy}{dt} + y = x$$

or

$$\frac{1}{\omega_n^2} \frac{d^2y}{dt^2} + \frac{2\zeta}{\omega_n} \frac{dy}{dt} + y = x$$

where

$$\omega_n = \left(\frac{k}{M}\right)^{\frac{1}{2}} \quad \text{and} \quad \zeta = \frac{c}{2(Mk)^{\frac{1}{2}}}$$

The transfer function is

$$\frac{Y(s)}{X(s)} = \frac{1}{\dfrac{s^2}{\omega_n^2} + \dfrac{2\zeta s}{\omega_n} + 1} \tag{7-7}$$

The block representing the spring-mass-damper system is shown in Fig. 7-5*b* and is the general form for all second-order elements.

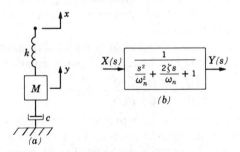

FIG. 7-5. A spring-mass-damper system and its block representation.

Four basic building blocks used in setting up block diagrams for linear control systems have been identified and illustrated. There are others (for example, a differentiating block), but the four discussed are by far the most common and will suffice for the present time.

7-4. The Block Diagram

A control system can be represented by a block diagram in which individual blocks are related so that the system equations are properly preserved. The justification for a block diagram is that it faithfully represents the system equations and their relationship to each other. Block diagrams can exist in any configuration consistent with this criterion.

Cascaded Blocks. In block diagrams, the output of one block is frequently the input of a second block; such blocks are said to be in series, or cascaded. Consider the two blocks shown in Fig. 7-6*a*; the output $B(s)$ of the first block is the input of the second. It is important to understand the meaning of the configuration. The implication is that the flow of information occurs only in the direction of the arrows; this in turn implies that the output of the first block is in no way affected by the presence of the second block. The same point is sometimes made by saying that the element being

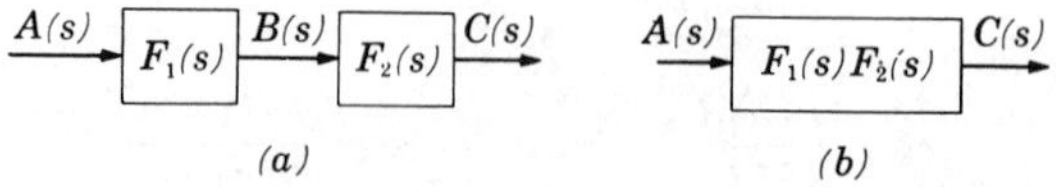

FIG. 7-6. Cascaded blocks.

represented by the second block must offer a high input impedance (impedance can be mechanical as well as electrical) so as not to *load* the element being represented by the first block. No matter how expressed, blocks can be shown in series only when the output of one block is not affected (or is only slightly affected) by the presence of the next following block. If loading effects (interaction between elements) are present in the system, the approach is to combine the elements in question into a single block before proceeding with a block diagram.

Any number of cascaded blocks can be replaced by a single block. For the configuration shown in Fig. 7-6*a*, the following equation can be written:

$$\frac{C(s)}{A(s)} = \frac{B(s)}{A(s)} \frac{C(s)}{B(s)} = F_1(s) F_2(s) \tag{7-8}$$

Therefore, the two blocks can be replaced by the single block shown in Fig. 7-6*b*. In general, the transfer function for a single block representing a number of cascaded blocks is simply the product of the individual transfer functions.

Summing Point. In order to construct a block diagram, some provision must be made for the algebraic summation of certain quantities. The device used is called a *summing point* and is drawn as a circle, as shown in Fig. 7-7. The mathematical signs indicate how the incoming quantities are to be summed to obtain the outgoing quantity. For example, the summing

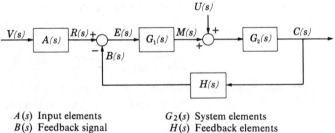

FIG. 7-7. Examples of summing points.

point shown in Fig. 7-7*a* is interpreted to mean that

$$C(s) = A(s) - B(s) \tag{7-9}$$

while that shown in Fig. 7-7*b* means that

$$G(s) = D(s) + E(s) + F(s) \tag{7-10}$$

A physical interpretation of the summing point will be given in connection with the example presented in Sec. 7-5.

Standard Notation. Various groups have attempted to standardize the symbols and terminology relating to feedback control systems. Figure 7-8 shows a block diagram of a typical feedback control system in which the symbols and terminology are patterned after those commonly used.

$A(s)$ Input elements	$G_2(s)$ System elements
$B(s)$ Feedback signal	$H(s)$ Feedback elements
$C(s)$ Controlled variable	$M(s)$ Manipulated variable
$E(s)$ Actuating signal	$R(s)$ Reference input
$G_1(s)$ Control elements	$U(s)$ Disturbance input

$V(s)$ Command

FIG. 7-8. Block diagram of typical feedback control system.

Closed-Loop Transfer Function. A transfer function which encompasses an entire control system is called a *closed-loop transfer function*. We will now investigate how to obtain these functions. For analytical purposes, the block diagrams for most control systems can be reduced to one of the two forms shown in Fig. 7-9. In Fig. 7-9*a*, the summing point is used to make the comparison between the controlled variable $C(s)$ and the reference input $R(s)$ so that an actuating signal $E(s)$ is provided. Note that the feedback portion of the diagram indicates that the controlled variable is measured and fed back for comparison without modification. The feed-

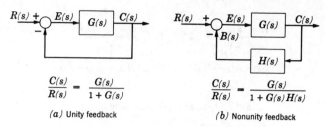

$$\frac{C(s)}{R(s)} = \frac{G(s)}{1 + G(s)}$$

(a) Unity feedback

$$\frac{C(s)}{R(s)} = \frac{G(s)}{1 + G(s)H(s)}$$

(b) Nonunity feedback

FIG. 7-9. Reduced forms of block diagrams and their closed-loop transfer functions.

back is handled on a one-to-one basis and is called *unity feedback*. The closed-loop transfer function for the configuration shown is obtained as follows. The relationship expressed by the block is

$$C(s) = G(s)E(s)$$

and that by the summing point is

$$E(s) = R(s) - C(s)$$

Substitution of the second expression into the first yields

$$C(s) = G(s)[R(s) - C(s)] = G(s)R(s) - G(s)C(s)$$
$$C(s) + G(s)C(s) = C(s)[1 + G(s)] = G(s)R(s)$$

The closed-loop transfer function is

$$\frac{C(s)}{R(s)} = \frac{G(s)}{1 + G(s)} \qquad (7\text{-}11)$$

The block diagram in Fig. 7-9*b* illustrates a system in which the feedback signal is modified before being compared with the reference input. This is called *nonunity feedback*. To obtain the closed-loop transfer function, one writes

$$C(s) = G(s)E(s)$$
$$E(s) = R(s) - B(s)$$
$$B(s) = H(s)C(s)$$

Simultaneous solution of these expressions yields the closed-loop transfer function

$$\frac{C(s)}{R(s)} = \frac{G(s)}{1 + G(s)H(s)} \qquad (7\text{-}12)$$

Equations 7-11 and 7-12 will be used frequently to obtain closed-loop transfer functions from block diagrams.

Block Diagram Manipulation. The initial block diagram for a system may well be more complicated than those shown in Fig. 7-9. In such cases, manipulation is required to achieve the appropriate form for use with Eqs. 7-11 and 7-12. The key idea is that a block diagram can exist

in any configuration as long as the system equations and their relationship to each other are preserved. Several examples will illustrate the
approach to block diagram manipulation.

Two summing-point configurations are shown in Fig. 7-10. The

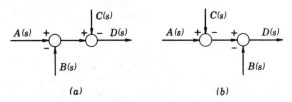

(a)　　　　　　　　　(b)

FIG. 7-10. Exchange of adjacent summing points.

difference between the two is that the summing points for Fig. 7-10*a* are
exchanged for Fig. 7-10*b*. Both configurations are equally valid because,
in each case,

$$D(s) = A(s) - B(s) - C(s)$$

Thus, if it serves a useful purpose, two adjacent summing points can be
exchanged.

Sometimes it is desirable to move a summing point around a block
as illustrated in Fig. 7-11. In Fig. 7-11*a*, one sees that the contribution

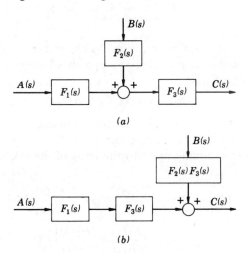

(a)

(b)

FIG. 7-11. Moving a summing point
around a block.

of $B(s)$ to $C(s)$ involves multiplication by both $F_2(s)$ and $F_3(s)$. For
the change of summing point shown in Fig. 7-11*b*, multiplication by both
transfer functions is accomplished by introducing $B(s)$ through a block
with a transfer function $F_2(s) F_3(s)$. The two configurations are equivalent
because, for each,

$$C(s) = F_1(s) F_3(s) A(s) + F_2(s) F_3(s) B(s)$$

The reader should be able to devise a scheme for moving the summing point of Fig. 7-11*a* to the left instead of to the right.

Figure 7-12 illustrates the movement of a take-off point around a

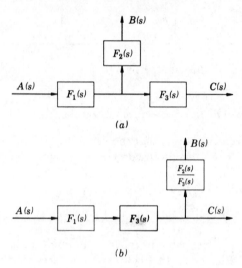

(a)

(b)

FIG. 7-12. Moving a take-off point around a block.

block. The new take-off point, Fig. 7-12*b*, includes multiplication by $F_3(s)$. Compensation is provided through division by $F_3(s)$ in the outgoing block. In both cases,

$$B(s) = F_1(s) F_2(s) A(s)$$

As a final example of block diagram manipulation, consider the diagram shown in Fig. 7-13*a*. The diagram of Fig. 7-13*b* is obtained by moving the inner summing point to the left and the take-off point to the right. One notes that

$$A(s) = R(s) - \frac{H_1(s)}{G_1(s)} C(s) - \frac{H_2(s)}{G_3(s)} C(s)$$

$$= R(s) - \left[\frac{H_1(s)}{G_1(s)} + \frac{H_2(s)}{G_3(s)} \right] C(s)$$

The second equation permits the system to be represented as shown in Fig. 7-13*c*. The closed-loop transfer function can now be obtained by the use of Eq. 7-12 with

$$G(s) = G_1(s) G_2(s) G_3(s)$$

and

$$H(s) = \frac{H_1(s)}{G_1(s)} + \frac{H_2(s)}{G_3(s)}$$

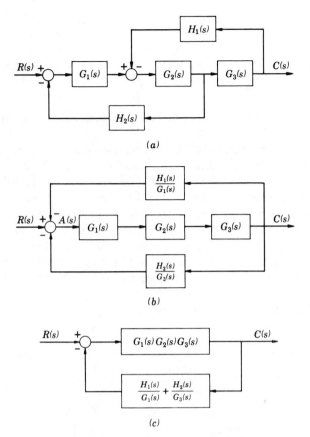

FIG. 7-13. Example of block diagram manipulation.

7-5. Example of the Use of Block Diagrams

The electromechanical positioning system introduced in Sec. 5-5 and shown in Fig. 7-14 will serve as a convenient example of the use of block

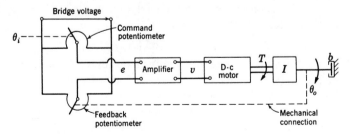

FIG. 7-14. Electromechanical positioning system.

diagrams. A summary of the system equations, previously determined in Sec. 5-5, is as follows:

$$e = K_1(\theta_i - \theta_o) = K_1\theta_i - K_1\theta_o \tag{7-13}$$

$$v = K_2 e \tag{7-14}$$

$$T = K_3 v \tag{7-15}$$

$$T = I\frac{d^2\theta_o}{dt^2} + b\frac{d\theta_o}{dt} \tag{7-16}$$

where K_1, K_2, and K_3 are the proportionality factors, or gains, for the various components.

To obtain the system block diagram, one first transforms the system equations and then represents them individually in block form. Block representations for Eqs. 7-13 through 7-16 are given in Fig. 7-15a through 7-15d, respectively. These blocks are then assembled to provide the system block diagram shown in Fig. 7-16a. Note in particular that the summing point indicates a comparison of two voltages from identical potentiometers —one corresponding to the input angular and the other to the output angular displacement. In this case, the summing point is a symbol representing the behavior of two potentiometers used in a bridge circuit.

The block diagram of Fig. 7-16a is the one most representative of the physical system. However, the diagram shown in Fig. 7-16b is also valid.

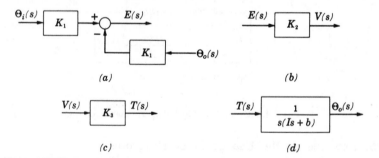

(a)   (b)

(c)   (d)

FIG. 7-15. Block representations for Eqs. 7-13 through 7-16.

It is based on the first form of Eq. 7-13 and the fact that the error $\epsilon = \theta_i - \theta_o$. The system equations and their relationship to each other remain unchanged because, from the diagram,

$$E(s) = K_1\epsilon(s) = K_1[\Theta_i(s) - \Theta_o(s)]$$

The most useful form of the block diagram for system analysis is shown in Fig. 7-16c, in which $K = K_1 K_2 K_3$. In this form, the *actuating signal* $E(s)$ is the actual system error.

The closed-loop transfer function will now be determined. The system block diagram (Fig. 7-16c) exhibits unity feedback; from Eq. 7-11,

$$\frac{\Theta_o(s)}{\Theta_i(s)} = \frac{G(s)}{1 + G(s)} = \frac{K/s(Is + b)}{1 + K/s(Is + b)} \qquad \frac{\Theta_o(s)}{\Theta_i(s)} = \frac{K}{Is^2 + bs + K} \tag{7-17}$$

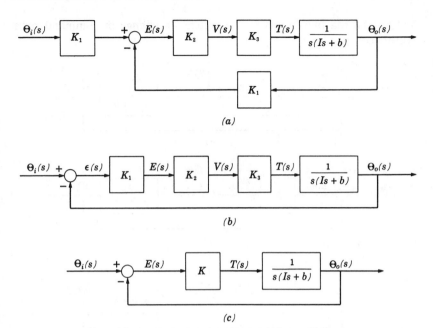

Fig. 7-16. Block diagrams for electromechanical positioning system.

The closed-loop transfer function can be used to obtain the system equation. Equation 7-17 can be written as

$$\Theta_o(s)[Is^2 + bs + K] = K\Theta_i(s)$$

Inverse transformation yields

$$I\frac{d^2\theta_o}{dt^2} + b\frac{d\theta_o}{dt} + K\theta_o = K\theta_i(t) \qquad (7\text{-}18)$$

This equation is, of course, the same as Eq. 5-23 but was obtained from a block-diagram approach. With Eq. 7-18, the response of the system to various inputs can be determined.

The use of block diagrams can be extended by considering a disturbance torque acting on the rotating mass in the system. The response of the system to such a torque was investigated in Sec. 5-5. In Fig. 7-14, assume a disturbance torque $T_d(t)$ acting on the mass in the positive direction. The only system equation requiring modification is Eq. 7-16 which becomes

$$T + T_d = I\frac{d^2\theta_o}{dt^2} + b\frac{d\theta_o}{dt} \qquad (7\text{-}19)$$

The use of a summing point to permit summation of the torques leads to the block diagram shown in Fig. 7-17a. This diagram illustrates the possibility of having two inputs to the system—a change in command or a disturbance.

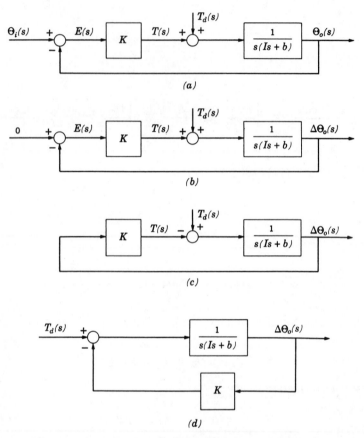

FIG. 7-17. Block diagrams for electromechanical positioning system with torque disturbance.

When considering a disturbance, it is convenient to assume the system initially at rest with zero error. In such an equilibrium condition, references are chosen making $\theta_i = \theta_o = 0$. The disturbance is then applied. Any deviation of the output will be designated as $\Delta\theta_o$ to indicate an unwanted change brought about by the disturbance.

The block diagram of Fig. 7-17a will be modified, in several steps, to obtain its most useful form. First, from the choice of references and the $\Delta\theta_o$ notation, the initial diagram can be represented as shown in Fig. 7-17b. Now, note that the only effect of the summing point with the zero input is to change the sign of the feedback signal as it passes through; that is,

$$E(s) = 0 - \Delta\Theta_o(s) = -\Delta\Theta_o(s)$$

However, the sign change can equally as well be taken into account by omitting the first summing point completely and changing the sign with which the torque $T(s)$ enters the second summing point. The resulting diagram is given in Fig. 7-17c. The final block diagram (Fig. 7-17d) is obtained by redrawing the diagram of Fig.7-17c in a more conventional form.

The block diagram of Fig. 7-17d exhibits nonunity feedback. Use of Eq. 7-12 yields

$$\frac{\Delta\Theta_o(s)}{T_d(s)} = \frac{G(s)}{1 + G(s)H(s)} = \frac{1/s(Is + b)}{1 + K/s(Is + b)}$$

$$\frac{\Delta\Theta_o(s)}{T_d(s)} = \frac{1}{Is^2 + bs + K}$$

The system equation is then

$$I\frac{d^2\Delta\theta_o}{dt^2} + b\frac{d\Delta\theta_o}{dt} + K\Delta\theta_o = T_d(t) \tag{7-20}$$

which is similar to Eq. 5-25.

The presentation of how to obtain system block diagrams and how to use them to determine closed-loop system equations is now complete. Other uses for the information contained in block diagrams will occur at later points in the book.

7-6. Signal Flow Graphs

A *signal flow graph* is a graphical representation of system equations and, in this sense, serves the same purpose as a block diagram. With a signal flow graph, system variables are represented as nodes with directed line segments between the nodes being used to show the relationships between the variables. Three examples will be presented to illustrate the basic concepts.

Figure 7-18 shows a simple block diagram and the equivalent signal

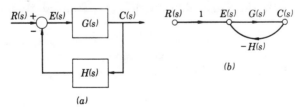

(a)

(b)

Fig. 7-18. A block diagram and the equivalent signal flow graph.

flow graph. In Fig. 7-18a, the three key variables are $R(s)$, $E(s)$, and $C(s)$. These variables are shown as nodes (the small circles) in the signal

flow graph of Fig. 7-18b. The directed line segments show the relationships between the variables including the appropriate transfer functions.
The expression for a variable at a node is obtained as the summation of
the incoming signals at that node. For example,

$$E(s) = 1 \cdot R(s) + [-H(s) \cdot C(s)]$$
$$= R(s) - H(s) C(s)$$

This expression is the same as that represented by the summing point
in the block diagram. Note that, with signal flow graphs, negative
summation is designated by the use of a negative sign with the appropriate transfer function. As with block diagrams, signals flow only
in the direction of the arrows.

A second example is shown in Fig. 7-19. The reader should be

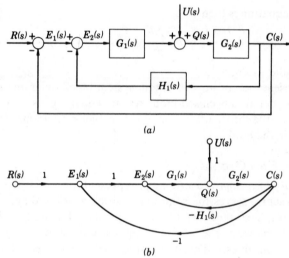

(a)

(b)

FIG. 7-19. Equivalent signal flow graph for a block
diagram.

able to see how the signal flow graph was obtained from the block
diagram.

An advantage of using a signal flow graph is that system transfer
functions are readily obtained without manipulation of the graph. The
equation used is

$$\frac{O(s)}{I(s)} = \sum_{k} \frac{M_k \Delta_k}{\Delta} \tag{7-21}$$

where $O(s)/I(s)$ = Transfer function, i.e., the ratio of the transforms of
the output and input variables.

M_k = Gain of the k-th forward path between $I(s)$ and $O(s)$.

Δ = 1 − (sum of all individual loop gains) + (sum of the gain product of all possible combinations of two non-touching loops) − (sum of the gain product of all possible combinations of three non-touching loops) + ···

Δ_k = Value of Δ for that part of the graph not touching the k-th path.

For use with Eq. 7-21, the term gain is interpreted to mean the product of the transfer functions.

The use of Eq. 7-21 can be illustrated by determining the closed-loop transfer function $C(s)/R(s)$ for the signal flow graph of Fig. 7-18b. There is only one forward path (direct path in the direction of the arrows) between $R(s)$ and $C(s)$. Thus,

$$M_1 = 1 \cdot G(s) = G(s)$$

There is only one loop; therefore,

$$\Delta = 1 - [-G(s) H(s)] = 1 + G(s) H(s)$$

The one loop touches the forward path; thus,

$$\Delta_1 = 1$$

With Eq. 7-21,

$$\frac{C(s)}{R(s)} = \frac{G(s)}{1 + G(s) H(s)}$$

The result is obviously correct.

Suppose that we wish the closed-loop transfer function $C(s)/U(s)$ (with $R(s) = 0$) for the block diagram of Fig. 7-19a. It should be obvious that block diagram manipulation will be required. However, with the signal flow graph of Fig. 7-19b and Eq. 7-21,

$$M_1 = G_2(s)$$
$$\Delta = 1 - [-G_1(s) G_2(s) H_1(s) - G_1(s) G_2(s)]$$
$$\Delta_1 = 1$$

Therefore,

$$\frac{C(s)}{U(s)} = \frac{G_2(s)}{1 + G_1(s) G_2(s) H_1(s) + G_1(s) G_2(s)}$$

Direct Use of Signal Flow Graphs. Although signal flow graphs were introduced in relation to block diagrams, there is no reason why one cannot make direct use of signal flow graphs. An example will illustrate the point.

Consider the liquid-level system shown in Fig. 7-20. The first defining equation is

$$C_1 \frac{dh_1}{dt} = q_i - q_1 - \frac{h_1}{R_1}$$

Transformation and solution for $H_1(s)$ yields

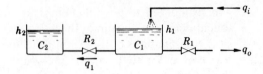

FIG. 7-20. Liquid-level system.

$$H_1(s) = \frac{R_1}{R_1 C_1 s + 1} [Q_i(s) - Q_1(s)]$$

$$= \frac{R_1}{R_1 C_1 s + 1} [\Sigma Q(s)]$$

This equation is represented by the partial signal flow graph of Fig. 7-21*a*.

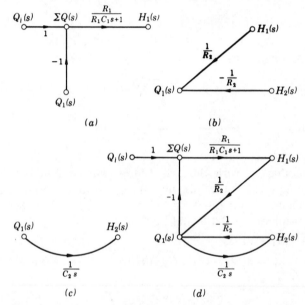

FIG. 7-21. Signal flow graphs for liquid-level system.

A second defining equation is

$$q_1 = \frac{h_1 - h_2}{R_2}$$

or

$$Q_1(s) = \frac{1}{R_2} H_1(s) - \frac{1}{R_2} H_2(s)$$

Figure 7-21*b* shows the corresponding partial signal flow graph.
The final defining equation is

$$C_2 \frac{dh_2}{dt} = q_1$$

or
$$H_2(s) = \frac{1}{C_2 s} Q_1(s)$$

The partial signal flow graph is given in Fig. 7-21c.

Combination of the partial signal flow graphs produces the system signal flow graph shown in Fig. 7-21d.

Equation 7-21 can be used to obtain the transfer function $H_1(s)/Q_i(s)$. There is one forward path and so

$$M_1 = \frac{R_1}{R_1 C_1 s + 1}$$

There are two loops and Δ becomes

$$\Delta = 1 - \left[-\frac{R_1}{R_2(R_1 C_1 s + 1)} - \frac{1}{R_2 C_2 s} \right]$$

One loop does not touch the forward path; therefore,

$$\Delta_1 = 1 - \left[-\frac{1}{R_2 C_2 s} \right]$$

The transfer function $H_1(s)/Q_i(s)$ is

$$\frac{H_1(s)}{Q_i(s)} = \frac{\dfrac{R_1}{R_1 C_1 s + 1} \left[1 + \dfrac{1}{R_2 C_2 s} \right]}{1 + \dfrac{R_1}{R_2(R_1 C_1 s + 1)} + \dfrac{1}{R_2 C_2 s}}$$

$$= \frac{R_1(R_2 C_2 s + 1)}{R_1 C_1 R_2 C_2 s^2 + (R_1 C_1 + R_2 C_2 + R_1 C_2)s + 1}$$

The transfer function $H_2(s)/Q_i(s)$ can be obtained in a similar manner.

With the introduction to signal flow graphs completed, attention is now turned to the development of transfer functions for certain control system components. These transfer functions will then be available for later use in problems which illustrate various aspects of control theory.

7-7. Transfer Functions for Transducers

The transducers to be considered are typically used as measurement devices when it is desired to obtain an electrical signal representative of some physical quantity; for example, displacement, speed, pressure, and so forth. Transfer functions for four commonly used transducers will be obtained beginning with the potentiometer.

Potentiometers. One type of potentiometer was discussed in Sec. 7-3. Its transfer function, based on Eq. 7-2, can be written as

$$\frac{E_o(s)}{X(s)} = K \tag{7-22a}$$

where $E_o(s)$ is the transformed output voltage and $X(s)$ is the transformed input displacement. For a rotary potentiometer,

$$\frac{E_o(s)}{\Theta(s)} = K \qquad (7\text{-}22b)$$

where $\Theta(s)$ is the transformed input angular displacement.[2]

The wire-wound potentiometer is a convenient and widely utilized transducer. It can be used with either a-c or d-c voltages. The output-input relationship is not precisely linear, and the wear produced by the sliding action of the wiper on the winding tends to impose some limitation on the useful life of the device.

Linear Variable Differential Transformers. The linear variable differential transformer (LVDT) is an electromagnetic a-c device typically used as a displacement transducer. A schematic drawing of an LVDT is shown in Fig. 7-22a. The device consists of a primary winding, energized with a fixed a-c voltage e, and two secondary windings. A moveable magnetic core

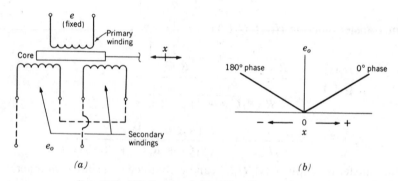

Fig. 7-22. The linear variable differential transformer.

provides magnetic coupling between the windings. The two secondaries can be wired opposing each other, as shown in Fig. 7-22a by dashed lines. The operation is as follows. With the core centered, the same voltage is induced in both secondaries because the flux path for each is identical. The output voltage e_o is then zero because the secondary windings are wired to oppose each other. A displacement of the core unbalances the flux paths and results in an increase of induced voltage in one secondary at the expense of the other. This in turn provides a net output voltage e_o of the same phase as that of the favored secondary.

The performance of the LVDT is shown graphically in Fig. 7-22b.

---

[2] In this section, the letter K will frequently be used to indicate the gain of a component. All gains are certainly not equal as might be implied. In working with a system in which there are several components, numerical subscripts (e.g. K_1, K_2, etc.) will be used to identify the gains of the various components.

Increasing displacements in the positive direction provide proportionately larger output voltages of 0° phase (compared with some reference voltage) while negative displacements produce output voltages of 180° phase. Thus an LVDT provides an output voltage proportional to displacement with a phase relationship dependent upon the direction of displacement. The transfer function then is

$$\frac{E_o(s)}{X(s)} = K \tag{7-23}$$

The LVDT is physically arranged in a tubular form. It is a rugged device in which wear is not a problem because the core need not make contact with the bore of the tube. LVDT's are linear only through their specified range of operation.

Tachometer Generators. The tachometer generator is a useful transducer for providing a voltage signal e_o proportional to rotating speed ω. The transfer function is

$$\frac{E_o(s)}{\Omega(s)} = K \tag{7-24}$$

Tachometer generators are small instrument generators. The output voltage can be either a-c or d-c depending on the type purchased. With an a-c unit, a reversal of rotation is indicated by a change in phase (for example, 180° phase instead of 0° phase); with a d-c unit, by a change of polarity.

Synchros. The term synchro is used for a large class of a-c electromechanical devices found extensively in transmission, computing, and controlling systems. Synchros look like small electric motors. However, they are essentially rotary differential transformers, shaft angular displacement being the mechanical input or output. Synchros are commonly used in pairs. In applications where a remote indication of angular displacement is desired, a synchro *transmitter* and a synchro *receiver* are used. The transmitter converts shaft position to an electrical signal which, when introduced to the receiver, causes its shaft to assume a corresponding position. When used in this manner, the combination of synchros has been described as an "electrical flexible shaft," the intermediate link being the electrical wiring between the two devices.

A synchro transmitter and a synchro *control transformer* may be combined to form a widely used error detection device in control systems. Figure 7-23 schematically represents such a combination in which the output e_o is an electrical signal proportional to the relative angle between the transmitter and control transformer rotor shafts.

Figure 7-24 shows a typical error voltage versus angular difference curve for the connection shown in Fig. 7-23. The output voltage is essentially proportional to the angular difference over a range of about 70°.

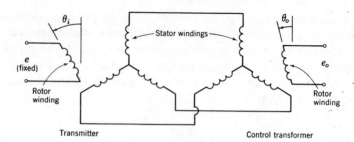

FIG. 7-23. Schematic drawing of a synchro transmitter and control transformer as an error detection device.

Thus,

$$e_o = K(\theta_i - \theta_o)$$

or

$$E_o(s) = K[\Theta_i(s) - \Theta_o(s)] \tag{7-25}$$

7-8. Error Detection Devices

One of the prime functions in a feedback control system, as discussed in Chapter 1, is the detection of differences between the system input and output. The physical nature of an error detector depends on the control equipment with which it is associated. It may be necessary, for example, to sense mechanical positions and produce another mechanical position proportional to their difference. On the other hand, it may be desirable to have an electrical output proportional to the difference between two mechanical measurements. Other examples might include producing an electrical or mechanical signal proportional to the difference in hydraulic or pneumatic pressures. Or, conversely, the difference between two voltages might be used to produce a pressure signal.

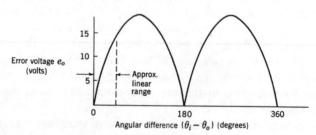

FIG. 7-24. Typical error voltage verus relative displacement curve.

Table 7-1 illustrates some typical error-detection devices found in control systems. The information presented permits rapid correlation of the schematic diagram of the device, the defining equation, and the block-diagram representation. With the background of earlier discussions, Table 7-1 should be self-explanatory.

It should be emphasized that many of the detection schemes of Table 7-1 may be modified to detect the error between other types of signals. For example, opposing LVDT's may be used to sense a pressure error by connecting the movable cores to diaphragms or spring-loaded pistons, such that the core displacements are proportional to pressure.

7-9. Transmission and Actuating Devices

Transfer functions for several hydraulic components will be developed starting with the pump which is the source of fluid power. Pumps used in fluid-power applications are of the *positive-displacement* variety which means that a fixed volume of oil is displaced (pumped) per revolution of the unit. Positive-displacement pumps can have a *fixed displacement*, in which case the displacement per revolution is constant and cannot be changed. On the other hand, a *variable-displacement* pump is one in which the displacement per revolution can be changed in some manner. Our concern will be with variable-displacement pumps. These units are frequently driven by constant-speed electric motors and, more often than not, utilize reciprocating pistons to pump the oil.

Variable-Displacement Pumps. A variable-displacement pump is shown schematically in Fig. 7-25a. The letters *PV* indicate that the pump displacement (and therefore the output flow rate) is variable. When the pump is driven at a constant speed, the output flow rate q can be varied by changing the displacement which is accomplished by varying the *pump stroke x*. Thus the fluid output from the pump is controlled by the pump stroke.

Operating characteristics are shown in Fig. 7-25b. With zero stroke ($x = 0$), no fluid is pumped. As the pump is stroked in the positive direction, the output flow rate increases in a linear manner. The same is true for negative strokes except that the direction of oil flow is reversed. Thus a variable displacement pump will supply oil from zero to some maximum

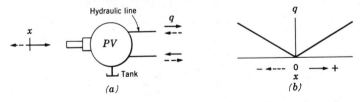

Fig. 7-25. The variable-displacement hydraulic pump.

flow rate in either direction depending on the stroke of the pump. The transfer function for a variable-displacement pump can be written as

$$\frac{Q(s)}{X(s)} = K \qquad (7\text{-}26)$$

TABLE 7-1. ERROR DETECTING DEVICES

Description	Schematic Drawing	Defining Equation	Block-Diagram Representation	Comments
Bevel-gear differential		$\phi = \frac{1}{2}(\theta_i - \theta_o)$	$\theta_i(s)$, $\theta_o(s)$ → $\frac{1}{2}$ → $\Phi(s)$	Can be used with angular displacements or velocities
Rack-and-gear differential		$z = \frac{1}{2}(x - y)$	$X(s)$, $Y(s)$ → $\frac{1}{2}$ → $Z(s)$	Used with linear displacements
Hydraulic valve		$\delta = x - y$ δ = valve opening	$X(s)$, $Y(s)$ → $\Delta(s)$	Used with linear displacements
Potentiometer bridge		$e_o = K_1 x - K_2 y$ e = fixed voltage	$X(s)$ → K_1; $Y(s)$ → K_2 → $E_o(s)$	Displacements can be linear or angular. Voltage e can be a-c or d-c. For identical potentiometers: $K_1 = K_2$
LVDT's wired opposing		$e_o = K_1 x - K_2 y$ e_1 = fixed a-c voltage e_2 = fixed a-c voltage	$X(s)$ → K_1; $Y(s)$ → K_2 → $E_o(s)$	Used with linear displacements. For identical LVDT's with $e_1 = e_2$: $K_1 = K_2$
Synchro pair		$e_o = K(\theta_i - \theta_o)$ e = fixed a-c voltage	$\theta_i(s)$, $\theta_o(s)$ → K → $E_o(s)$	Used with angular displacements
Potentiometer-tachometer combination for speed control		$e_o = K_1\theta - K_2\omega$ e = fixed voltage	$\theta(s)$ → K_1; $\Omega(s)$ → K_2 → $E_o(s)$	Illustrates combining different transducers

Equation 7-26 does not take into account the internal leakage (called *slip*) of the pump, which is a function of pressure. Slip is usually considered with the unit being supplied by the pump and suitable over-all system loss factors are introduced here.

Hydraulic Cylinders. The hydraulic cylinder is a frequently used component. One is shown in Fig. 7-26 with an inertial load; the assumption is made that only viscous friction exists in the system. The input is the flow rate q of oil and the output is the displacement y of the mass. The cylinder pressure p acts on the piston area A to provide the force necessary to move the mass. The volume V of oil being compressed (subjected to pressure) is considered constant.[3] Oil from the cylinder is returned to the pump tank, a point of zero gage pressure.

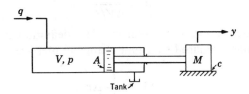

FIG. 7-26. Hydraulic cylinder with load.

In Sec. 7-3, the cylinder was treated as an integrating device which it is basically. However, there are other factors involved which must be considered in getting a more descriptive transfer function. These factors will be discussed as a more sophisticated transfer function is developed.

Referring to Fig. 7-26 and using Newton's second law, one writes

$$M \frac{d^2y}{dt^2} = -c \frac{dy}{dt} + Ap$$

or
$$M \frac{d^2y}{dt^2} + c \frac{dy}{dt} = Ap(t) \qquad (7\text{-}27)$$

Equation 7-27 is one of two major equations required.

A second equation, related to flow rates, is

$$q = q_y + q_l + q_c \qquad (7\text{-}28)$$

where q_y is the flow rate utilized in displacing the mass, q_l is the flow rate

---

[3] Treating V as a constant means that the resulting transfer function will be valid for only small displacements about that displacement defined by the particular value chosen for V. Of course, various values for V can be introduced, and the response of the system investigated at different positions of piston travel. You ask, why not let $V = f(y)$? A nonlinear differential equation results if this is done, and we are attempting to avoid these for the time being.

lost through leakage, and q_c is the flow rate lost in compression of the oil. Each quantity will now be examimed separately.

The flow rate used in displacing the mass is

$$q_y = A\frac{dy}{dt} \tag{7-29a}$$

The loss through leakage is proportional to system pressure; therefore

$$q_l = Lp(t) \tag{7-29b}$$

where L is the leakage coefficient.[4]

The flow rate lost in compression of the oil will require more explanation than in the previous cases. One property of a hydraulic fluid is its *bulk modulus*, defined as

$$\beta = \frac{\Delta p}{\Delta V/V}$$

where p is pressure and V is volume. The definition involves the unit change in volume produced by a unit change in pressure. The expression can be written as

$$\Delta V = \frac{V}{\beta}\Delta p$$

Introducing a time increment Δt and taking the limit as $\Delta t \rightarrow 0$ yields

$$q_c = \frac{dV}{dt} = \frac{V}{\beta}\frac{dp}{dt} \tag{7-29c}[5]$$

By substituting Eqs. 7-29a,b,c into Eq. 7-28,

$$q = A\frac{dy}{dt} + Lp(t) + \frac{V}{\beta}\frac{dp}{dt}$$

Transformation, assuming all initial conditions zero, yields

$$Q(s) = AsY(s) + LP(s) + \frac{V}{\beta}sP(s) \tag{7-30}$$

Equation 7-27 can be transformed and solved for $P(s)$ as follows:

$$Ms^2Y(s) + csY(s) = AP(s)$$

$$P(s) = \left(\frac{Ms^2 + cs}{A}\right)Y(s) \tag{7-31}$$

Combination of Eqs. 7-30 and 7-31 provides the transfer function for the cylinder:

---

[4]For the reader interested in analogies, L is *hydraulic conductance*, that is, the reciprocal of hydraulic resistance.

[5]The expression defining β can be written as $V/\beta = \Delta V/\Delta p$. Thus V/β is *hydraulic capacitance* because it is the change in quantity contained (ΔV) per unit change of some reference variable (Δp).

$$\frac{Y(s)}{Q(s)} = \frac{A}{s\left[\dfrac{VM}{\beta} s^2 + \left(LM + \dfrac{Vc}{\beta}\right)s + (A^2 + Lc)\right]} \qquad (7\text{-}32)$$

The transfer function given in Eq. 7-32 can be checked by supposing that $M = c = 0$; then

$$\frac{Y(s)}{Q(s)} = \frac{A}{s(A^2)} = \frac{1/A}{s}$$

which is identical to Eq. 7-5.

Hydraulic Motors. A hydraulic motor is used when the required motion of the output is rotation. A *fixed-displacement motor* is one whose theoretical behavior is defined by the relationship

$$q = D_m \frac{d\theta}{dt} \qquad (7\text{-}33)$$

in which θ is the angular position in radians and D_m is the motor displacement, expressed in volumetric units/radian, which cannot be changed for any given motor. The output torque T for an ideal motor is

$$T = D_m p \qquad (7\text{-}34)$$

in which p is the pressure drop across the motor.

Figure 7-27 shows a fixed-displacement motor driving a load consisting of inertia and damping. The flow rate q of oil is the input, and the angular

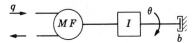

FIG. 7-27. Hydraulic motor with load.

position θ is the output. The transfer function is obtained in a manner similar to that used for the hydraulic cylinder. The result is

$$\frac{\Theta(s)}{Q(s)} = \frac{D_m}{s\left[\dfrac{VI}{\beta} s^2 + \left(LI + \dfrac{Vb}{\beta}\right)s + (D_m^2 + Lb)\right]} \qquad (7\text{-}35)$$

The inertia I is taken as the combined effect of the motor and load inertias with the assumption that the connecting coupling is rigid. The same is true for the damping coefficient b. In the case of the hydraulic motor, the volume V of oil subjected to pressure is essentially constant in contrast with the variable volume associated with a cylinder.

A transfer function involving rotating speed ω can be obtained from Eq. 7-35 by recalling that

$$\Omega(s) = s\Theta(s) \qquad (7\text{-}36)$$

Hydraulic Transmissions. A hydraulic transmission is the combination of a variable-displacement pump and a fixed-displacement motor. Its transfer function is obtained from Eqs. 7-26 and 7-35.

$$\frac{\Theta(s)}{X(s)} = \frac{Q(s)}{X(s)}\frac{\Theta(s)}{Q(s)} \tag{7-37}$$

Electrohydraulic Servo Valves. The electrohydraulic servo valve is a very important component in an electrohydraulic control system, being the primary junction point between the electrical and the hydraulic portions of the system. (Electrohydraulic systems are used to combine the advantages of electrical control with those of hydraulic power.) The electrohydraulic servo valve utilizes an electrical signal, usually provided by an amplifier, to control the flow rate of hydraulic fluid. A schematic drawing of an amplifier-driven servo valve is given in Fig. 7-28a. A voltage e (which can be positive or negative) to the amplifier causes an actuator displacement x which controls the spool position in a valve similar to those discussed in Sec. 2-5. The valve is supplied with fluid at a constant pressure P; the spool displacement x then controls the flow rate q of fluid through the valve. If the pressure drop across the valve is constant, the valve output is as shown in Fig. 7-28b (see Eq. 2-28).

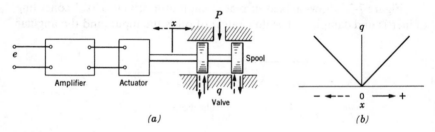

FIG. 7-28. Amplifier-driven electrohydraulic servo valve.

Amplifiers and electrohydraulic servo valves differ widely in design, but a practical transfer function for an amplifier-actuator combination is

$$\frac{X(s)}{E(s)} = \frac{K_1}{\tau s + 1} \tag{7-38}$$

The valve transfer function, from Eq. 2-28, is

$$\frac{Q(s)}{X(s)} = K_v \tag{7-39}$$

Thus, the transfer function for an amplifier-driven servo valve, with the assumption of a constant pressure drop across the valve, is[6]

$$\frac{Q(s)}{E(s)} = \frac{K}{\tau s + 1} \qquad (7\text{-}40)$$

Armature Controlled D-C Motors. The d-c motor of Fig. 7-29 with a variable d-c voltage applied to the armature and constant field excitation

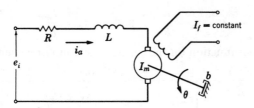

FIG. 7-29. D-c motor with constant field excitation.

may be used as a positioning device in control systems. The transfer function relates the output position θ to the applied armature voltage e_i. The total voltage drop in the armature circuit is the sum of the voltage drops in the series resistance, inductance, and the back emf which is proportional to the rotor speed. Thus the transformed voltage equation is

$$E_i(s) = RI_a(s) + LsI_a(s) + K_e s\Theta(s) \qquad (7\text{-}41)$$

The motor torque may be assumed proportional to the armature current; thus,

$$T(s) = K_t I_a(s) \qquad (7\text{-}42)$$

---

[6]If the pressure drop across the valve cannot be assumed constant, Eq. 2-27, in which the flow rate q is seen to be a function of both the spool displacement x and the pressure drop p across the valve, must be used as a basis for determining valve flow. Since the supply pressure P to a servo valve is held constant, the pressure drop across the valve is a function of the load pressure p_L; specifically,

$$p = P - p_L$$

Therefore,

$$q = f(x, p_L)$$

An approximate equation for valve flow can be obtained as

$$\Delta q \cong dq = \frac{\partial q}{\partial x} \Delta x + \frac{\partial q}{\partial p_L} \Delta p_L$$

Based on selected operating reference points

$$q = A_1 x + A_2 p_L$$

where A_1 and A_2 are constants which can be obtained from performance curves for a particular valve at particular displacement and load-pressure operating reference points. Thus the equation, while useful, does have distinct limitations.

The motor torque is available to accelerate the rotor inertia and overcome any rotor friction (assumed viscous); thus,

$$T(s) = I_m s^2 \Theta(s) + bs\,\Theta(s) \tag{7-43}$$

Combining Eqs. 7-41, 7-42, and 7-43 algebraically yields

$$E_i(s) = \frac{Rb}{K_t}[s(\tau_a s + 1)(\tau_m s + 1)]\Theta(s) + K_e s\Theta(s) \tag{7-44}$$

where $\qquad \tau_a = \dfrac{L}{R} \qquad$ and $\qquad \tau_m = \dfrac{I_m}{b} \tag{7-45}$

The transfer function for the constant field motor is then

$$\frac{\Theta(s)}{E_i(s)} = \frac{K_t/Rb}{s\left[\tau_m \tau_a s^2 + (\tau_a + \tau_m)s + \left(\dfrac{K_e K_t}{Rb} + 1\right)\right]} \tag{7-46}$$

If the armature time constant τ_a is small compared with the motor time constant τ_m (that is, $L/R \ll I_m/b$), the transfer relation is of the form

$$\frac{\Theta(s)}{E_i(s)} = \frac{K}{s(\tau s + 1)} \tag{7-47}$$

Two-Phase A-C Servomotors. A two-phase a-c servomotor is shown schematically in Fig. 7-30a. It is composed of a rotor and two independently wound fields, a control field and a fixed reference field positioned at 90° to each other. A constant a-c voltage is applied to the fixed reference field through a capacitor or a 90° phase-shift network. An a-c voltage of variable magnitude, either in phase or 180° out of phase with the a-c source, is applied to the control field. The phase relationship determines the direction of shaft rotation.

For steady-state operation, typical torque speed curves may be approximated by Fig. 7-30b. The assumption generally made in writing the torque equation is that the instantaneous dynamic behavior does not differ

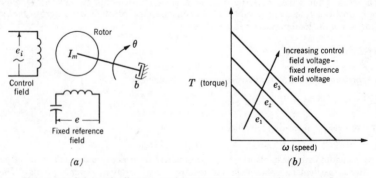

Fig. 7-30. Schematic diagram and torque speed curves for a two-phase a-c servomotor.

appreciably from the steady-state behavior. Thus, from Fig. 7-30*b*, the torque equation is

$$T = K_t' e_i - K_b \omega \qquad (7\text{-}48)$$

where K_t' and K_b are constants with units such as $K_t' = $ oz-in./volt and $K_b = $ oz-in./(radians/sec). In transformed form, Eq. 7-48 becomes

$$T(s) = K_t' E_i(s) - K_b s \Theta(s) \qquad (7\text{-}49)$$

The torque produced in the motor is available to accelerate the rotor inertia and to overcome friction (assumed viscous). Thus, in transformed form

$$T(s) = I_m s^2 \Theta(s) + b s \Theta(s) \qquad (7\text{-}50)$$

Equating Eqs. 7-49 and 7-50 yields

$$\frac{\Theta(s)}{E_i(s)} = \frac{K_t'/(b + K_b)}{s(\tau_m s + 1)} \qquad (7\text{-}51)$$

where

$$\tau_m = \frac{I_m}{b + K_b}$$

Although Eq. 7-51 is quite usable for most servomotors in the range of their application, actual tests reveal that a more complex transfer function better describes their dynamic behavior. The relationship generally proposed is

$$\frac{\Theta(s)}{E_i(s)} = \frac{K}{s(\tau_m s + 1)(\tau_e s + 1)} \qquad (7\text{-}52)$$

where τ_e may be interpreted as the electrical time constant of the motor. When servomotors are used with other mechanical and electromechanical devices, the frequencies at which the servomotor is driven are low and the electrical time constant is justifiably neglected.

7-10. Closure

In closing, three comments will be made. The first has to do with block diagrams. Some references provide tabulations of equivalent block-diagram arrangements and call this approach *block-diagram algebra* [14]. Such tabulations can prove helpful. However, remember the basic underlying concept—block diagrams can be arranged in any form as long as the system equations and their relationship to each other are maintained.

The second point is related to initial conditions. Recall that transfer functions are obtained by transformation with the assumption that all initial conditions are zero. Should there be nonzero initial conditions, they can be introduced into the over-all system equation (obtained from the block diagram) when transforming it to effect a solution.

Finally, it should be noted that the predicted response of a control system is only as good as the assumptions made in obtaining the transfer functions for the components making up the system. However, the fewer the assumptions, the more complex are the transfer functions. Some compromise must be made between the simplicity of the mathematical model and the accuracy of the prediction. In this latter view, computers play a major role, for more complicated mathematical models may be used without exhausting one's ability to study the system.

Problems

7-1. Determine the transfer function for each of the systems shown in parts *a* to *i* of the accompanying illustration and represent each in block form.

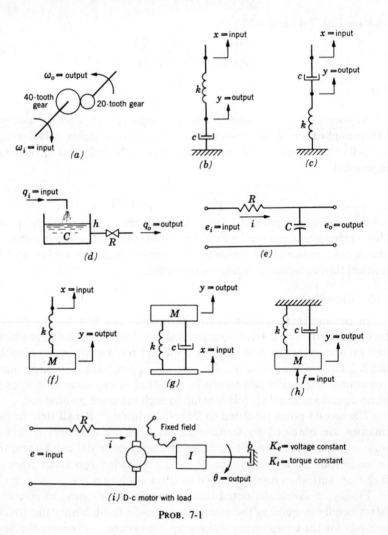

PROB. 7-1

7-2. Represent the systems shown in parts *a* and *b* of the accompanying illustration in cascaded block form and then as a single block.

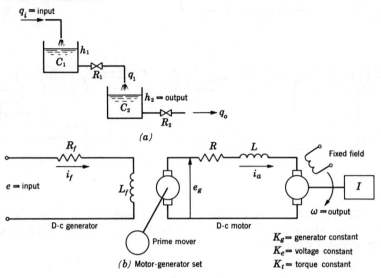

(a)

(b) Motor-generator set

K_g = generator constant
K_e = voltage constant
K_t = torque constant

PROB. 7-2

7-3. Represent the following equations with summing points:

a) $E(s) = R(s) - C(s)$

b) $E(s) = R(s) - B_1(s) - B_2(s)$

7-4. Occasionally in setting up block diagrams, a configuration of the type shown in the accompanying illustration will occur. Derive the expression for the closed-loop transfer function. It will be similar to Eq. 7-11 and can be used whenever the reduced form of a block diagram is as shown.

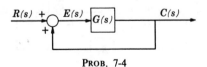

PROB. 7-4

7-5. Show each of the block diagrams given in parts *a* to *d* of the accompanying illustration in reduced form. Then determine the closed-loop transfer function for each by using either Eq. 7-11 or Eq. 7-12. HINT: In parts *c* and *d*, first replace the inner loop by a single block.

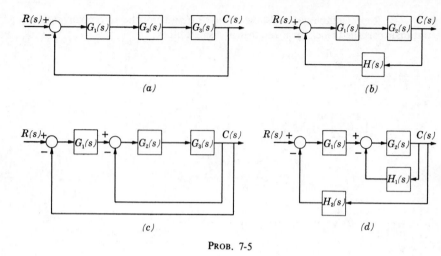

(a)   *(b)*

(c)   *(d)*

PROB. 7-5

7-6. For each of the block diagrams shown in parts *a* to *c* of the accompanying illustration, determine the closed-loop transfer function $C(s)/R(s)$ and from it, the over-all system differential equation.

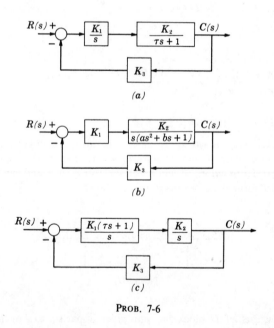

(a)

(b)

(c)

PROB. 7-6

7-7. For each of the diagrams in parts *a* to *c* of the accompanying illustration, redraw the block diagram in the form used for investigating a disturbance $U(s)$. Assume that $R(s)$ remains unchanged. Then determine the transfer function

$\Delta C(s)/U(s)$. Finally, obtain the appropriate system differential equation from the transfer function.

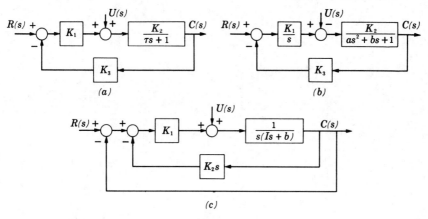

PROB. 7-7

7-8. Determine the transfer function $H_2(s)/Q_i(s)$ from the signal flow graph shown in Fig. 7-21d.

7-9. The transfer function for a hydraulic motor with load is given in Eq. 7-35 as

$$\frac{\Theta(s)}{Q(s)} = \frac{D_m}{s\left[\frac{VI}{\beta}s^2 + \left(LI + \frac{Vb}{\beta}\right)s + (D_m^2 + Lb)\right]}$$

Derive this transfer function using Eqs. 7-33 and 7-34 and following the approach used with the hydraulic cylinder in Sec. 7-9.

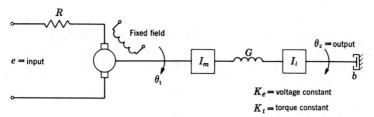

PROB. 7-10

7-10. In presenting the transfer function for a hydraulic motor with load (Eq. 7-35), the assumption of a rigid coupling between motor and load inertias was made. The same assumption is implied in part i of the figure for Prob. 7-1 in which the inertia I is taken to be the combined effect of the motor inertia I_m and the load inertia I_l, that is, $I = I_m + I_l$. The accompanying sketch represents a system in which the coupling is not rigid but has a torsional spring rate G.

a) Determine the transfer function $\Theta_2(s)/E(s)$.

b) Check the solution to part a) by letting $G \rightarrow \infty$ and comparing the result with the solution for part i of Prob. 7-1.

Chapter 8

Control Actions
and System Types

8-1. Introduction

In feedback control systems, an actuating signal (proportional to error) is obtained from a comparison of the reference input and the feedback signal. The actuating signal is then used to make any correction necessary to bring the output into correspondence with the input. The term *control action* refers to the manner in which the actuating signal is utilized by the control portions of the system (i.e., the control elements) to achieve the correction. A knowledge of the characteristics of the various control actions is helpful in predicting the response of control systems, especially in those cases where transfer functions for the parts of the system being controlled (i.e., the system elements) are either unknown or not well defined. An example is the comfort control for a commercial building; heating and cooling equipment, ductwork, and the building itself are all in the control loop. A knowledge of control actions also permits the control engineer to select the one best suited to a particular application.

A control system, for which all transfer functions are known, can be classified by *system type*. The system-type number is obtained from information concerning not only the control elements but the system elements as well. As with control actions, a knowledge of the characteristics of the various system types is useful for the prediction of control system performance.

The reader should recognize the fact that many control systems can be classified either by control action or system type. In some industries, control engineers speak almost exclusively of control actions while in other industries, the system-type classification is used. The purpose of this chapter is to present both viewpoints.

This chapter presents four basic control actions and certain commonly encountered combinations of them. Each action is illustrated by an example of a specific control system. Three basic system types are then discussed; the chapter closes with control applications illustrating two widely used types.

It is worth remarking at this point that the control systems presented were chosen to serve a double purpose—first, to illustrate the subject matter and second, to acquaint the reader with some actual control systems.

8-2. Proportional Control

With proportional control, corrective action is taken which is proportional to the error. Expressed mathematically

$$\delta m = K_p e \tag{8-1}$$

where δm is a *change* of the manipulated variable (the quantity varied to achieve correction), K_p is the gain of the proportional controller, and e is the error, or the actuating signal.

One important characteristic of proportional control action is revealed by Eq. 8-1. Should a sustained correction (brought about by a sustained disturbance to the system) be required, an accompanying steady-state error will exist, that is, $e_{ss} = \delta m_{ss}/K_p$. This point will be further developed in considering a specific control system.

An application involving the speed control of a gas turbine will serve to illustrate the use of proportional control action. Attention is first directed to the speed governor.

Proportional Governor. A schematic drawing of a centrifugal governor is given in Fig. 8-1.[1] The input shaft is driven at turbine speed ω. Through

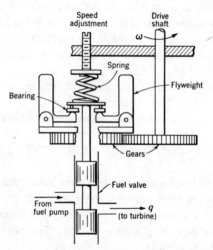

FIG. 8-1. Speed governor exhibiting proportional control.

[1] It should be pointed out that the governor is represented in its most elementary form. A typical gas-turbine governor would also provide an adjustable bypass flow arrangement so that fuel to the turbine can never be completely shut off during operation. Furthermore, some means of limiting fuel flow during acceleration is required so that safe operating temperatures are not exceeded. Finally, a differential pressure relief valve might well be used to maintain a constant pressure drop across the fuel valve.

the gears shown, the two flyweights are rotated, resulting in a centrifugal force which is opposed by the spring force. The spring preload is a measure of the desired speed setting. Under steady-state operating conditions, the position of the spool in the fuel valve, which controls the flow rate q of fuel to the turbine, is determined by the spring compression required to attain force equilibrium. Thus, the fuel flow rate is related to the spring preload and the driven speed. With the spring preload fixed (the usual operating condition), it can be seen that a speed change is required in order to vary q. The simplest equation which can be written defining the behavior of the governor is

$$\delta q = K_p e \qquad (8\text{-}2)$$

where δq is the *change* in fuel flow rate, K_p is the gain of the governor, and e is the actual speed error expressed in radians per second.[2] Equation 8-2 is that of proportional control action.

Gas Turbine. The purpose of a gas turbine is to convert the energy in the supplied fuel to useful power. With the turbine operating at some base speed, and assuming small variations of speed during operation, the change in power developed can be considered to be proportional to the change of fuel flow rate, δq. Since power is proportional to the product of torque and speed, and speed is assumed to be essentially constant, one writes

$$\delta T = K_1 \delta q \qquad (8\text{-}3)$$

where δT is the change in developed torque and K_1 is the appropriate proportionality factor, or gain, which would be obtained from the performance curves of the turbine.

A gas turbine will exhibit rotating inertia and damping. However, if the connection between the turbine and the driven load can be assumed rigid, turbine inertia and damping can be conveniently combined with that of the load. Thus, the turbine will be treated as a torque producer with Eq. 8-3 being the defining equation.

Load. Driven loads frequently exhibit inertia and damping and, in addition, involve a load torque. Figure 8-2 shows such a load in which I and b represent the combined effects of both turbine and load inertia and damping. The driving torque is $T(t)$, the opposing load torque is $T_l(t)$ while $\omega_c(t)$ is the rotating speed (controlled variable). The appropriate load

[2]One might well question the validity of Eq. 8-2 when it is recalled that centrifugal force is a function of ω^2. However, if we assume operation at some established speed setting and that variations from this base speed are small, we can, with reasonable accuracy, work with the tangent to the force-speed curve. This technique is known as *linearization* and is one way of handling nonlinearities. (Linearization is discussed in detail in Chapter 14.) A more subtle assumption inherent in Eq. 8-2 is that the natural frequency of the governor is high compared with the frequency of any speed oscillations encountered. If such is not the case, the governor itself, being a spring-mass-damper system, may oscillate.

equation is

$$T(t) - T_l(t) = I \frac{d\omega_c(t)}{dt} + b\omega_c(t) \qquad (8\text{-}4)$$

The components making up the control system are now described mathematically. Attention can be turned to obtaining a system block diagram.

System Block Diagram. Before proceeding with a system block diagram, some necessary preliminary remarks will be made having to do with the choice of references.

At the start of an analysis involving a linear control system, it is convenient to assume the system in some equilibrium condition with zero error and to choose references such that the input, controlled variable, and any other system variables are all zero. The problem is then worked with incremental changes such as $\delta\omega_c$, δT, δq, and so forth. However, by choice of

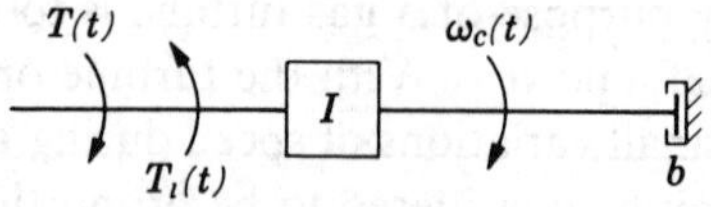

Fig. 8-2. The load.

references, $\delta\omega_c = \omega_c$ (a deviation of ω_c from zero reference is simply ω_c), $\delta T = T$, and so forth. Therefore, the delta notation can be omitted. This is the basis for drawing system block diagrams. The approach will be illustrated in obtaining a suitable block diagram for the speed control system.

First, consider the load shown in Fig. 8-2. The driving torque $T(t)$ will be assumed to be composed of two parts, a constant part T and a variable part $\delta T(t)$:

$$T(t) = T + \delta T(t) \qquad (8\text{-}5a)$$

The constant portion is the value of the driving torque when the system is in some equilibrium condition. The variable portion permits $T(t)$ to change as the system responds to some input.

In a similar manner, the controlled speed $\omega_c(t)$ is thought of as being made up of two parts:

$$\omega_c(t) = \Omega_c + \delta\omega_c(t) \qquad (8\text{-}5b)$$

where Ω_c is the equilibrium speed value and $\delta\omega_c(t)$ is any speed change occurring because of some input to the system.

The load torque $T_l(t)$ is separated into two parts to yield

$$T_l(t) = T_L + T_d(t) \qquad (8\text{-}5c)$$

where T_L is the constant equilibrium value of load torque and $T_d(t)$ is any change in load torque, that is, a torque disturbance.

Substitution of Eqs. 8-5a, b,c into Eq. 8-4 yields

$$T + \delta T(t) - T_L - T_d(t) = I\frac{d\Omega_c}{dt} + I\frac{d\delta\omega_c(t)}{dt} + b\Omega_c + b\delta\omega_c(t) \quad (8\text{-}6)$$

At the equilibrium condition

$$T - T_L = I\frac{d\Omega_c}{dt} + b\Omega_c \quad (8\text{-}7)$$

Subtracting Eq. 8-7 from Eq. 8-6,

$$\delta T(t) - T_d(t) = I\frac{d\delta\omega_c(t)}{dt} + b\delta\omega_c(t) \quad (8\text{-}8)$$

If references are chosen so that, at the initial equilibrium condition (before any torque disturbance has occurred), $T = 0$, $T_L = 0$, and $\Omega_c = 0$, the delta notation can be omitted. Equation 8-8 becomes

$$T(t) - T_d(t) = I\frac{d\omega_c(t)}{dt} + b\omega_c(t) \quad (8\text{-}9)$$

Equations 8-2, 8-3, and 8-9 can be utilized to obtain block representations for the system components. The blocks are shown in Fig. 8-3. The

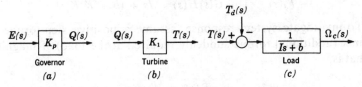

FIG. 8-3. Block representations of system components.

delta notation of Eqs. 8-2 and 8-3 was omitted because of the choice of references.

The last relationship required to complete the block diagram for the speed control system is

$$e = \omega_r - \omega_c \quad (8\text{-}10)$$

where ω_r is the speed reference input.

The completed diagram is given in Fig. 8-4. It permits two types of input—a reference change or a torque disturbance. A disturbance input will now be considered to illustrate an important characteristic of proportional control.

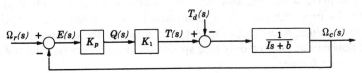

FIG. 8-4. System block diagram.

System Response to Disturbance Torque. The block diagram of Fig. 8-4 can be modified, in the manner described in Sec. 7-5, to the form shown in Fig. 8-5 for convenience in investigating the response to a disturbance torque. The output is given as $\Delta\Omega_c(s)$. Recall that this delta notation was introduced in Sec. 7-5 to identify an unwanted change of the controlled variable brought about by a disturbance input to the system. Note also that

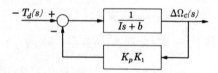

FIG. 8-5. Block diagram for use in
investigating a torque disturbance.

the sign of $T_d(s)$ was changed so that the signs of the inputs to the summing point would appear in the usual fashion. The closed-loop transfer function is

$$\frac{\Delta\Omega_c(s)}{-T_d(s)} = \frac{G(s)}{1 + G(s)H(s)} = \frac{1}{Is + (b + K_pK_1)} \tag{8-11}$$

To investigate system response, assume a positive step torque disturbance (additional torque suddenly required at the load) of magnitude T_D, that is,

$$T_d(s) = \frac{T_D}{s} \tag{8-12}$$

Substitution of Eq. 8-12 into Eq. 8-11 yields

$$\Delta\Omega_c(s) = \frac{-T_D}{s[Is + (b + K_pK_1)]} \tag{8-13}$$

The steady-state speed deviation can be obtained from Eq. 8-13 by using the final value theorem:

$$\Delta\omega_{css}(t) = \lim_{t \to \infty} \Delta\omega_c(t) = \lim_{s \to 0} s\,\Delta\Omega_c(s)$$

$$\Delta\omega_{css}(t) = \frac{-T_D}{(b + K_pK_1)} \tag{8-14}$$

From Eq. 8-14, it can be seen that the additional torque required at the load results in a reduction of system speed. The speed loss is entirely plausible because, referring to Fig. 8-1, one sees that the necessary increase in fuel flow can be secured only by a speed reduction which permits the fuel valve to open further.

The speed control application presented illustrates the use of proportional control and an important characteristic of it—when a sustained correction is necessary, a sustained (or steady-state) error results.

8-3. Integral Control

A second control action is that provided by integral control. Here a correction is made which is proportional to the time integral of the error. Expressed mathematically

$$\delta m = K_i \int e \, dt \tag{8-15a}$$

or

$$\frac{d \delta m}{dt} = K_i e \tag{8-15b}$$

where δm is a change of the manipulated variable, K_i is the gain of the integral controller, and e is the error, or the actuating signal.

Equation 8-15b reveals that correction takes place at a *rate* proportional to error and so an integral controller will continue to correct until the error is zero. This tendency to eliminate any steady-state system error is the prime benefit obtained from integral control action. But, there is a weakness also; integral control tends to overshoot, thereby producing an oscillatory response and, in some cases, instability.

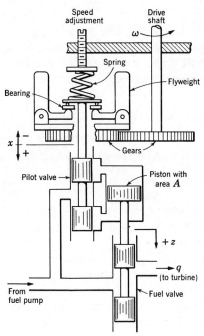

FIG. 8-6. Speed governor exhibiting integral control.

The use and characteristics of integral control action will be illustrated with the speed control application introduced in the previous section.

Integral Governor. Figure 8-6 shows a schematic drawing of a speed governor exhibiting integral control. In many respects, it is the same as the proportional governor given in Fig. 8-1. However, in the integral governor, the flyweights control the position of the pilot-valve spool which, in turn, controls the output displacement of a small cylinder. The fuel-valve spool then meters fuel flow to the turbine.

Making the same assumptions as with the proportional governor,

$$x = k_1 e \qquad (8\text{-}16a)$$

in which x is the pilot-valve spool displacement, k_1 is the appropriate gain, and e is the actual speed error in radians per second. Note that the reference $x = 0$ is taken with the spool centered in the pilot valve, as shown. With $x = 0$, the piston does not move.

A second equation relates the displacement x with the *change* of fuel-valve spool position δz; the previous spool displacement is chosen as a reference and is considered to be zero. The relationship obtained by equating the flow rate through the pilot valve to that utilized by the cylinder is

$$k_2 x = -A \frac{d\delta z}{dt} \qquad (8\text{-}16b)$$

where k_2 is the pilot-valve gain. The negative sign is required because a positive value of x produces velocity in the negative direction. Equation 8-16b also assumes that the piston-rod area is negligible.

The equation for the fuel valve is

$$\delta q = -k_3 \delta z \qquad (8\text{-}16c)$$

in which δq is the *change* in fuel flow rate, and k_3 is the fuel-valve gain.

Equations 8-16a, b, and c are transformed and combined to yield the transfer function for the governor.

$$\frac{Q(s)}{E(s)} = \frac{K_i}{s} \qquad (8\text{-}17)$$

where the gain of the integral governor $K_i = k_1 k_2 k_3 / A$. The delta notation is omitted in Eq. 8-17 because of the reference assumptions made in setting up a system block diagram.

It is worth remarking that the governor of Fig. 8-6, in addition to providing integral control action, also includes a *hydraulic amplifier* composed of the pilot valve and the cylinder. This amplifier permits the relatively low force required to move the pilot-valve spool to produce a relatively high force at the fuel-valve spool by virtue of controlling the system fuel pressure acting on the piston area. With hydraulic amplification, the integral governor would be able to control higher fuel flow rates than is possible with the simple configuration of Fig. 8-1. For this reason, many proportional governors also include a hydraulic amplifier.

System Block Diagram. The remainder of the speed control system is the same as in Sec. 8-2. Therefore, the complete system block diagram illustrating integral control action is as given in Fig. 8-7a. The modified diagram for use in investigating a torque disturbance is shown in Fig. 8-7b.

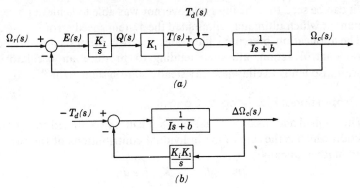

(a)

(b)

FIG. 8-7. Block diagrams illustrating the use of integral control action.

System Response to Disturbance Torque. The closed-loop transfer function, from Fig. 8-7b, is

$$\frac{\Delta\Omega_c(s)}{-T_d(s)} = \frac{s}{Is^2 + bs + K_iK_1} \quad (8\text{-}18)$$

A comparison of Eqs. 8-18 and 8-11 shows that the use of integral control will result in a higher-order system differential equation and the possibility of an oscillatory response. This fact illustrates the tendency of integral control action to produce oscillations.

To investigate system response, assume a positive step torque disturbance with

$$T_d(s) = \frac{T_D}{s} \quad (8\text{-}19)$$

Equation 8-18 can be written as

$$\Delta\Omega_c(s) = -\frac{sT_d(s)}{Is^2 + bs + K_iK_1} \quad (8\text{-}20)$$

Substitution of Eq. 8-19 yields[3]

[3]The reader should recall that transfer functions are obtained with the assumption of zero initial conditions. If there are *arbitrary* non-zero initial conditions, the system differential equation must be obtained from the closed-loop transfer function and then transformed to include the initial conditions. In the case being considered, the step change of torque produces an instantaneous acceleration—a non-zero initial condition. However, this initial condition is not arbitrary and so the approach being followed is acceptable.

$$\Delta\Omega_c(s) = \frac{-T_D}{Is^2 + bs + K_iK_1} \qquad (8\text{-}21)$$

The steady-state speed deviation can be obtained from Eq. 8-21 by using the final value theorem:

$$\Delta\omega_{css}(t) = \lim_{s \to 0} s\Delta\Omega_c(s) = 0 \qquad (8\text{-}22)$$

Thus it can be seen that the integral governor was able to achieve correction in a manner which ultimately eliminated the system speed error.

The application presented illustrates the two prime characteristics of integral control action; although tending to produce an oscillatory response, it also tends to eliminate any steady-state system error.[4]

8-4. Proportional Plus Integral Control

The control actions of Secs. 8-2 and 8-3 can be combined to produce a correction which is the sum of the individual contributions of the two. Expressed mathematically

$$\delta m = K_pe + K_i \int e\, dt \qquad (8\text{-}23)$$

The combination defined by Eq. 8-23 is a "cure" for certain control system "diseases." To illustrate proportional plus integral control, a "disease" must first be developed so that the "cure" can be applied.

Consider a speed control application in which neither the prime mover nor the load exhibits significant damping, that is, $b = 0$.[5] An integral governor is to be used. A block diagram for the system is shown in Fig. 8-8. Assuming no torque disturbance $[T_d(s) = 0]$, the closed-loop transfer

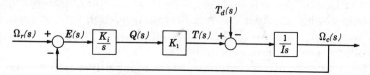

FIG. 8-8. Speed control system.

function is

$$\frac{\Omega_c(s)}{\Omega_r(s)} = \frac{K_iK_1}{Is^2 + K_iK_1} \qquad (8\text{-}24)$$

[4] The reader may be puzzled by the statement that integral control action *tends to* eliminate any system error when, by Eq. 8-22, the steady-state speed deviation is zero. The reason for the statement is that imperfections existing in physical components will not permit the *complete* elimination of errors. Again we are faced with the difference between mathematical models and actual hardware.

[5] The question might well arise at this stage as to how to establish a value for damping. Because, by general definition, $b = dT/d\omega$, torque-speed curves for the equipment involved provide one means for obtaining damping coefficients. The definition suggests using a tangent to the curve at the operating speed.

Equation 8-24 can be written as

$$\Omega_c(s)[Is^2 + K_iK_1] = K_iK_1\Omega_r(s)$$

From Chapter 6, the characteristic equation in s is seen to be

$$Is^2 + K_iK_1 = 0 \qquad (8\text{-}25)$$

The roots of Eq. 8-25 are complex conjugates with zero real parts. Thus, the undesirable condition of limited stability (persistent oscillation in response to a step input) exists. This is the required "disease."

A system block diagram with proportional plus integral control is given in Fig. 8-9.[6] Here the closed-loop transfer function is

$$\frac{\Omega_c(s)}{\Omega_r(s)} = \frac{K_pK_1s + K_iK_1}{Is^2 + K_pK_1s + K_iK_1} \qquad (8\text{-}26)$$

and the characteristic equation is

$$Is^2 + K_pK_1s + K_iK_1 = 0 \qquad (8\text{-}27)$$

A comparison of Eqs. 8-27 and 8-25 shows that the addition of proportional control action has made the system stable by adding "damping."

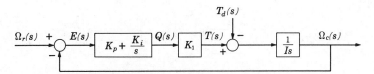

FIG. 8-9. System with proportional plus integral control.

Also, further investigation would reveal that the integral control action will ultimately eliminate any steady-state system error.

The application presented illustrates a case where proportional plus integral control provides the best features of each—the stability of proportional control and the error elimination of integral control.

8-5. Derivative Control

With derivative control, a correction is made which is proportional to the time derivative of the error. Expressed mathematically

$$\delta m = K_d \frac{de}{dt} \qquad (8\text{-}28)$$

where K_d is the gain of the derivative controller.

---

[6]The relationship expressed by the first block in the diagram can be written as

$$\frac{Q(s)}{E(s)} = K_p + \frac{K_i}{s} = \frac{K_ps + K_i}{s} = \frac{K_i(\tau s + 1)}{s}$$

where $\tau = K_p/K_i$. The latter form of the transfer function is more familiar to control engineers who do not generally think in terms of control actions.

Derivative control is useful because it responds to the rate of change of error and can produce a significant correction before the actual magnitude of the error is great. For this reason, it is sometimes said that derivative control *anticipates* the error and thus initiates an early corrective action. However, despite its usefulness, derivative control cannot be used alone because it will not respond to a steady-state error. Therefore, it must be used in combination with other control actions.

Consider the block diagram, Fig. 8-10, of a positioning control system utilizing proportional plus derivative control actions. With $T_d(s) = 0$, the closed-loop transfer function is

$$\frac{\Theta_o(s)}{\Theta_i(s)} = \frac{G(s)}{1 + G(s)} = \frac{K_p + K_d s}{Is^2 + (b + K_d)s + K_p} \tag{8-29}$$

and the characteristic equation is

$$Is^2 + (b + K_d)s + K_p = 0 \tag{8-30}$$

from which it can be seen that one effect of the derivative control action was to increase the damping in the system.

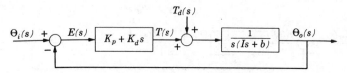

FIG. 8-10. Positioning system with proportional plus derivative control.

The combination of control actions indicated in Fig. 8-10 may not be easy to obtain for the positioning system. However, a derivative action in a feedback loop is relatively easily obtained by using a tachometer generator. Derivative feedback will be used in connection with the illustrative system which follows.

The effects of derivative control will be further illustrated by comparing the responses of a system with and without the addition of this control action. The electromechanical positioning system, discussed in Sec. 7-5 and shown in Fig. 8-11, will be used. The system equation, Eq. 7-20, previously

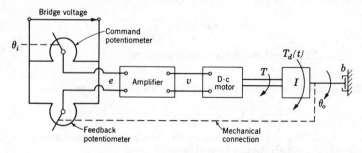

FIG. 8-11. Electromechanical positioning system.

developed for investigating a torque disturbance acting on the rotating mass, is

$$I \frac{d^2 \Delta \theta_o}{dt^2} + b \frac{d \Delta \theta_o}{dt} + K \Delta \theta_o = T_d(t) \qquad (8\text{-}31)$$

From Eq. 8-31, the steady-state response to a step torque disturbance $T_D u(t)$ is

$$\Delta \theta_{oss} = \frac{T_D}{K} \qquad (8\text{-}32)$$

and the system damping ratio is

$$\zeta = \frac{b}{2(IK)^{1/2}} \qquad (8\text{-}33)$$

It is seen that by increasing the gain K, the unwanted deviation (and system error) can be made smaller. However, the greater value of K will reduce the damping ratio which means that a more oscillatory response will result.

Now consider the addition of one type of derivative action in which a tachometer generator is suitably attached to the rotating mass. The transfer function for the generator, based on Eq. 7-24, is

$$\frac{E_1(s)}{\Omega(s)} = K_1 = \frac{E_1(s)}{s \Theta_o(s)}$$

and

$$\frac{E_1(s)}{\Theta_o(s)} = K_1 s \qquad (8\text{-}34)$$

Equation 8-34 defines the behavior of a device which provides a voltage output proportional to the derivative of the input. Although, in this case, the input is the controlled variable instead of the actuating signal, the end result is the same as will be seen. The voltage output from the tachometer generator is fed back as indicated in the system block diagram given in Fig. 8-12a.

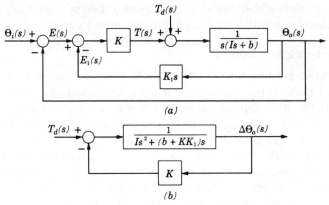

(a)

(b)

FIG. 8-12. Block diagrams for system using derivative feedback.

To investigate the response to a disturbance torque, the block diagram of Fig. 8-12a can be modified (in several steps) to that shown in Fig. 8-12b. (Of course, other forms for the diagram are equally as valid, the form being dependent on the manipulative approach followed.) The closed-loop transfer function is

$$\frac{\Delta\Theta_o(s)}{T_d(s)} = \frac{1}{Is^2 + (b + KK_1)s + K}$$

and the system equation is

$$I\frac{d^2\Delta\theta_o}{dt^2} + (b + KK_1)\frac{d\Delta\theta_o}{dt} + K\Delta\theta_o = T_d(t) \tag{8-35}$$

It can be seen that the derivative feedback produced additional system damping, the same result as obtained by adding derivative control action in the system of Fig. 8-10 (see Eq. 8-30). From Eq. 8-35, the steady-state response to a step torque disturbance $T_D u(t)$ and the system damping ratio are

$$\Delta\theta_{oss} = \frac{T_D}{K} \tag{8-36}$$

$$\zeta = \frac{b + KK_1}{2(IK)^{1/2}} \tag{8-37}$$

Comparing these expressions with Eqs. 8-32 and 8-33 reveals that the derivative action does not affect the steady-state deviation directly, but does permit the use of a larger value of K. In Eq. 8-37 the numerator of ζ is increased by K as well as the denominator, and K_1 may be adjusted to maintain the same value of ζ. Thus, with derivative control, an acceptable damping ratio can be maintained while keeping the proportional gain K high enough to reduce the system error to some tolerable value.

The application presented illustrates a case in which one form of derivative action added damping to the system thereby permitting a higher proportional gain with a resulting improvement in accuracy.

8-6. Two-Position Control

In a two-position control system, the final controlling element has only two fixed positions (or outputs)—in many cases, *on* and *off*. This is in contrast with the other control actions in which the controlling device is capable of continuous variation of the output over a specific range. Being limited to two positions, the two-position controller either supplies too much or too little correction to the system. Thus, the controlled variable must continuously move between the two limits required to cause the controlling element to move from one fixed position to the other. The range through which the controlled variable must move is called the *differential* of the controller. The "oscillation" of the controlled variable between two limits is one important characteristic of two-position control and one which sometimes limits its usefulness. However, two-position control is relatively simple and inexpensive and, for this reason, is widely used.

Home Heating System. A common use of two-position control is in home heating systems, one of which will now be described.

A schematic drawing of a gas-fired, forced-warm-air furnace is given in Fig. 8-13. Gas flow to the burner is controlled by an electrically oper-

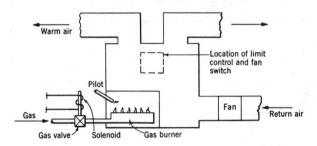

FIG. 8-13. Schematic drawing of gas-fired, forced-warm-air furnace.

ated gas valve which is open when the valve solenoid is energized and closed when the solenoid is de-energized. The valve is the final controlling element and has two fixed positions—open or closed. A pilot is also provided which supplies a continuous flame for igniting the gas supplied to the burner. Return air from the dwelling is forced through the furnace by a fan, the warm air leaving the furnace being used to heat the house. The location of the limit control and fan switch, to be described later, is indicated.

The electrical circuits for the system are shown in Fig. 8-14. The control of the fan motor is the simpler of the two circuits and will be discussed first. The purpose of the fan is to circulate air through the house, but only if the air is sufficiently warm. Thus, the furnace air temperature is sensed by some means, illustrated by a bellows in this case, which closes the fan switch when the air temperature is high enough to be circulated and opens the switch when the air temperature reaches some lower limit.

The output of the gas burner is controlled by the solenoid-actuated gas valve. The solenoid does not require the use of line voltage (115 volts) and so a transformer is used (Fig. 8-14) to provide a low-voltage circuit in which the wiring need not be placed in conduit. The essential control devices in the low-voltage circuit are the thermostat and the gas valve. The thermostat, illustrated as a bimetal blade which warps with temperature change, is a thermal switch which closes the electrical circuit at some lower temperature limit and opens it at some higher limit. (The difference between these temperatures is called the *differential* of the thermostat.) Thus, when the room temperature drops, the thermostat will close the circuit, thereby energizing the gas-valve solenoid which, in turn, opens the valve and supplies gas to the burner. When the room temperature has risen sufficiently, the thermostat will cause the gas valve to close.

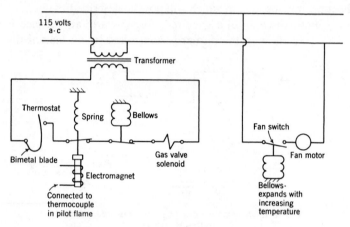

FIG. 8-14. Electrical circuits for heating system.

Room temperature variations with time are depicted in Fig. 8-15. The temperature variations exceed the thermostat differential because there are various heat storage elements in the system. The system is composed of

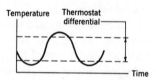

FIG. 8-15. Room   temperature variation with time.

several thermal time-constant elements in series which slow its response. The figure suggests that temperature variations could be reduced by decreasing the differential. While this is true, the result would also include shorter periods of burner operation and shorter periods between them. Such a condition reduces component life and may lessen the efficiency of the furnace. Thus, practical considerations limit how small the differential can be made.

Although the thermostat and gas valve are necessary for system performance, the other two elements shown in the low-voltage circuit of Fig. 8-14 are essential for safe operation. The first guards against the opening of the gas valve in the event of a pilot-flame failure. A spring-loaded switch is held closed by an electromagnet which is powered from a thermocouple heated by the pilot flame. Should the pilot become extinguished, the switch will open making the heating system inoperative. A second safety control guards against having too high an air temperature within the furnace. A bellows, located in the top of the furnace, controls the switch shown which is opened if the furnace air temperature exceeds a safe limit.

Liquid-Level System. A second illustration of two-position control is given in conjunction with the liquid-level application shown in Fig. 8-16a. The head h of the tank is measured by means of a float, which, through a

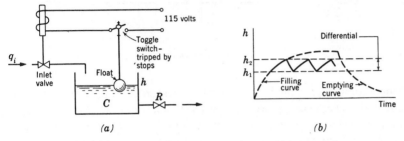

(a)   *(b)*

Fɪɢ. 8-16. Liquid-level application using two-position control.

switch, controls the inlet-valve solenoid so as to permit either full flow or zero flow into the tank.

The differential equation which defines the behavior of the tank without control is

$$RC \frac{dh}{dt} + h = R\, q_i(t) \qquad (8\text{-}38)$$

With two-position control, $q_i(t)$ must be either some constant flow rate Q (valve open) or zero (valve closed). The response of the system is shown in Fig. 8-16b with solid lines. In going from h_1 to h_2 (through the *differential*) the response follows the filling curve for the tank—the curve resulting from the solution of Eq. 8-38 with $q_i(t) = Qu(t)$. From h_2 to h_1, the response follows that of the emptying curve obtained from the solution of Eq. 8-38 with $q_i(t) = 0$. Thus, the response is obtained on a step-by-step basis, the first step involving a solution of Eq. 8-38 with h_1 as an initial condition and $q_i(t) = Qu(t)$ as the forcing function. This solution is valid until $h = h_2$ at which time the inlet valve closes and a new expression for the response is required. The second step involves a solution of Eq. 8-38 with h_2 as an initial condition and $q_i(t) = 0$ as the forcing function. This solution provides the response until $h = h_1$ at which time a typical cycle is complete. It should be noted that the response does not exceed the limits of the differential when the controlled portion of the system is a single time-constant element.

The discussion of control actions is now complete and attention can be turned to the classification of control systems by system type.

8-7. System Types

Classification of control systems by *system type* is based on information concerning not only the control elements (the basis for control actions) but the system elements as well. In the reduced form of block diagram, shown in Fig. 8-17, the *forward transfer function* $G(s)$ is used to obtain the system type number. A constant other than unity can exist in the feedback

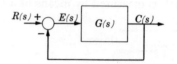

FIG. 8-17. Block diagram.

loop without changing the type number because the character of $G(s)$ is not affected, the constant merely being a scale factor in the system. Classification by system type is particularly useful for follow-up systems, that is, systems where the major function is to keep the controlled variable in close correspondence with the reference variable.

In introducing the concepts of system types, it is convenient to investigate the steady-state errors which result from various inputs. To facilitate the study, the transfer function $E(s)/R(s)$ is first obtained. With the block diagram of Fig. 8-17,

$$C(s) = G(s) E(s) \quad \text{and} \quad E(s) = R(s) - C(s)$$

Combination of these equations, to eliminate $C(s)$, yields

$$\frac{E(s)}{R(s)} = \frac{1}{1 + G(s)} \tag{8-39}$$

The forward transfer function $G(s)$ can be expressed, in general form, as the ratio of two polynomials:

$$G(s) = \frac{N(s)}{D(s)} = \frac{a_i s^i + a_{i-1}s^{i-1} + \cdots + a_1 s + a_0}{s^m(b_n s^n + b_{n-1}s^{n-1} + \cdots + b_1 s + 1)} \tag{8-40}$$

With $G(s)$ in this form, m is the *system type number*. In terms of a physical system, m is the number of integrations in the forward loop.

Substitution of Eq. 8-40 into Eq. 8-39, and solution for $E(s)$ yields

$$E(s) = \frac{1}{1 + \dfrac{N(s)}{D(s)}} R(s) = \frac{D(s)}{D(s) + N(s)} R(s)$$

$$= \frac{s^m(b_n s^n + \cdots + 1)}{s^m(b_n s^n + \cdots + 1) + (a_i s^i + \cdots + a_0)} R(s) \tag{8-41}$$

As an example, consider the response of a type 0 system to a step input, that is, $m = 0$ and $R(s) = 1/s$. With Eq. 8-41,

$$E(s) = \frac{b_n s^n + \cdots + 1}{(b_n s^n + \cdots + 1) + (a_i s^i + \cdots + a_0)} \frac{1}{s}$$

The steady-state error is obtained, by use of the final value theorem, as

$$e_{ss}(t) = \lim_{s \to 0} sE(s) = \frac{1}{1 + a_0}$$

There is a steady-state error and it is finite.

Consider next the response of a type 0 system to a ramp input, that is, $R(s) = 1/s^2$. With Eq. 8-41,

$$E(s) = \frac{b_n s^n + \cdots + 1}{(b_n s^n + \cdots + 1) + (a_i s^i + \cdots + a_0)} \frac{1}{s^2}$$

The use of the final value theorem yields

$$e_{ss}(t) = \infty \tag{8-42}$$

Equation 8-42 reveals that a type 0 system will not follow a ramp input.

As a final example, let us consider the response of a type 1 system to a ramp input. Substitution of $m = 1$ and $R(s) = 1/s^2$ in Eq. 8-41 yields

$$E(s) = \frac{s(b_n s^n + \cdots + 1)}{s(b_n s^n + \cdots + 1) + (a_i s^i + \cdots + a_0)} \frac{1}{s^2}$$

With the final value theorem,

$$e_{ss}(t) = \frac{1}{a_0}$$

Thus it can be seen that a type 1 system will follow a ramp input but with a finite error.

Further analyses of the steady-state errors resulting from various combinations of system type number and input can be made by following the above procedure. A summary is given in Table 8-1. This table is

TABLE 8-1

STEADY-STATE ERRORS FOR VARIOUS COMBINATIONS
OF SYSTEM TYPE NUMBER AND INPUT

System Type	Input		
Number	Step	Ramp	Parabolic
0	Finite	∞	∞
1	0	Finite	∞
2	0	0	Finite

quite useful. For example, if one wished to design a follow-up system with zero steady-state error to a ramp input, the table shows that a type 2 system is required.

A comparison of the performances of the three system types shows that, as the type number is increased, accuracy is improved. However, each additional integration in the forward loop aggravates the stability problem. Some compromise is therefore necessary.

Type 0 and type 1 control systems are very common. An application illustrating each is given in Sec. 8-8.

8-8. Applications

Two applications will now be presented, a hydraulic system to illustrate a type 0 system and an electrohydraulic system to illustrate a type 1 system. The applications will also serve as further examples of setting up system block diagrams.

Pressure-Compensated Pump System. In fluid power systems, some type of pump is used as the power source for a circuit in which cylinders or motors perform the required work and suitable valving arrangements exist to direct the fluid flow to the proper point at the proper time. The application to be presented involves a pressure-compensated, variable-displacement pump used in a hydraulic circuit. A schematic drawing of such a pump is given in Fig. 8-18. The spring force acts to increase the stroke x which, in turn, provides a greater flow rate q_1. System pressure

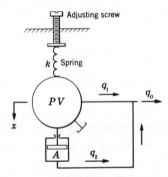

FIG. 8-18. Schematic drawing of pressure-compensated, variable-displacement pump.

acting on the piston (with area A) creates a force which opposes that of the spring and acts to reduce the pump stroke. A force balance produces a fixed stroke position and a constant output flow rate q_o ($q_2 = 0$ with x constant). In the pump being described, x can vary from zero (no fluid pumped) to some maximum positive value (maximum output). The adjusting screw is used to vary the spring force and thereby to change the system operating pressure.

The operation of a pressure-compensated pump in a circuit can be described as follows. If additional flow to the circuit is required, system pressure will decrease permitting the spring to increase the pump stroke by the amount necessary to provide the additional flow. If, on the other hand, less flow is required, the rising system pressure will reduce the stroke by an appropriate amount. Thus, a pressure-compensated pump acts to provide the flow rate required by the circuit at nearly a constant system pressure. Not only does the pump provide pressure control but a saving in power as well —only the fluid required by the circuit at any particular time is pumped.

The first logical step in setting up a system block diagram is to obtain a transfer function for the pump. The internal mechanism, which controls the pump displacement, involves mass M and friction c which can be assumed to be viscous (as a first approximation). A net force f_e (the actuating signal as it turns out to be) acting on the mass produces motion. The differential equation is

$$M \frac{d^2x}{dt^2} + c \frac{dx}{dt} + kx = f_e(t) \tag{8-43}$$

Transforming,

$$X(s)[Ms^2 + cs + k] = F_e(s) \tag{8-44}$$

The pump output flow rate is

$$q_o = q_1 + q_2 = K_1 x + A \frac{dx}{dt} \tag{8-45}$$

where K_1 is the gain of the pump (see Eq. 7-26). Thus,

$$Q_o(s) = X(s)[As + K_1] \tag{8-46}$$

Equations 8-44 and 8-46 can be combined to give the transfer function for the pump:

$$\frac{Q_o(s)}{F_e(s)} = \frac{As + K_1}{Ms^2 + cs + k} \tag{8-47}$$

The net force acting on the mass is the difference between the spring force f_r (actually the reference input obtained by turning the adjustment screw) and the pressure force. Thus,

$$f_e = f_r - Ap \tag{8-48}$$

where p is the system pressure. Transforming,

$$F_e(s) = F_r(s) - AP(s) \tag{8-49}$$

The equation relating the flow rates in the circuit is

$$q_o = q_L + q_c + q_l \tag{8-50}$$

where q_L is the flow rate required to drive some load, q_c is the flow rate involved in the compressibility of the fluid, and q_l is the flow rate lost through leakage in the system. Substitution of Eqs. 7-29b and 7-29c yields

$$q_o = q_L + \frac{V}{\beta} \frac{dp}{dt} + Lp \tag{8-51}$$

Therefore,

$$Q_o(s) - Q_L(s) = P(s)\left(\frac{V}{\beta}s + L\right) \qquad (8\text{-}52)$$

Equations 8-47, 8-49, and 8-52 define the behavior of the system and can be represented by the block diagram shown in Fig. 8-19. The diagram can be used to investigate the system pressure response to either a change in reference or a change in load flow rate (a disturbance). If the diagram is

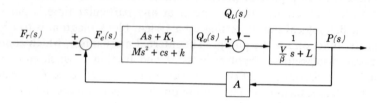

FIG. 8-19. Block diagram for hydraulic system.

drawn in reduced form, assuming no disturbance, the forward transfer function is

$$G(s) = \frac{As + K_1}{(Ms^2 + cs + k)\left(\dfrac{V}{\beta}s + L\right)} \qquad (8\text{-}53)$$

which is that of a type 0 system.

Electrohydraulic Ship Steering System. Ship steering can be accomplished in various ways, but the approach presented here involves a system in which some components are electrical and others are hydraulic. Electrohydraulic systems are widely used because they combine the advantages of electrical control with those of hydraulic power.

A schematic drawing of the ship steering system is shown in Fig. 8-20. The wheel position is sensed by a linear variable differential transformer (LVDT) which supplies an electrical signal to the input of the amplifier.

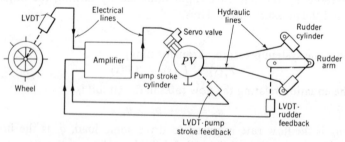

FIG. 8-20. Electrohydraulic ship steering system.

The amplifier output drives the electrohydraulic servo valve which, by using a cylinder, controls the stroke of the variable-displacement pump. The pump permits both positive and negative stroke positions, meaning that the output flow can be reversed in direction. Fluid from the pump enters one or the other of the two rudder cylinders which, acting on the rudder arm, control the rudder position. A second LVDT senses the rudder position and provides the primary feedback for the system. An additional LVDT supplies pump-stroke information to the amplifier; this feedback is necessary to achieve absolute stability.

A system block diagram can now be obtained using transfer functions previously determined. The transfer function for the amplifier-driven servo valve, from Eq. 7-40, can be written as

$$\frac{Q_1(s)}{E_1(s)} = \frac{K_1}{\tau s + 1} \tag{8-54}$$

The pump-stroke cylinder can be represented, with reasonable accuracy, as an integrating device. Thus,

$$\frac{X(s)}{Q_1(s)} = \frac{K_2}{s} \tag{8-55}$$

The transfer function for the pump, from Eq. 7-26, is

$$\frac{Q_2(s)}{X(s)} = K_3 \tag{8-56}$$

For simplicity, the rudder cylinders will be treated as integrators although Eq. 7-32 could be used. Therefore,

$$\frac{C(s)}{Q_2(s)} = \frac{K_4}{s} \tag{8-57}$$

where $C(s)$ is the controlled variable and is a measure of the position of the rudder. Equation 7-23 reveals that the LVDT's are proportional elements.

The system block diagram of Fig. 8-21 can now be drawn. The reference input is the signal from the wheel LVDT. In order to determine the system type number, the inner loop must first be replaced with a single block. The appropriate transfer function is

$$\frac{X(s)}{E(s)} = \frac{K_1 K_2}{\tau s^2 + s + K_1 K_2 K_5} \tag{8-58}[7]$$

With this transfer function substituted into the block diagram, the forward transfer function is

$$G(s) = \frac{K_1 K_2 K_3 K_4}{s(\tau s^2 + s + K_1 K_2 K_5)} \tag{8-59}$$

which is that of a type 1 system.

[7]The comment has been made that each additional integration in a system aggravates the stability problem. Equation 8-58 points out the fact that, in feeding back around an integrating element, the integration is eliminated. This is one way of improving system performance.

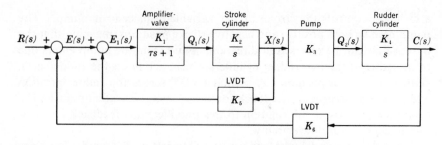

FIG. 8-21. Block diagram for ship steering system.

8-9. Closure

In this, and in previous chapters, the approach to control system analysis has been to study system response using a step input as the forcing function. Such an approach is called the *transient-response method* because transients are considered. The method, while valuable, can become unwieldy for more complicated systems which are defined by higher-order differential equations. An alternative approach involves the steady-state response to a sinusoidal input and is known as the *frequency-response method*. The next three chapters deal with this method and its application to control system analysis and design.

Problems

8-1. The use of proportional control action for controlling the liquid level of a tank is illustrated schematically in the accompanying figure. The inlet valve has a stem travel of 1 in. and a capacity from 0 to 10 gpm (depending on the position of the stem). Determine the numerical value for the proportional gain K_p and state the units.

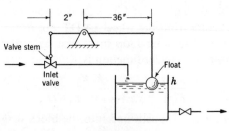

PROB. 8-1

8-2. A block diagram for a speed control system is shown in the accompanying illustration. a) Obtain the expression for the response $\omega_c(t)$ if a step reference input $\Omega_R u(t)$ is applied. Assume that the disturbance torque $T_d(s)$ is zero. b) Obtain the expression for the deviation $\Delta\omega_c(t)$ if a step disturbance torque $T_D u(t)$ is applied. Assume that the reference input is zero.

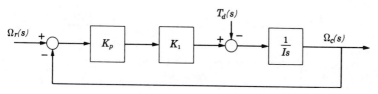

PROB. 8-2

8-3. a) Show that the block representation for the tank shown in the accompanying illustration is correct. The manipulated variable is q_m, and q_u is a disturbance flow rate. Liquid level h_c is the controlled variable. b) The liquid level is to be controlled by proportional control action as indicated in the block diagram. Determine the minimum proportional gain K_p such that, with a sustained disturbance q_u of 0.5 ft³/sec, the deviation of liquid level will not exceed 0.1 ft. Assume $C = 9$ ft² and $R = 10$ sec/ft².

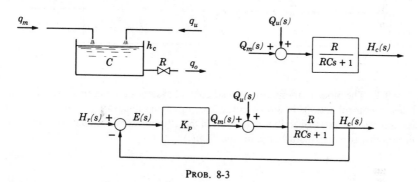

PROB. 8-3

8-4. A block diagram for a speed control system is shown in the accompanying illustration. Determine the minimum gain $K_p K_1$ such that, with sustained torque disturbance of 1000 lb-in., the speed deviation will not exceed 2 radians/sec. The time constant $\tau = I/b = 1.9$ sec. The value for I is 130 lb-in.-sec².

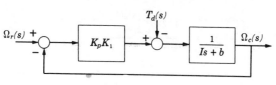

PROB. 8-4

8-5. A block diagram for a liquid-level system is shown in the accompanying illustration. Determine the steady-state response $h_{css}(t)$ if a step reference input $H_R u(t)$ is applied.

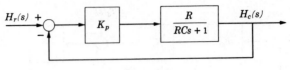

PROB. 8-5

8-6. Integral control action is used in a speed control system as indicated in the block diagram shown in the accompanying illustration. a) Obtain the expression for the response $\omega_c(t)$ if a step reference input $\Omega_R\,u(t)$ is applied. Assume that $T_d(s) = 0$ and that the response is oscillatory. b) Obtain the expression for the deviation $\Delta\omega_c(t)$ if a positive step disturbance torque $T_D\,u(t)$ is applied. Assume that the reference input is zero and that the response is oscillatory.

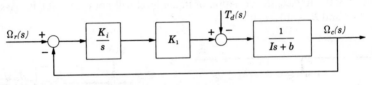

PROB. 8-6

8-7. The block diagram for a speed control system utilizing proportional plus integral control actions is shown in the accompanying illustration. Show that this combination of control actions will result in the ultimate elimination of system error in response to a step disturbance torque $T_D\,u(t)$.

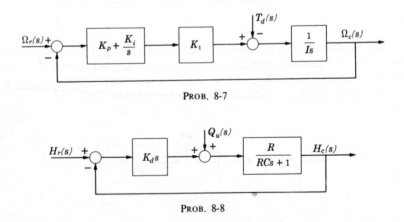

PROB. 8-7

PROB. 8-8

8-8. This problem will illustrate the fact that derivative control action should not be used alone. For the liquid-level system shown in the accompanying illustration: a) Obtain the steady-state response $h_{css}(t)$ for a step reference input $H_R\,u(t)$. Assume $Q_u(s) = 0$. Is this a desirable response? b) Obtain the steady-state deviation $\Delta h_{css}(t)$ for a step disturbance input $Q_U u(t)$. Assume $H_r(s) = 0$. Compare the response with that resulting from having no control at all.

8-9. Assume that the block diagram shown in the accompanying illustration is that of a system already built. The response of the system is considered to be too oscillatory. A suggestion was made to add derivative control action to improve the response, that is, use derivative plus integral control actions. Investigate this addition and then recommend whether or not derivative control should be used.

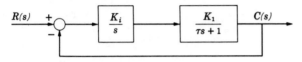

PROB. 8-9

8-10. Two-position control action will be used to control the liquid level in a tank as shown in Fig. 8-16. The head h is to be 20 ± 1 ft. The outlet valve is set so that, at $h = 20$ ft, the outflow is 0.25 ft³/sec. Tank capacitance $C = 9$ ft². The inlet valve permits a flow rate of 0.5 ft³/sec when open and 0 ft³/sec when closed. Determine the period (in seconds) for oscillations of liquid level. Assume the resistance of the outlet valve to be linear.

8-11. This problem will illustrate the fact that a type 1 system, subjected to a disturbance input, may or may not eliminate the system error depending on the location of the integration. For each of the block diagrams shown in the accompanying illustration, determine the steady-state deviation $\Delta c_{ss}(t)$ for a step disturbance input $U u(t)$.

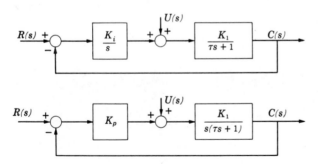

PROB. 8-11

8-12. A hydraulic system involving a pressure-compensated pump was presented in Sec. 8-8. The block diagram is given in Fig. 8-19. Determine the steady-state deviation $\Delta p_{ss}(t)$ for a step load flow rate disturbance $Q_L u(t)$.

8-13. For the hydraulic system of Fig. 8-19, determine the relationship between the system parameters required for absolute stability. Use Routh's criterion.

8-14. The block diagram for the electrohydraulic ship steering system of Sec. 8-8 is given in Fig. 8-21. Determine the steady-state response $c_{ss}(t)$ for a step change of reference input $R\,u(t)$.

8-15. The block diagram for the electrohydraulic ship steering system of Sec. 8-8 is given in Fig. 8-21. If the pump stroke feedback were eliminated, show that the system would be unstable.

Chapter 9

Frequency Response

9-1. Introduction

The term *frequency response* refers to the steady-state response of a system subjected to a sinusoidal input of fixed amplitude but with the frequency varied over some range. The concept is illustrated in Fig. 9-1a in which a linear system is forced by an input $a \sin \omega t$; the output is $b \sin (\omega t + \phi)$.[1] The input and output wave forms are shown in Fig. 9-1b. Of interest is the ratio of the amplitudes b/a and the phase angle ϕ both of which are functions of frequency. The ratio b/a is commonly called the *magnitude ratio* and will be designated as $M(\omega)$. The phase angle (or phase shift as it is sometimes called) will be denoted by $\phi(\omega)$.

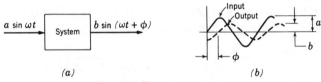

(a) (b)

FIG. 9-1. Frequency response.

This chapter is concerned primarily with determining frequency-response information analytically although such data can be obtained experimentally if the system exists. The material presented is important in that it provides a convenient means for getting the steady-state response for any linear system subjected to a sinusoidal input. It is also intimately related to the method of control-system analysis to be covered in Chapter 10.

The procedure for obtaining frequency-response data analytically is relatively simple and is embodied in four steps.

Step 1. Obtain the transfer function for the element or combination of elements involved, that is, $O(s)/I(s) = F(s)$ where $O(s)$ and $I(s)$ are the transforms of the output and input respectively. Any initial conditions are ignored because these do not affect the steady-state response.

Step 2. In the transfer function, replace each s by $j\omega$. The justification for this substitution will follow shortly.

---

[1] The expression for the output is given in general form in which $+\phi$ indicates the addition of an angle which may be either positive or negative.

Step 3. For various values of frequency ω, determine the magnitude ratio M and the phase angle ϕ.

Step 4. Plot the results from step 3 in polar-coordinate or rectangular-coordinate form. These plots are not only convenient means for presenting frequency-response data but are also the basis for analytical and design methods which will be covered in later chapters.

9-2. Justification for Replacing s with $j\omega$

Step 2 in Sec. 9-1 called for replacing each s in the transfer function by $j\omega$. This substitution will now be justified. The procedure will be to work with a general transfer function, first obtaining the steady-state response to a sinusoidal input by using Laplace transforms and then by making the $j\omega$-for-s substitution. If the responses prove to be identical, the substitution can be considered valid. The presentation which follows not only serves the purpose at hand but will also provide a review of some of the material covered in Chapter 6.

Assume a general transfer function with numerator $N(s)$ and denominator $D(s)$. Then

$$\frac{O(s)}{I(s)} = F(s) = \frac{N(s)}{D(s)} \tag{9-1}$$

Let the input be a sinusoid with unity amplitude ($\sin \omega t$). The transform of the input $I(s)$ is $\omega/(s^2 + \omega^2)$. Therefore,

$$O(s) = \frac{\omega}{s^2 + \omega^2} \frac{N(s)}{D(s)} = \frac{\omega}{(s + j\omega)(s - j\omega)} \frac{N(s)}{D(s)}$$

Recall that any initial conditions can be ignored because these do not affect the steady-state response.

By partial fraction expansion,

$$O(s) = \frac{C_1}{s + j\omega} + \frac{C_2}{s - j\omega} + \frac{C_3}{s + r_1} + \frac{C_4}{s + r_2} + \cdots + \frac{C_n}{s + r_{n-2}}$$

where $s + r_1, s + r_2, \ldots$ are factors of $D(s)$. For a stable system, the transients die out [the inverse transforms of $C_3/(s + r_1), \cdots$ go to zero as $t \to \infty$] and the steady-state response is

$$O_{ss}(t) = C_1 e^{-j\omega t} + C_2 e^{j\omega t} \tag{9-2}$$

The constants C_1 and C_2 are determined by the method introduced in Chapter 6.

$$C_1 = \frac{\omega}{s - j\omega} \frac{N(s)}{D(s)} \bigg|_{s = -j\omega} = -\frac{1}{2j} \frac{N(-j\omega)}{D(-j\omega)}$$

$$C_2 = \frac{\omega}{s + j\omega} \frac{N(s)}{D(s)} \bigg|_{s = j\omega} = \frac{1}{2j} \frac{N(j\omega)}{D(j\omega)}$$

The fraction $N(j\omega)/D(j\omega)$ can be expressed in alternative form as[2]

$$\frac{N(j\omega)}{D(j\omega)} = A(\omega) + jB(\omega) \tag{9-3}$$

and

$$\frac{N(-j\omega)}{D(-j\omega)} = A(\omega) - jB(\omega) \tag{9-4}$$

where $A(\omega)$ and $B(\omega)$ are real numbers and are functions of frequency. Therefore,

$$C_1 = -\frac{1}{2j}[A(\omega) - jB(\omega)] \quad \text{and} \quad C_2 = \frac{1}{2j}[A(\omega) + jB(\omega)]$$

Substitution in Eq. 9-2 yields

$$\begin{aligned}
o_{ss}(t) &= -\frac{1}{2j}[A(\omega) - jB(\omega)]e^{-j\omega t} + \frac{1}{2j}[A(\omega) + jB(\omega)]e^{j\omega t} \\
&= -\frac{A(\omega)}{2j}e^{-j\omega t} + \frac{jB(\omega)}{2j}e^{-j\omega t} + \frac{A(\omega)}{2j}e^{j\omega t} + \frac{jB(\omega)}{2j}e^{j\omega t} \\
&= A(\omega)\left[\frac{e^{j\omega t} - e^{-j\omega t}}{2j}\right] + B(\omega)\left[\frac{e^{j\omega t} + e^{-j\omega t}}{2}\right] \\
&= A(\omega)\sin\omega t + B(\omega)\cos\omega t
\end{aligned} \tag{9-5}$$

The steady-state response is seen to be composed of the sum of a sine and a cosine wave. The summation of these two waves is accomplished by vector addition as shown in Fig. 9-2. Here a vector of magnitude $A(\omega)$ is considered rotating counterclockwise at some frequency ω. The projection

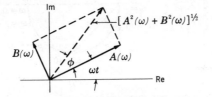

FIG. 9-2. Vector addition for combining sine and cosine waves.

of this vector on the imaginary axis produces the required sine wave. A vector of magnitude $B(\omega)$ is shown 90° ahead of the first vector; its projection on the imaginary axis produces the appropriate cosine wave. The two vectors can be replaced by a single vector of magnitude $[A^2(\omega) + B^2(\omega)]^{1/2}$ and phase angle $\phi = \tan^{-1} B(\omega)/A(\omega)$. Therefore, Eq. 9-5 can be expressed as

$$o_{ss}(t) = [A^2(\omega) + B^2(\omega)]^{1/2}\sin(\omega t + \phi) \tag{9-6}$$

where $\phi = \tan^{-1} B(\omega)/A(\omega)$.

[2] See Prob. 9-1 for a specific example.

Recall that the input was a sine wave with unity amplitude. Therefore, the magnitude ratio, that is, the ratio of the output to input amplitudes, is

$$M(\omega) = [A^2(\omega) + B^2(\omega)]^{\frac{1}{2}}$$

The phase angle is

$$\phi(\omega) = \tan^{-1}\frac{B(\omega)}{A(\omega)}$$

It should be recognized that the phase angle may well be negative.

Having completed the solution using Laplace transforms, it remains to be shown that the same result can also be obtained directly from Eq. 9-1. Making the substitution $s = j\omega$ and from Eq. 9-3

$$\frac{O(j\omega)}{I(j\omega)} = \frac{N(j\omega)}{D(j\omega)} = A(\omega) + jB(\omega)$$

This expression yields

$$M(\omega) = [A^2(\omega) + B^2(\omega)]^{\frac{1}{2}} \quad \text{and} \quad \phi(\omega) = \tan^{-1}\frac{B(\omega)}{A(\omega)}$$

the same result as previously obtained and with far less effort.

9-3. The Polar Plot

Attention is now turned to two illustrative systems for which frequency-response information will be presented in polar-coordinate form. The alternate form, the rectangular plot, will be discussed in Sec. 9-4.

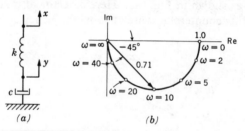

FIG. 9-3. Polar plot for system shown.

Consider the system shown in Fig. 9-3a. The differential equation is

$$\frac{c}{k}\frac{dy}{dt} + y = x(t) \tag{9-7}$$

and the transfer function is

$$\frac{Y(s)}{X(s)} = \frac{1}{\tau s + 1} \tag{9-8}$$

where the time constant $\tau = c/k$. Making the substitution $s = j\omega$ and assuming $\tau = 0.1$ sec, yields

$$\frac{Y(j\omega)}{X(j\omega)} = \frac{1}{1 + j\omega\tau} = \frac{1}{1 + j0.1\omega} \tag{9-9}$$

A table is now constructed giving the magnitude ratio $M(\omega)$ and the phase angle $\phi(\omega)$ for various values of ω. To illustrate the calculations involved, take $\omega = 10$ as an example. Then,

$$\frac{Y}{X} = \frac{1}{1 + j1} = \frac{1}{1.41\,\underline{/45°}} = 0.71\,\underline{/-45°}$$

The student should check other values. The table of $1 + j\omega\tau = A\,\underline{/\theta°}$ given in Appendix E will simplify making the calculations.

TABLE 9-1

FREQUENCY RESPONSE DATA

ω (radians/sec)	$M(\omega)$	$\phi(\omega)$ (degrees)
0	1.00	0.0
2	0.98	-11.3
5	0.89	-26.6
10	0.71	-45.0
20	0.45	-63.4
40	0.24	-76.0
∞	0.00	-90.0

The data from Table 9-1 are plotted on the complex plane with ω as a parameter. The result is shown in Fig. 9-3*b*.

As a second example, consider the system shown in Fig. 9-4*a*. The differential equation is

$$M\frac{d^2y}{dt^2} + c\frac{dy}{dt} + ky = kx(t) \tag{9-10}$$

and the transfer function is

$$\frac{Y(s)}{X(s)} = \frac{k}{Ms^2 + cs + k} = \frac{1}{\dfrac{M}{k}s^2 + \dfrac{c}{k}s + 1} = \frac{1}{\dfrac{s^2}{\omega_n^2} + \dfrac{2\zeta s}{\omega_n} + 1} \tag{9-11}$$

Assuming a natural frequency of 10 radians/sec and a damping ratio of 0.5, and making the substitution $s = j\omega$, yields

$$\frac{Y(j\omega)}{X(j\omega)} = \frac{1}{0.01(j\omega)^2 + 0.1j\omega + 1} = \frac{1}{(1 - 0.01\omega^2) + j0.1\omega} \tag{9-12}$$

A table is now made giving the magnitude ratio and phase angle for various values of ω. To illustrate the calculations involved, take $\omega = 5$ as an example. Then,

$$\frac{Y}{X} = \frac{1}{(1 - 0.25) + j0.5} = \frac{1}{0.9\,\underline{/33.7°}} = 1.11\,\underline{/-33.7°}$$

The student should check other values. A polar plot for the information contained in Table 9-2 is shown in Fig. 9-4*b*.

The method for calculating frequency-response data from transfer

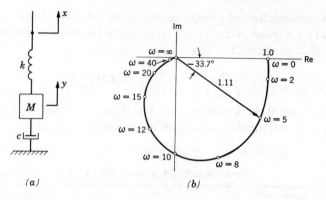

FIG. 9-4. Polar plot for system shown.

functions and presenting this information in polar form has been sufficiently illustrated. More complicated transfer functions can be handled in
a similar manner. As a guide, Fig. 9-5 shows sketches of polar plots for
some typical transfer functions.

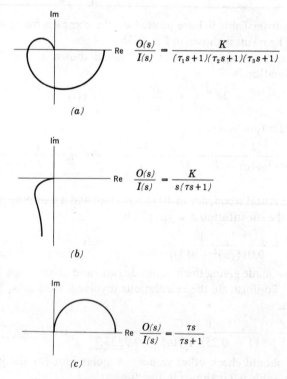

$$\frac{O(s)}{I(s)} = \frac{K}{(\tau_1 s+1)(\tau_2 s+1)(\tau_3 s+1)}$$

(a)

$$\frac{O(s)}{I(s)} = \frac{K}{s(\tau s+1)}$$

(b)

$$\frac{O(s)}{I(s)} = \frac{\tau s}{\tau s+1}$$

(c)

FIG. 9-5. Sketches of polar plots for some typical
transfer functions.

TABLE 9-2
FREQUENCY RESPONSE DATA

ω (radians/sec)	$M(\omega)$	$\phi(\omega)$ (degrees)
0	1.00	0.0
2	1.02	−11.8
5	1.11	−33.7
8	1.14	−65.8
10	1.00	−90.0
12	0.78	−110.1
15	0.51	−129.8
20	0.28	−146.3
40	0.06	−165.1
70	0.02	−171.7
∞	0.00	−180.0

9-4. The Rectangular Plot

Frequency-response data can be presented in rectangular-coordinate form. The magnitude ratio and the phase angle are plotted against frequency. It is customary and convenient to make the plots against $\log_{10} \omega$. In the case of phase angles, a linear scale is used and so the $\phi(\omega)$ vs. ω plot is made on semilog paper. The magnitude ratio can be handled in two ways. It can be plotted as $\log_{10} M(\omega)$ in which case the $M(\omega)$ vs. ω plot is made on log-log paper, or it can be expressed in *decibels*, a logarithmic unit, and plotted on a linear scale. In the latter case, semilog paper is used. The decibel approach will be employed in this book because it is widely used and also permits the plotting of both $M(\omega)$ and $\phi(\omega)$ on the same sheet, both plots requiring semilog paper.

If M is a magnitude and m is that magnitude expressed in decibels (db), then

$$m = 20 \log_{10} M \qquad (9\text{-}13)$$

To illustrate:

If $M = 1$:　　$m = 20(0) = 0$ db
If $M = 10$:　　$m = 20(1) = 20$ db
If $M = 100$:　$m = 20(2) = 40$ db
If $M = 0.1$:　$m = 20(9.0 - 10) = -20$ db
If $M = 0.2$:　$m = 20(9.301 - 10) = -13.98$ db

The conversion of magnitude ratios to decibels is greatly facilitated by the decibel conversion table given in Appendix F.

The data given in Tables 9-1 and 9-2, corresponding to the systems shown in Figs. 9-3a and 9-4a, are plotted in Figs. 9-6 and 9-7 in rectangular form. As a guide, sketches of rectangular plots for some typical transfer functions are shown in Fig. 9-8.

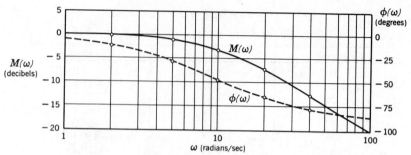

FIG. 9-6. Rectangular plot for system shown in Fig. 9-3a.

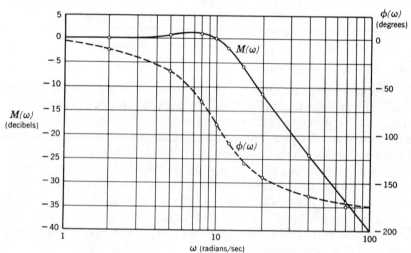

FIG. 9-7. Rectangular plot for system shown in Fig. 9-4a.

9-5. The Graphical Viewpoint

Additional insight into frequency response can be obtained by considering a semigraphical method for calculating frequency-response data. The graphical viewpoint will also be used to develop certain material to be presented later in the book, making its introduction here doubly worth while.

Consider the simple transfer function

$$\frac{O(s)}{I(s)} = \frac{K}{s + r} \tag{9-14}$$

As a matter of terminology, $-r$ is called a *pole* of $O(s)/I(s)$; a pole is, by definition, a value of s for which $O(s)/I(s)$ is infinite. The pole is shown plotted on the complex plane of Fig. 9-9a and is designated by a small cross. A vector from the origin to $-r$ is the $-r$ vector. In addition, a vector from the origin to $j\omega_1$ is the $j\omega_1$ vector.

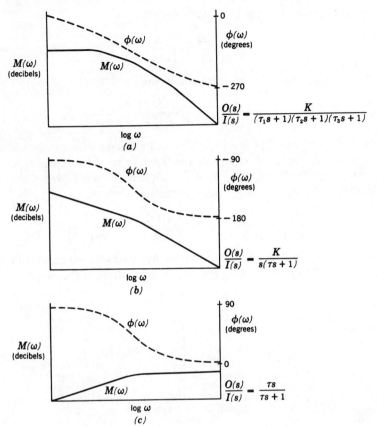

FIG. 9-8. Sketches of rectangular plots for some typical transfer functions.

A vector is now drawn between $-r$ and $j\omega_1$ as shown in Fig. 9-9b. Based on the graphical subtraction of vectors, this *difference vector* is

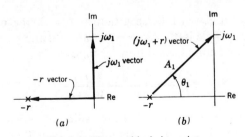

FIG. 9-9. The graphical viewpoint.

$$j\omega_1 - (-r) = j\omega_1 + r \tag{9-15}$$

The vector can be scaled for magnitude A_1 and angle θ_1.

Substitution of $j\omega$ for s in Eq. 9-14 yields

$$\frac{O(j\omega)}{I(j\omega)} = \frac{K}{j\omega + r} \tag{9-16}$$

With the aid of Fig. 9-9b, one determines that

$$\left.\frac{O}{I}\right|_{\omega=\omega_1} = \frac{K}{j\omega_1 + r} = \frac{K}{A_1\,\underline{/\,\theta_1}} = \frac{K}{A_1}\,\underline{/-\theta_1}$$

Magnitude ratios and phase angles for other values of ω can be obtained in a similar manner by drawing vectors from $-r$ to other values of $j\omega$.

As a specific illustration, let it be required to determine the magnitude ratio and phase angle for the transfer function

$$\frac{O(s)}{I(s)} = \frac{1}{s + 1} \quad \text{or} \quad \frac{O(j\omega)}{I(j\omega)} = \frac{1}{j\omega + 1}$$

for $\omega = 1$. The pole, -1, is plotted on the complex plane as shown in Fig. 9-10. A vector is drawn from the pole location to the point $j1$. The

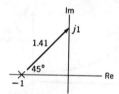

FIG. 9-10. Specific illustration of the graphical approach.

length of the vector is scaled as 1.41, and the angle with the real axis is measured as 45°. Therefore,

$$\left.\frac{O}{I}\right|_{\omega=1} = \frac{1}{1.41\,\underline{/\,45}} = 0.71\,\underline{/-45°}$$

From such a plot it can also be seen that at $\omega = 0$, the magnitude ratio is 1, and the phase angle is 0°. As $\omega \to \infty$, $M \to 0$ and $\phi \to -90°$.

A slightly more complicated transfer function is

$$\frac{O(s)}{I(s)} = \frac{K(s + r_1)}{(s + r_2)(s + r_3)} \tag{9-17}$$

Here, $-r_1$ is a *zero* of $O(s)/I(s)$, that is, a value of s for which the transfer function is zero. The zero is designated by a small circle on the complex plane of Fig. 9-11. The complex poles $-r_2$ and $-r_3$ are also shown.

Vectors are drawn from the zero and pole locations to the point $j\omega_1$. These difference vectors represent the factors of the transfer function

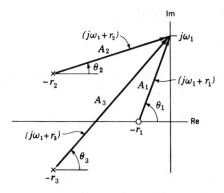

FIG. 9-11. The graphical approach with a zero and complex poles.

with $s = j\omega_1$. The magnitudes and angles of the vectors are scaled. With Eq. 9-17,

$$\left.\frac{O}{I}\right|_{\omega=\omega_1} = \frac{KA_1\underline{/\theta_1}}{(A_2\underline{/\theta_2})(A_3\underline{/\theta_3})} = \frac{KA_1}{A_2A_3}\underline{/\theta_1 - \theta_2 - \theta_3}$$

9-6. Experimental Determination of Frequency Response

Thus far the discussion of frequency response has assumed that the transfer function of the physical device was known. Unfortunately, this is not always the case in practice. In such circumstances, it is necessary to obtain frequency-response information experimentally. Such data may then be used to establish the transfer function of the device because, as has been illustrated in this chapter, the character of a transfer function is portrayed graphically by its frequency-response plots. This concept will be illustrated further in Chapters 10 and 11 through the Nyquist and Bode plots. Experimental frequency-response data can also be used to verify a transfer function which has been obtained analytically.

The experimental method requires that the actual component or system be subjected to a sinusoidal input and that the output be observed and compared in both magnitude and phase with the input. Unfortunately, this is not always a simple matter, particularly if the input and output of the device are mechanical in nature. It is perhaps advantageous to discuss the procedure with electrical components or, for that matter, any combination of physical elements which is driven by a voltage input and results in a voltage output. In this case, there are two possibilities for electrical signals—a system controlled by a varying d-c voltage or a system controlled by an a-c voltage of variable amplitude.

Figure 9-12 illustrates a simple setup for sinusoidally testing d-c control elements. The source of voltage input is a sine oscillator and the measuring device is a dual-beam oscilloscope. A sine wave is fed into the control

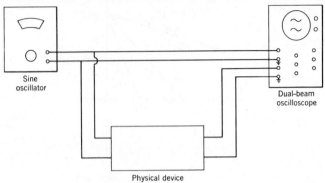

Sine oscillator

Dual-beam oscilloscope

Physical device

FIG. 9-12. Schematic for sinusoidal testing of d-c control components.

component and directly into one beam of the oscilloscope. The output of the test component is fed into the other beam of the oscilloscope. The amplitudes and phase relationship of the waveforms may be measured directly on the scope screen. (A convenient technique for phase measurement is by the use of Lissajous figures.)

If the elements under test are a-c devices, the input and output signals are amplitude modulated sine waves. A wide variety of special test equipment is commercially available for the accurate and rapid determination of magnitude and phase relationships. These servo analyzers are particularly useful in analyzing a-c equipment.

For elements with mechanical inputs and outputs, it is easy to appreciate the difficulties that may be encountered in testing. If equipment is added to convert information to electrical signals, care must be taken not to choose electrical equipment which will effect the transfer relationships of the equipment under test.

9-7. Closure

Obtaining frequency-response information from a known transfer function and plotting the results is simple in concept as has been illustrated. However, the ideas involved are very important because they turn out to be the basis for the frequency-response approach to control system analysis and design—the polar plot becomes a Nyquist diagram and the rectangular plot becomes a Bode plot. No more need be said about these now because they are the subjects of the next two chapters.

Problems

9-1. Show that

$$\frac{N(j\omega)}{D(j\omega)} = \frac{j\omega}{1 + j\omega} = A(\omega) + jB(\omega)$$

and that
$$\frac{N(-j\omega)}{D(-j\omega)} = A(\omega) - jB(\omega)$$

where $A(\omega) = \omega^2/(1 + \omega^2)$ and $B(\omega) = \omega/(1 + \omega^2)$. HINT: Multiply both numerator and denominator by the conjugate of the denominator.

9-2. From the accompanying illustration, make polar plots for the following: a) The system shown in view (a). Assume $c/k = 0.1$ sec. b) The system shown in view (b). Assume $RC = 0.05$ sec.

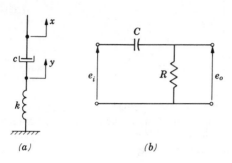

(a) (b)

PROB. 9-2

9-3. Make polar plots for the systems with the following transfer functions:

a) $\dfrac{O(s)}{I(s)} = \dfrac{5}{(0.1s + 1)(0.5s + 1)(2s + 1)}$ b) $\dfrac{O(s)}{I(s)} = \dfrac{2}{s(0.1s + 1)}$

9-4. Make rectangular plots for a) Prob. 9-2(a), b) Prob. 9-3(a), and c) Prob. 9-3(b).

9-5. Referring to Fig. 9-4a and Eq. 9-12, check the results given in Table 9-2 for $\omega = 0, 5, 10$. Use the graphical approach. In making the plot, keep the same scale on the real and imaginary axes. HINT: Work with the transfer function in the form
$$\frac{Y(s)}{X(s)} = \frac{100}{s^2 + 10s + 100} = \frac{100}{(s + r_1)(s + r_2)}$$

9-6. Write a digital computer program for determining the magnitude ratio and the phase angle for the system represented by Eq. 9-12. Run the program for $\omega = 0, 5, 10, 15, 20$. Compare the results with those given in Table 9-2.

9-7. The system shown in the accompanying illustration is given in most books on mechanical vibrations. These books derive the following expressions:
$$M.F. = \frac{X}{X_{st}} = \frac{1}{[(1 - \omega^2/\omega_n^2)^2 + (2\zeta\omega/\omega_n)^2]^{1/2}}$$
$$\phi = -\tan^{-1}\frac{2\zeta\omega/\omega_n}{1 - (\omega/\omega_n)^2}$$

where $M.F.$ is the magnification factor and X_{st} is the static deflection which equals F_o/k. The phase angle is ϕ.

Determine the transfer function for the system and show that the equations given can be obtained by making the $s = j\omega$ substitution.

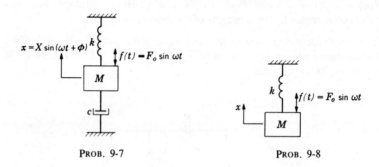

PROB. 9-7   PROB. 9-8

9-8. Following the graphical approach, verify that, for the undamped system shown in the accompanying illustration, the phase angle between the input forcing function $f(t)$ and the output displacement x is $0°$ for $\omega < \omega_n$ and $-180°$ for $\omega > \omega_n$. HINT: The root locations are $s = \pm j\omega_n$.

9-9. Write a digital computer program to obtain the magnitude A and the angle θ associated with the expression $1 + j\omega\tau$. Let $\omega\tau$ range from 1 to 10 in increments of one. Compare the results with the table given in Appendix E.

Chapter 10

System Analysis Using Polar Plots (Nyquist Diagrams)

10-1. Introduction

Chapter 9 introduced the polar plot as a means of graphically representing (on the complex plane) the steady-state response of a system subjected to a simple harmonic excitation. The plot provides the magnitude and phase relationships between the input and the response for any excitation frequency ω. Although previously illustrated with simple mechanical systems, the polar plot contains information of far greater value when used in connection with feedback control systems. In particular, it can be used to predict feedback control system stability.

This chapter will further the use of the polar plot by developing techniques for the determination of open-loop (the control system without feedback) and closed-loop (the system with feedback) responses to simple harmonic inputs. The concepts of magnitude ratio $M(\omega)$ and phase angle $\phi(\omega)$ are again applicable. For a harmonically forced linear system the response is related to the excitation by these parameters no matter how complex or involved the system.

One might logically question why the analysis of system response with a sinusoidal input is so important when, in use, control systems are seldom subjected to harmonic inputs. The answer is that information gained from sinusoidal analysis may be used to establish the nature of the response to a large class of inputs. Further, the analysis is convenient to handle mathematically and experimentally.

The prediction of control system stability by application of the *Nyquist criterion* is one widely accepted use of the polar plot. If a linear system is unstable when subjected to a harmonic input, it is unstable for any other input. The use of the polar plot bypasses the need for finding the roots of the characteristic equation of the system and eliminates the need for taking the inverse transform as in the transient response approach to system response analysis. Thus, the polar plot may be considered a labor-saving technique in the analysis of dynamic behavior.

The polar plot, or *Nyquist diagram*, is quite simple to use for most linear control systems. However, the concepts underlying the approach are more subtle. These will be developed first.

10-2. The Characteristic Equation and Criterion for Stability

The typical closed-loop control system may be represented by the simplified block diagram of Fig. 10-1. The *closed-loop transfer function*, or

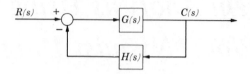

FIG. 10-1. Simplified system block diagram.

relationship between output and input of the system, was shown in Eq. 7-12 to be

$$\frac{C(s)}{R(s)} = \frac{G(s)}{1 + G(s)H(s)} \qquad (10\text{-}1)$$

The *open-loop transfer function* $G(s)H(s)$ (the transfer function with the feedback loop broken at the summing point) is typically a ratio of polynomials in s, or $N(s)/D(s)$. Thus,

$$1 + G(s)H(s) = 1 + \frac{N(s)}{D(s)} = \frac{D(s) + N(s)}{D(s)} \qquad (10\text{-}2)$$

The quantity $D(s) + N(s)$ is called the *characteristic function*, and if set equal to zero, becomes the characteristic equation

$$D(s) + N(s) = 0 \qquad (10\text{-}3)$$

EXAMPLE 10-1. Show that the characteristic equation for a system with $G(s) = 1/(\tau s + 1)$ and $H(s) = 1/s$ is defined by Eq. 10-3.

Solution:

$$G(s)H(s) = \frac{N(s)}{D(s)} = \frac{1}{s(\tau s + 1)}$$

Therefore, $N(s) = 1$, $D(s) = s(\tau s + 1)$, and by Eq. 10-3, the characteristic equation is

$$\tau s^2 + s + 1 = 0$$

This may be verified by expanding $C(s)/R(s)$ directly to give

$$\frac{C(s)}{R(s)} = \frac{\dfrac{1}{\tau s + 1}}{1 + \dfrac{1}{s(\tau s + 1)}} = \frac{s}{s(\tau s + 1) + 1}$$

Rearranging,

$$(\tau s^2 + s + 1)C(s) = sR(s)$$

It should be recalled from Chapter 6 that

$$\tau s^2 + s + 1 = 0$$

is the characteristic equation in s. *Ans.*

Equation 10-2 may be modified by representing $1 + G(s)H(s)$ as the ratio of two factored polynomials in s.

$$1 + G(s)H(s) = \frac{(s + r_1)(s + r_2)(s + r_3)\cdots}{(s + r_a)(s + r_b)(s + r_c)\cdots} \tag{10-4}$$

where $(s + r_1)$, $(s + r_2)$, and $(s + r_3)$ are the factors of the numerator $D(s) + N(s)$, and $(s + r_a)$, $(s + r_b)$, and $(s + r_c)$ are the factors of the denominator $D(s)$.

The characteristic equation may then be represented in the general case by

$$(s + r_1)(s + r_2)\cdots(s + r_n) = 0 \tag{10-5}$$

Here $-r_1, -r_2, \cdots, -r_n$ are the roots of the characteristic equation. If s takes on one of these values, $1 + G(s)H(s)$ becomes zero (and $C(s)/R(s)$ becomes infinite). Thus the roots of the characteristic equation are called the *zeros* of $1 + G(s)H(s)$. To complete the terminology, the zeros of $D(s)$, that is, $-r_a$, $-r_b$, and $-r_c$ from Eq. 10-4 are called the *poles* of $1 + G(s)H(s)$—the values of s for which $1 + G(s)H(s)$ is infinite.

Criterion for Stability. For stable operation of the control system of Fig. 10-1, *all* roots of the characteristic equation must be negative real numbers or complex numbers with negative real parts. The roots of the characteristic equation, zeros of $1 + G(s)H(s)$, may be plotted on the complex plane, or s plane, of Fig. 10-2. The imaginary axis of Fig. 10-2 divides the complex plane into two regions—the right half plane and the left half plane. Roots of the characteristic equation lying on the right half plane lead to instability of the system (for example, $+2 \pm j1$), and roots lying on the left half plane lead to a stable system (for example, $-3 \pm j2$).

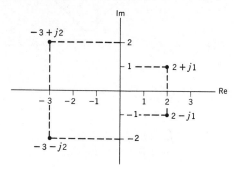

Fig. 10-2. Stability on the complex plane.

The procedure for investigating the stability of a system then is to search the right half plane for zeros of $1 + G(s)H(s)$. If there are no zeros of $1 + G(s)H(s)$ in the right half plane, then there are no roots of the characteristic equation in the right-half plane, and the system is stable. Obviously, it is impractical, if not impossible, to investigate every point on the right half of the s plane and so it is necessary to devise some shorter method. The procedure for searching the right half of the s plane and the interpretation of this procedure on the polar plot is the subject of the following section.

10-3. Stability and the Polar Plot (Nyquist Criterion)

In order to investigate stability on the polar plot,[1] it is first necessary to correlate the region of instability on the s plane with identification of instability on the polar plot, or $1 + GH$ plane. The $1 + GH$ plane is frequently the name given to the plane where $1 + G(s)H(s)$ is plotted in complex coordinates with s replaced by $j\omega$. Likewise, the plot of $G(s)H(s)$ with s replaced by $j\omega$ is often termed the GH plane. This terminology will be adopted in the remainder of the text. Recall from Chapter 6 that s is a complex variable, $s = \sigma + j\omega$. Although s may take on any value, in general, the values of $s = 0 \pm j\omega$ have particular significance.

The characteristic function $1 + G(s)H(s)$ may be plotted for specific values of s on a complex coordinate system. This coordinate system will be termed the $1 + G(s)H(s)$ plane. Likewise, the rectangular coordinate system on which $G(s)H(s)$ is plotted is called the $G(s)H(s)$ plane.

Equation 10-4 is the link between the $1 + G(s)H(s)$ plane and the s plane. If some line is chosen on the s plane and values of s along this line are substituted in Eq. 10-4, a corresponding curve on the $1 + G(s)H(s)$

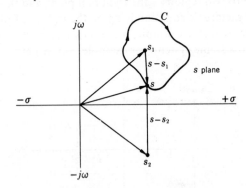

FIG. 10-3. Stability on the s plane.
$(s = \sigma + j\omega)$

[1]The polar plot is a plot of frequency response obtained by replacing s by $j\omega$ as discussed in Secs. 9-2 and 9-3.

plane is generated. If a path C is chosen on the s plane so as to enclose a zero s_1, as shown in Fig. 10-3, what curve is generated on the $1 + G(s)H(s)$ plane?

Each point on the s plane may be considered as the tip of a vector. (On Fig. 10-3 the point s_1 is the tip of the vector s_1, the point s_2 is the tip of the vector s_2, and the variable point s is the tip of a variable vector s.) The vectors $s - s_1$ and $s - s_2$ are obtained by vector subtraction and are shown in Fig. 10-3. Each of the difference vectors is a factor $(s + r_1)$ of Eq. 10-4. Equation 10-4 may be rewritten for the specific number of poles and zeros not all shown in Fig. 10-3.

$$1 + G(s)H(s) = \frac{(s - s_1)(s - s_2)(s + s_3)}{(s - s_a)(s + s_b)(s - s_c)} \tag{10-6}$$

If values of s are chosen so that path C is a closed contour traversed in a clockwise direction and enclosing the zero s_1, then $s - s_1$ makes one complete clockwise revolution. Correspondingly, as all other difference vectors are external to the path enclosing s_1, they make no net revolutions on the s plane as C is traversed. Thus, the right-hand side of Eq. 10-6 experiences a net phase change of 360° clockwise. As a result, $1 + G(s) H(s)$, which is a vector, must also experience a 360° clockwise phase change. Figure 10-4 illustrates the path followed on the $1 + G(s) H(s)$ plane corresponding to one net encirclement of the zero s_1 on the s plane. The result which should be realized here is that encirclement of a zero on the s plane corresponds to encirclement of the origin on the $1 + G(s) H(s)$ plane.

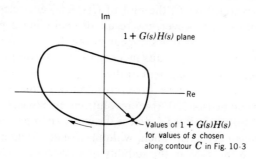

Fig. 10-4. The $1 + G(s)H(s)$ plane.

The $1 + G(s)H(s)$ plane may easily be converted to the $G(s)H(s)$ plane, or open-loop transfer function plane, by subtracting one from each side of Eq. 10-6.

$$G(s)H(s) = \frac{(s - s_1)(s - s_2)(s + s_3)}{(s - s_a)(s + s_b)(s - s_c)} - 1 \tag{10-7}$$

The plot is identical to $1 + G(s)H(s)$ except that the origin for $1 +$

$G(s)H(s)$ is now shifted to the left by one, specifically it is at the point $-1 + j0$. This process is illustrated in Fig. 10-5.

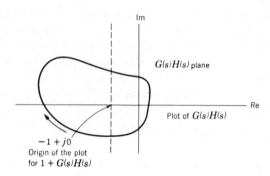

Fig. 10-5. The $G(s)H(s)$, or open-loop, transfer function plane.

The basic criterion for closed-loop system instability, determined from the open-loop transfer function plotted on the $G(s)H(s)$ plane, is that the point $-1 + j0$ be encircled when a zero on the right half of the s plane is encircled. Suppose that the contour C is enlarged to encircle two zeros, s_1 and s_2. Then there are two difference vectors, $s - s_1$ and $s - s_2$, which rotate $360°$ as the path is traversed for one revolution. Thus, $G(s)H(s)$ undergoes two complete revolutions about the $-1 + j0$ point. For each zero encircled by C, the plot of $G(s)H(s)$ encircles the $-1 + j0$ point once. Consider, however, the result of encircling a pole s_a on the right half of the s plane. As poles appear in the denominator of Eq. 10-7, clockwise encirclement of s_a results in a negative $360°$ rotation of $G(s)H(s)$ or a counterclockwise encirclement of the $-1 + j0$ point on the $G(s)H(s)$ plane. Thus if C should enclose three zeros and one pole, the net rotation on the $G(s)H(s)$ plane is $+720°$ (two revolutions clockwise).

The key to this analysis is to enclose the entire right half of the s plane, including the imaginary axis. Then, without actually finding all the poles and zeros lying in this region, the stability of a system may be ascertained on the $G(s)H(s)$ plane by the nature of the encirclement of the point $-1 + j0$. Figure 10-6 illustrates the path followed to enclose the entire right half plane.[2] The plot starts at 0 and proceeds to $+\infty$ along the imaginary axis. The outskirts of the plane are then enclosed by a clockwise circle at an infinite radius from $+\infty$ to $-\infty$. Finally, the region is completely enclosed by returning to the origin from $-\infty$ along the imaginary axis. In general,

---

[2]It is necessary to modify the curve along the imaginary axis to either include or exclude poles which lie on the axis; particularly common are poles at the origin. This modification does not appreciably complicate the actual procedure discussed later.

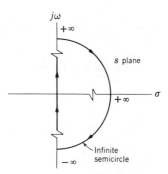

FIG. 10-6. Enclosure of the right half of the s plane.

values to be used are $s = \pm j\omega$ since in traversing the imaginary axis the real part of s is zero ($\sigma = 0$). For all points along the infinite circle on the s plane, $s = \infty$. For $s = \infty$, $G(s)H(s)$ takes on a finite value or infinity. Commonly, when the denominator of $G(s)H(s)$ is of higher degree than the numerator, $G(s)H(s)$ takes on the value of zero. $G(s)H(s)$ is said to map into the origin of the $G(s)H(s)$ plane. The process of translating a plot on the s plane to the $G(s)H(s)$ plane is called *conformal mapping*.

The points on the s plane lying on the imaginary axis from 0 to $+\infty$ are mapped onto the $G(s)H(s)$ plane by substituting specific values of ω into $G(j\omega)H(j\omega)$ after replacing s by $j\omega$. This information corresponds to the polar plot of the open-loop frequency response of the system.

Conformal mapping of the negative half of the imaginary axis is simplified by the fact that it is the mirror image of the positive half. The resulting polar plot of the open-loop transfer function $G(s)H(s)$ is generally known as a *Nyquist diagram*.

EXAMPLE 10-2. Obtain the Nyquist diagram for the system shown in Fig. 10-7a and ascertain system stability.

Solution: In this case $G(s) = K/(\tau s + 1)$ and $H(s) = 1$. Therefore,

$$G(s)H(s) = \frac{K}{\tau s + 1}$$

and it is seen that the single pole of $G(s)H(s)$, $-1/\tau$, does not lie in the right half of the s plane. Substituting $s = j\omega$.

$$G(j\omega)H(j\omega) = \frac{K}{j\omega\tau + 1}$$

At $\omega = 0$, $G(j\omega)H(j\omega) = K$; and at $\omega = +\infty$, $G(j\omega)H(j\omega) = 0$. Between $\omega = 0$ and $\omega = +\infty$, $G(j\omega)H(j\omega)$ takes on values on the polar plot of magnitude $M(\omega) = K\{1/[(\tau\omega)^2 + 1]^{1/2}\}$ and phase angle $\phi(\omega) = -\tan^{-1}\omega\tau$. The complete

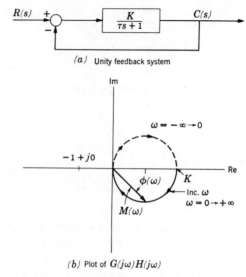

(a) Unity feedback system

(b) Plot of $G(j\omega)H(j\omega)$

FIG. 10-7. Figures for Example 10-2.

Nyquist diagram is given in Fig. 10-7b. For identification purposes the solid curve is the map of the positive half of the imaginary axis while the dashed curve is the mirror image obtained by mapping the negative half of the imaginary axis.

System stability is quickly ascertained. Since the $-1 + j0$ point is not enclosed in the closed contour and there are no poles in the right half plane, there are no zeros in the right-hand half plane: The system is definitely stable.   *Ans.*

EXAMPLE 10-3. Plot the Nyquist diagram for the system of Fig. 10-8a and ascertain the stability of the system.

Solution: In this instance,

$$G(s) = \frac{K}{s^2(\tau s + 1)} \quad \text{and} \quad H(s) = s$$

Thus,

$$G(s)H(s) = \frac{K}{s(\tau s + 1)}$$

The presence of s in the denominator indicates that there is a pole at the origin of the s plane and that the enclosure illustrated in Fig. 10-6 should be modified to circumvent it. The choice here is to eliminate the pole from the right half-plane enclosure by traveling along a small circular arc to the right of the origin as shown in Fig. 10-8b. The solution proceeds by substituting $j\omega$ for s.

$$G(j\omega)H(j\omega) = \frac{K}{j\omega(j\tau\omega + 1)}$$

As ω approaches $\pm\infty$, $G(j\omega)H(j\omega)$ tends toward zero. The magnitude and phase of the $G(j\omega)H(j\omega)$ vector is given by

$$M(\omega) = \frac{K}{\omega} \frac{1}{[(\omega\tau)^2 + 1]^{1/2}}$$

and
$$\phi(\omega) = -\frac{\pi}{2} - \tan^{-1}(\omega\tau)$$

Therefore, as ω varies from a finite positive value toward $+\infty$, $M(\omega)$ decreases in magnitude and $\phi(\omega)$ varies from $-\pi/2$ to $-\pi$. Thus the solid portion of the curve is as shown in Fig. 10-8c. In the actual problem the shape of the curve may be readily determined by plotting several points for specific values of ω.

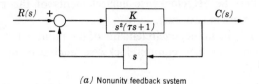

(a) Nonunity feedback system

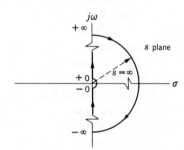

(b) s-plane enclosure for Ex. 10-3

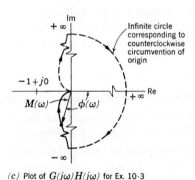

(c) Plot of $G(j\omega)H(j\omega)$ for Ex. 10-3

FIG. 10-8. Figures for Example 10-3.

In the region of $\omega = 0$, the s vector in Fig. 10-8b goes from $-\pi/2$ to $+\pi/2$ (counterclockwise rotation) and it has some infinitesimally small magnitude γ. Thus, let $s = \gamma e^{j\theta}$, where $\gamma \to 0$ and $-\pi/2 \leq \theta \leq +\pi/2$. Since s appears in the denominator of $G(s)H(s)$, the $G(j\omega)H(j\omega)$ vector is dependent on $(1/\gamma)e^{-j\theta}$. The result is that $M(\omega) \to \infty$ and $\phi(\omega)$ varies from $+\pi/2$ to $-\pi/2$ in a clockwise direction. The infinite circle on the Nyquist diagram can only go in the direction indicated in Fig. 10-8c.

From inspection, it is apparent that the system is stable, there being no poles in the right half plane. The $-1 + j0$ point can never be enclosed by the plot of $G(s)H(s)$.   *Ans.*

In summary, the polar plot, or Nyquist diagram, may yield three general results, depending on the nature of the open-loop transfer function of the system being investigated. These results are:

1. There may be net clockwise encirclements of the point $-1 + j0$ (one or more may occur).
2. There may be no net encirclements of the point $-1 + j0$.
3. There may be net counterclockwise encirclements of the point $-1 + j0$.

In case 1 the system *must* be unstable because the plot shows that there are an excess of zeros on the right half of the s plane. For case 2, the system *may* be stable if there are no zeros and no poles in the right half plane. However, it is possible that there are an equal number of poles and zeros with positive real parts in which case the system would be unstable (there cannot be any zeros with positive real parts for stable operation). In case 3, there are poles with positive real parts, but there might also be zeros, only fewer in number. In both case 2 and case 3, further tests are necessary to ascertain definitely if the system is stable or unstable. One approach is to determine the number of poles in the right half plane. With many systems there are none.

The analysis presented here, utilizing the encirclement of the point $-1 + j0$ on the polar plot of $G(j\omega)H(j\omega)$ to ascertain stability, is known as the *Nyquist stability criterion*. The criterion may be expressed mathematically as

$$Z = N + P \tag{10-8}$$

where Z = number of zeros on the right half of the s plane;

N = net clockwise encirclements of the point $-1 + j0$ (a counterclockwise encirclement has a negative sign);

P = number of poles on the right half of the s plane.

If the number of net clockwise (+) encirclements and the number of poles with positive real parts are known, the number of zeros with positive real parts may quickly be determined.[3]

EXAMPLE 10-4. The open-loop transfer function of a system is known to possess one pole on the right-hand side of the s plane. The source of this pole is an

---

[3]If it should happen that the net counterclockwise encirclement is greater than P (Z is negative), there is an error in the evaluation of N or P.

element which, by itself, is highly unstable. The Nyquist diagram for this transfer function is sketched in Fig. 10-9. Determine the stability of the closed-loop system.

Solution: The quantity $G(s)H(s)$ possesses one pole on the right-hand half of the s plane. Thus, $P = 1$. From inspection of the Nyquist diagram there are two counterclockwise encirclements of $-1 + j0$ and one clockwise encirclement of $-1 + j0$. Thus $N = -1$. The result from Eq. 10-8 is that $Z = -1 + 1 = 0$, and the system is actually stable.   *Ans.*

Example 10-4 illustrates a very important point. The stability of a closed-loop control system is not only dependent on the individual components that make up that system, but on the interaction between components and in some cases on the location of the components. It is entirely possible to have a combination of stable elements combine to create an un-

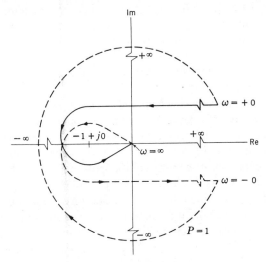

FIG. 10-9. Nyquist diagram for Example 10-4.

stable system (this is a basic reason for stability analysis). Conversely, an unstable element may interact with stable components to result in an over-all system that is stable (as was the case in Example 10-4).

10-4. Simplified Nyquist Analysis

One goal of the control designer, as discussed in Sec. 1-4, is to formulate a system that is stable. It is reasonable that an over-all stable system is more readily attainable when constructed of stable elements than when one or more of the elements is unstable by itself. Often an element that tends to be unstable is modified to have a stable operation before it is incorporated into a complete system.

The fact that most practical control systems are not composed of unstable components reduces the possibility of a pole of $1 + G(s)H(s)$ existing on the right half of the s plane. For example, if the system of Fig. 10-1 is composed of two stable elements $G_1(s)$ and $H_2(s)$, where

$$G_1(s) = \frac{N_1(s)}{D_1(s)} \quad \text{and} \quad H_2(s) = \frac{N_2(s)}{D_2(s)}$$

the zeros of $D_1(s)$ and $D_2(s)$ must have negative real parts because $D_1(s)$ and $D_2(s)$ are actually the characteristic functions for these elements. Because the zeros of $D_1(s)$ and $D_2(s)$ are the sources of the poles of $1 + G_1(s)H_2(s)$, there can be no poles with positive real parts. The result is that Eq. 10-8 reduces to

$$Z = N \tag{10-9}$$

The Nyquist diagram for most practical control systems will then take two general forms.

1. There is no encirclement of the point $-1 + j0$; thus the system is *stable.*

2. There is a net clockwise encirclement of the point $-1 + j0$, indicating an *unstable system.*

Survey of Commonly Encountered Nyquist Diagrams. Most linear physical devices found in control systems have open-loop transfer functions that are represented by a linear gain K, an integration K/s, a time constant $K/(\tau s + 1)$, or a quadratic $K/(as^2 + bs + 1)$, or combinations of these basic dynamic elements. Because many components are defined by similar transfer functions, it is valuable to have a set of s-plane plots and corresponding Nyquist diagrams which illustrate the stability possibilities and indicate the general shapes of plots for commonly encountered functions. A series of plots for increasingly complex transfer functions is given in Fig. 10-10a–f. Plots for many other functions may be found in the literature [11]. The s-plane plots in Fig. 10-10 illustrate the general locations of the poles (identified by ×) of $1 + G(s)H(s)$ and the nature of the enclosure of the right-hand half plane. The Nyquist diagrams shown are all drawn for the transfer function producing a stable condition where possible. Those transfer functions which may become unstable by altering the parameters of the system are identified. It should be recognized that the basic shape of the polar plot is determined by the form of the transfer function, whereas the proximity of the plot to the point $-1 + j0$ is a function of the actual values of the system parameters. Increasing the system gain K, for example, expands the plot but does not change its shape. Conversely, reducing the system gain shrinks the entire plot. The nearness of an open-loop plot to the point $-1 + j0$ can be used as a measure of the margin of stability. The plot also provides valuable information concerning the transient response of the system. In general, the closer a plot comes to encircling $-1 + j0$, the more oscillatory the behavior of the system. These concepts are discussed in detail in the next section.

10-5. Graphical Relationships on the GH Plane

The polar plot of $G(j\omega)H(j\omega)$ may be used for system analysis in several ways other than the strict determination of absolute stability. The plot of $G(j\omega)H(j\omega)$ is in itself a graphical picture of the open-loop frequency response of the system. By a simple graphical construction on the GH plane it is possible to obtain the closed-loop frequency response. Also,

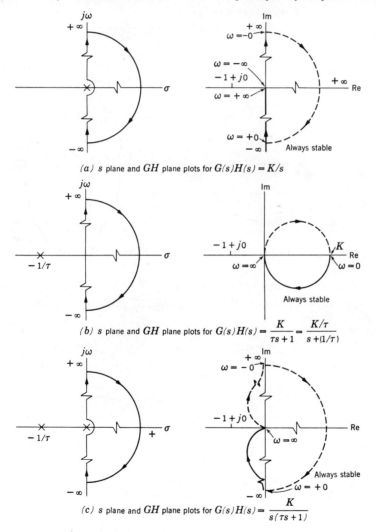

(a) s plane and GH plane plots for $G(s)H(s) = K/s$

(b) s plane and GH plane plots for $G(s)H(s) = \dfrac{K}{\tau s + 1} = \dfrac{K/\tau}{s + (1/\tau)}$

(c) s plane and GH plane plots for $G(s)H(s) = \dfrac{K}{s(\tau s + 1)}$

FIG. 10-10. Summary of common Nyquist diagrams.

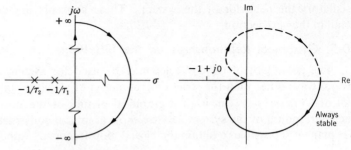

(d) s plane and **GH** plane plots for

$$G(s)H(s) = \frac{K/\tau_1\tau_2}{[s+(1/\tau_1)][s+(1/\tau_2)]}$$

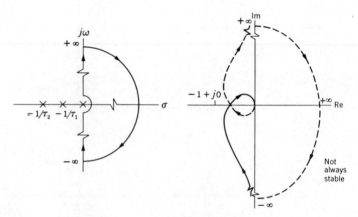

(e) s plane and **GH** plane plots for

$$G(s)H(s) = \frac{K/\tau_1\tau_2}{s[s+(1/\tau_1)][s+(1/\tau_2)]}$$

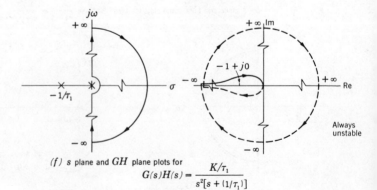

(f) s plane and **GH** plane plots for

$$G(s)H(s) = \frac{K/\tau_1}{s^2[s+(1/\tau_1)]}$$

Fig. 10-10. *Continued.*

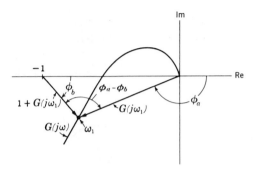

FIG. 10-11. Obtaining the closed-loop response (unity feedback).

the margin of stability of the control system may be determined. The plot of $G(j\omega)H(j\omega)$ may be further used as a means of establishing the most desirable setting of the control system's variable parameters; for example, the gain of amplifiers, positions of linkages, or the value of resistances.

Obtaining the Closed-Loop Response from the Open-Loop Response. If a system has unity feedback $[H(j\omega) = 1]$, it is a simple matter to determine the closed-loop system response from the open-loop response. The quantity plotted on the GH plane is the open-loop response, specifically, $G(j\omega)$. The closed-loop frequency response is then

$$\frac{C(j\omega)}{R(j\omega)} = \frac{G(j\omega)}{1 + G(j\omega)} \tag{10-10}$$

The closed-loop frequency response at any particular value of ω may be specified in polar form by the magnitude and phase quantities $M(\omega)$ and $\phi(\omega)$ introduced in Chapter 9. Thus,

$$M(\omega) = \left|\frac{C(j\omega)}{R(j\omega)}\right| = \frac{|G(j\omega)|}{|1 + G(j\omega)|} \tag{10-11}$$

and $\qquad \phi(\omega) = \underline{/\text{angle of } G(j\omega)} - \underline{/\text{angle of } 1 + G(j\omega)} \qquad (10\text{-}12)$

Figure 10-11 shows the vectors on the GH plane that represent the terms $G(j\omega_1)$ and $1 + G(j\omega_1)$ at the particular frequency ω_1. The ratio of the magnitude of these vectors is $M(\omega_1)$.

$$M(\omega_1) = \frac{|G(j\omega_1)|}{|1 + G(j\omega_1)|}$$

The phase angle between the excitation and response at this frequency is then

$$\phi(\omega_1) = \underline{/\phi_a - \phi_b}$$

It should be noted that for the particular case shown in Fig. 10-11, ϕ_a

and ϕ_b are negative angles, and the net angle $\phi(\omega_1)$ would be negative, indicating that the response lags the excitation.

In the general case where there is a nonunity element in the feedback path, the plot on the GH plane (Fig. 10-12) is for the quantity $G(j\omega)H(j\omega)$.

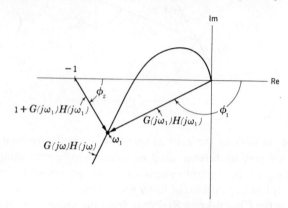

FIG. 10-12. Obtaining the closed-loop response
(nonunity feedback).

The ratio of the two vectors is

$$\frac{|G(j\omega_1)H(j\omega_1)|}{|1 + G(j\omega_1)H(j\omega_1)|} \qquad (10\text{-}13)$$

which obviously is not the magnitude of the closed-loop frequency response $|C(j\omega)/R(j\omega)|$. If, however, one multiplies Eq. 10-13 by $|1/H(j\omega)|$, $|C(j\omega)/R(j\omega)|$ may be obtained. Thus,

$$\left|\frac{C(j\omega)}{R(j\omega)}\right| = \frac{|G(j\omega)H(j\omega)|}{|1 + G(j\omega)H(j\omega)|}\left|\frac{1}{H(j\omega)}\right| \qquad (10\text{-}14)$$

At a particular value of frequency ω_1, the closed-loop frequency response is, in polar form,

$$\frac{C(\omega_1)}{R(\omega_1)} = \frac{|G(j\omega_1)H(j\omega_1)|\ \underline{/\phi_1 - \phi_2 - \phi_H}}{|1 + G(j\omega_1)H(j\omega_1)|\ |H(j\omega_1)|} = M(\omega_1)\ \underline{/\phi(\omega_1)}$$

Loci of Constant Magnitude M. The magnitude of the frequency response ratio $M(\omega)$ is a useful measure of system performance and may be utilized in several ways. If, for example, it is desirable to have some maximum value M_m not exceeded over the complete range of frequencies encountered, Eq. 10-11 could be tested for each value of ω. Then the curve of $G(j\omega)$ could be adjusted and the process repeated. It is possible, however, to superimpose the loci of constant values of M on the GH plane because there is a definite value of M for each point on the GH plane. Such loci can be used in the process of design and synthesis.

For the case of unity feedback, consider representing $G(j\omega)$ on the GH plane by a complex variable, specifically,

$$G(j\omega) = u + jv \tag{10-15}$$

Then, substituting this variable into Eq. 10-11 yields

$$M = \left(\frac{u^2 + v^2}{1 + 2u + u^2 + v^2}\right)^{\frac{1}{2}} \tag{10-16}$$

Squaring both sides of Eq. 10-16 and clearing fractions gives

$$M^2 + 2uM^2 + u^2(M^2 - 1) + v^2(M^2 - 1) = 0 \tag{10-17}$$

Dividing by $M^2 - 1$, rearranging, and adding $M^2/(M^2 - 1)^2$ to both sides yields

$$u^2 + 2u\,\frac{M^2}{M^2 - 1} + \frac{M^4}{(M^2 - 1)^2} + v^2 = \frac{M^2}{(M^2 - 1)^2} \tag{10-18}$$

It may then be noted that the first three terms on the left form a perfect square, thus,

$$\left(u + \frac{M^2}{M^2 - 1}\right)^2 + v^2 = \frac{M^2}{(M^2 - 1)^2} \tag{10-19}$$

Equation 10-19 is recognized as the equation of a circle with center at

$$u = -\frac{M^2}{M^2 - 1} \qquad v = 0 \tag{10-20}$$

and with radius

$$R = \left|\frac{M}{M^2 - 1}\right|$$

A family of these circles (called M circles) for specified values of M are shown in Fig. 10-13. If a particular $G(j\omega)$ were plotted, the value of $M(\omega)$ at any frequency could be rapidly established. In particular, the maximum magnitude ratio M_m could be determined. In a similar fashion, loci of constant $\phi(\omega)$ may be plotted on the GH plane. These loci are commonly called N circles [3].

Gain and Phase Margins. The *gain* and *phase margins* of a control system are quantities which may be used as design criteria. Although they are perhaps more directly useful in the method of analysis presented in Chapter 11, it is valuable to look briefly at their interpretation on the GH plane. These design parameters, taken together, give a measure of how close a system is to instability.

The *gain margin m_G* is defined as the reciprocal of $|G(j\omega)|$ when the phase angle $\phi(\omega)$ is $-180°$. This quantity is pictured graphically in Fig. 10-14. The *phase margin* is defined as the angle between the negative real axis and the $G(j\omega)$ vector where $|G(j\omega)|$ is unity. In Fig. 10-14 this angle

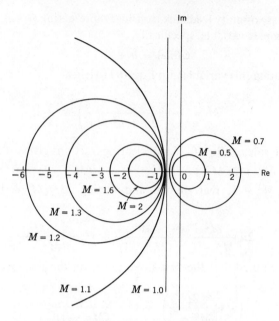

FIG. 10-13. Loci of constant M (M circles).

is given as α. Both criteria indicate the margin of stability or, in a sense, the factor of safety.

Typical design criteria may specify a gain margin from 2 to 10 and a phase margin from 30° to 60°. For a system with oscillatory tendencies, as would be that of Fig. 10-14, a phase margin of 45° would normally insure satisfactory operation. If α were made to approach 60°, the equivalent damping in the system would be increased reducing the oscillation. Further physical interpretation may be given to the angle α for the transient behavior of the system. As α is increased, the transient behavior becomes

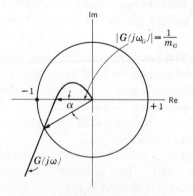

FIG. 10-14. Gain and phase margin.

less oscillatory; while as α is decreased, the transient response becomes more oscillatory (i.e., relative damping is decreased). For $\alpha = 0$, a system with a $G(j\omega)$ plot as shown in Fig. 10-14 would be on the threshold of instability, that is, the Nyquist diagram would be on the verge of enclosing the point $-1 + j0$.

Setting the System Gain. It was pointed out in Sec. 10-4 that the gain K of a system alters the magnitude of the plot on the GH plane. Therefore, changing the system gain alters the gain and phase margins. The over-all system gain, and thus the individual component gains, has considerable control over the entire system behavior. It is perhaps fortunate that the variation of gain is, physically, a relatively simple matter. Where electrical signals appear in a system, the gain may be altered by a potentiometer or a variable-gain amplifier. With mechanical elements, including pneumatic and hydraulic devices, the gain is typically changed by varying linkage ratios. Thus, in optimizing the performance of a system that has been designed and assembled, variation of gain is usually the first consideration.

Even though most practical control systems provide a simple and rapid adjustment of the gain constant, it is necessary to calculate a theoretically optimum gain in the design stage. This provides a base from which variations can be made to compensate for differences between the physical and mathematical models. The criterion for setting the system gain may be based on a desirable value for the maximum magnitude ratio M_m, a prescribed gain margin m_G, or some allowable phase margin α. Any of these figures of merit are rather loose in the actual final result they produce, thus there is more justification for providing a suitable range of variation for the system gain K.

Of the three suggested criteria, the simplest to utilize for some classes of systems is the gain margin. The absolute magnitude of $G(j\omega)$ is proportional to K as seen by the relationship

$$|G(j\omega)| = |KG_N(j\omega)| = K|G_N(j\omega)| \qquad (10\text{-}21)$$

where $G_N(j\omega)$ is a normalized function of $j\omega$. For example, if

$$G(j\omega) = \frac{10}{j\omega(j\omega + 2)} = \frac{K}{j\omega(0.5j\omega + 1)}$$

then $K = 5$ and

$$G_N(j\omega) = \frac{1}{j\omega(0.5j\omega + 1)}$$

The gain margin is defined by

$$m_G = \frac{1}{|G(j\omega)|} = \frac{1}{K|G_N(j\omega)|}$$

when the phase angle of $G(j\omega)$ is $-180°$. Thus, the procedure for specifying the over-all system gain is as follows:

1. Choose an acceptable gain margin $m_{G \text{ design}}$.
2. Plot $G_N(j\omega)$ on the GH plane.
3. Obtain the gain margin of $G_N(j\omega)$, defined as m_G'.
4. Then, by a simple proportion, $K_{\text{design}} = m_G'/m_{G \text{ design}}$.

For some systems it may be more desirable to base the setting of gain on a specified maximum value of magnitude ratio M_m. In this case, superposition of the loci of constant magnitude (M circles) on the $G(j\omega)$ plot provides a useful technique. The proper gain may be obtained by trial and error using a set of M circles, or a direct method may be applied, again utilizing the proportionality expressed by Eq. 10-21.

If a locus for a constant M_m greater than unity is plotted on the GH plane, a circle of radius $|M_m/(M_m^2 - 1)|$ with its center at the point $-M_m^2/(M_m^2 - 1)$ results. This circle is shown in Fig. 10-15 with a line drawn

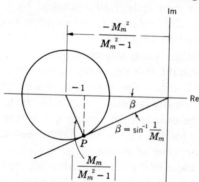

FIG. 10-15. Geometric relations of the constant M_m locus and tangent line.

from the origin of the GH plane tangent to the circle at some point P. It may be verified from the geometry of the figure that a line drawn from P, perpendicular to the real axis, intersects the real axis at the point -1 and that the angle $\beta = \sin^{-1} 1/M_m$. The procedure for setting the gain of a unity feedback system is then as follows:

1. Choose a maximum M as a design criterion, say M_m.
2. Plot the normalized forward transfer function $G_N(j\omega)$ on the GH plane. (This process may be followed on Fig. 10-16.)
3. Draw in a tangent line for M_m at an angle of $\beta = \sin^{-1}(1/M_m)$.
4. Draw in a circle choosing the center of the circle along the axis and with a radius such that the circle is tangent to both the plot of $G_N(j\omega)$ and the line at angle β.
5. Construct a perpendicular from the tangent point P to the real axis. This intersects at some value $-x \neq -1$. If the circle had been constructed from the correct center with the proper radius, tangent to

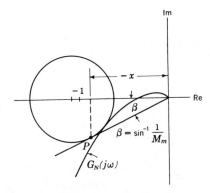

FIG. 10-16. Determination of system gain for specified M_m.

$G(j\omega)$ with $M(\omega) = M_m$, the point would be -1 as illustrated in Fig. 10-15. Because the gain K enlarges the $G(j\omega)$ plot in a direct proportion at all frequencies, inserting a gain K greater than unity would shift P to the left. Specifically, the circle must be shifted to the left until $-x$ corresponds to -1.

6. The proper design gain K_{design} for the specified M_m is then the value of K required to place the proper M_m circle at its proper location on the real axis. Thus

$$K_{\text{design}} = \frac{1}{x} \tag{10-22}$$

A specific example in the following section will illustrate the application of this method for setting the system gain.

10-6. An Example of System Analysis on the GH Plane

The concepts and techniques of this chapter may now be consolidated and applied to the analysis and design of a realistic control system. It is valuable to look at a specific example and to correlate the meaning of various design parameters, such as phase margin and maximum M_m criteria. To this end, this section will be devoted to the study of a typical closed-loop positioning system.

EXAMPLE 10-5. Analyze the simplified positioning system of Fig. 10-17. In the practical application of this system, the load would normally be driven through a gear train to allow the motor to function at a higher speed and lower torque. In that case, the load inertia must be reflected to the motor shaft. Further, the control transformer would likely be geared to the output shaft with a greater than 1:1 correspondence in order to detect the output shaft angle more precisely. In the block diagram, the gear ratios appear as gain constants. a) Determine the stability of the closed-loop system. b) For the specific values of the parameters given, establish the amplifier gain required for the system to have a maximum $M(\omega)$ of 1.2.

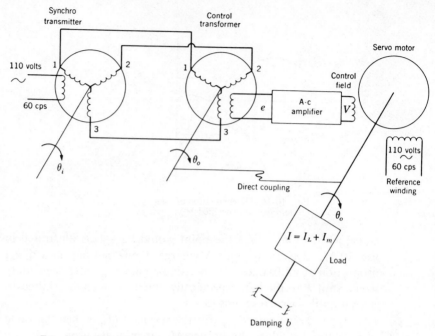

Fig. 10-17. Schematic of a simplified two-phase servomotor positioning system.

c) For the gain established in part b, determine the phase margin of the system.

The system parameters are given as

K_1 = synchro constant = 8.6 volts/radian

K_2 = amplifier gain − variable = volts/volt

K_3 = motor torque constant = 0.02 oz-in./volt

K_4 = motor damping = 10^{-2} oz-in.-sec/radian

$I = I_m + I_L = 14 \times 10^{-4}$ oz-in.-sec^2

$b = 70 \times 10^{-4}$ oz-in.-sec/radian

Solution: The first step in the analysis is to obtain the equations defining the various components in the system. The transformed error signal from the synchros is obtained from Eq. 7-25 as

$$E(s) = K_1[\Theta_i(s) - \Theta_o(s)]$$

The amplifier relationship is simply

$$V(s) = K_2 E(s)$$

The torque produced by the servomotor is

$$T(s) = K_3 V(s) - K_4 s \Theta_o(s)$$

which is the torque available to accelerate the combined inertia and overcome friction. Thus,

$$T(s) = Is^2 \Theta_o(s) + bs \Theta_o(s)$$

Therefore,

$$K_3 V(s) = (Is^2 + bs + K_4 s)\Theta_o(s)$$

or

$$\frac{\Theta_o(s)}{V(s)} = \frac{K_3/(K_4 + b)}{s(\tau s + 1)}$$

where $\tau = I/(K_4 + b)$.

The block diagram of the system is given in Fig. 10-18a with each of the blocks labeled. The diagram may be simplified by lumping the blocks to give Fig. 10-18b.

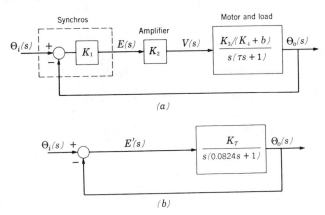

FIG. 10-18. Block diagrams for positioning system.

The total gain constant is

$$K_T = \frac{K_1 K_2 K_3}{K_4 + b}$$

and the time constant is

$$\tau = \frac{I}{K_4 + b} = \frac{14 \times 10^{-4}}{170 \times 10^{-4}} = 0.0824 \text{ sec}$$

The Nyquist diagram for the system represented by Fig. 10-18b is of the form shown in Fig. 10-19. It is apparent that the system could never be unstable (the point $-1 + j0$ will never be encircled, regardless of gain).

Next, it is desired to establish what over-all gain must be used to obtain a maximum magnitude ratio M_m of 1.2. The normalized transfer function

$$G_N(j\omega) = \frac{1}{j\omega(0.0824j\omega + 1)}$$

is plotted in Fig. 10-20a. The values of a few points are given in Table 10-1.

For $M_m = 1.2$, the tangent line is constructed at an angle

$$\beta = \sin^{-1} \frac{1}{1.2} = 56.5°$$

An M circle is then drawn tangent to this line and the plot of $G_N(j\omega)$. A perpendicular is constructed from point P and the design gain is established as

TABLE 10-1

Values for Polar Plot of $G_N(j\omega)$

ω radian/sec	$|G_N(j\omega)|$	Angle
10	0.0772	$-129.5°$
15	0.042	$-141°$
20	0.026	$-148.8°$

$$K_{design} = K_T = \frac{1}{x} = \frac{1}{0.071} = 14.1$$

Thus

$$\frac{K_1 K_2 K_3}{K_4 + b} = 14.1$$

Substituting the known constants, the amplifier gain is

$$K_2 = \frac{14.1(1.7 \times 10^{-2})}{8.6 \times 0.02} = 1.39$$

Thus, the system should be designed with an amplifier voltage gain of approximately 1.4.

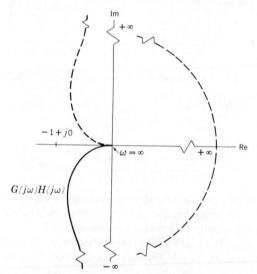

Fig. 10-19. Typical polar plot for
$$G(j\omega)H(j\omega) = \frac{K}{j\omega(j\omega\tau + 1)}.$$

The data for the polar plot of $G(j\omega)$ is obtained easily by multiplying each magnitude $|G_N(j\omega)|$ in Table 10-1 by the design gain $K_{design} = 14.1$. As pointed out in Sec. 10-5, at each value of ω, the angle is not changed, but the magnitude is proportional to K. This procedure yields the plot of Fig. 10-20b. The phase margin is obtained by constructing a unit circle and is found to be $\alpha = 48.5°$.   *Ans.*

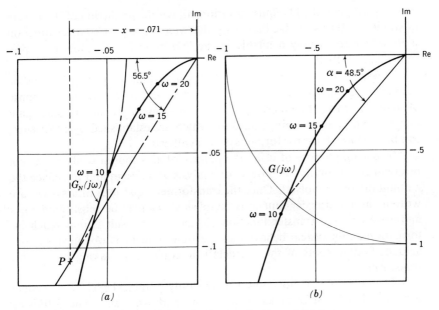

FIG. 10-20. Nyquist diagrams for Example 10-5.

The reader should recognize that the system of this example is a second-order system. As the over-all transfer function is

$$\frac{C(s)}{R(s)} = \frac{K_T}{\tau s^2 + s + K_T}$$

the characteristic equation is

$$s^2 + \frac{1}{\tau} s + \frac{K_T}{\tau} = 0$$

The results of the frequency analysis may be easily correlated to the transient behavior of the system by recalling the work of previous chapters. In this case the dimensionless damping ratio ζ is

$$\zeta = \frac{1}{2(\tau K_T)^{\frac{1}{2}}} = 0.465$$

Thus, if θ_i is a step input, the output θ_o will have some overshoot since there is less than critical damping. The actual response to a step input may be further appreciated by reviewing the curves of Fig. 5-6.

10-7. Closure

The polar plot of a system, representing the open-loop frequency response, has been used in this chapter as a means of analysis and design. Of paramount importance is the use of the polar plot, or Nyquist diagram, for the prediction of system stability. The Nyquist stability criterion, and particularly the simplified Nyquist criterion, allows the prediction of the nature of stability. Its use is facilitated by the wide variety of transfer-function plots which are readily available in the literature; several plots were presented in Fig. 10-10.

It was further demonstrated in this chapter that the polar plot may be used in the design process to set system gains and to predict the response of stable systems. Design criteria, such as maximum magnitude ratio, gain margin, and phase margin, values for which are established by experience, can be used as a basis for further system synthesis.

The main purpose of using the polar plot as a tool in studying system behavior is to simplify the process of analysis and design and to reduce the amount of labor involved. Once the transformed equations of a system are written and the transfer functions established, $j\omega$ may be substituted for s and the open-loop transfer function plotted. The result is that much information about system behavior in the time domain is obtained without actually finding roots of the characteristic equation or taking the inverse transform.

The polar plot is not, however, the only graphical picture of system behavior that is useful in analysis, design, and synthesis. The following chapters will deal with other methods, including frequency response on logarithmic coordinates and the root-locus method.

Problems

10-1. Determine whether the accompanying Nyquist diagrams for $G(j\omega) \cdot H(j\omega)$ represent stable or unstable closed-loop systems. (The r's are positive numbers.)

10-2. Plot Nyquist diagrams and ascertain the nature of stability for each of the following:

a) $G(s)H(s) = \dfrac{100}{(s - 0.5)(s + 10)}$

b) $G(s)H(s) = \dfrac{10.5}{(s + 0.2)(s + 0.8)(s + 10)}$

c) $G(s)H(s) = \dfrac{275}{s^4 + 8s^3 + 17s^2 + 10s}$

d) $G(s)H(s) = \dfrac{2.5(s + 1)}{s^2(s + 2)}$

e) $G(s)H(s) = \dfrac{1.5(s + 2)}{s^2 + 3s + 2}$

f) $G(s)H(s) = \dfrac{200}{s(0.001s + 1)}$

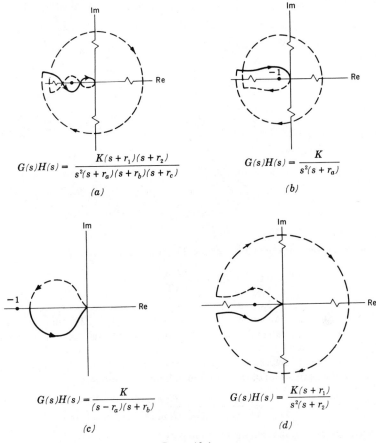

$$G(s)H(s) = \frac{K(s + r_1)(s + r_2)}{s^2(s + r_a)(s + r_b)(s + r_c)}$$

(a)

$$G(s)H(s) = \frac{K}{s^2(s + r_a)}$$

(b)

$$G(s)H(s) = \frac{K}{(s - r_a)(s + r_b)}$$

(c)

$$G(s)H(s) = \frac{K(s + r_1)}{s^2(s + r_2)}$$

(d)

Prob. 10-1

10-3. Using the Nyquist stability criterion, establish the maximum gain K that just maintains stability of a unity feedback closed-loop system with an open-loop transfer function defined by

$$G(s) = \frac{K}{(0.1s + 1)(0.3s + 1)(3s + 1)(0.05s + 1)}$$

Verify the calculation by Routh's criterion.

10-4. Adjust the gain K in the transfer function of Prob. 10-3 so that the system has a phase margin of 35°. What value of K satisfies this condition?

10-5. The Nyquist plot shown in the accompanying illustration is for a stable unity feedback system. The plot is for the normalized transfer function ($K = 1$). Establish the design gain that should be specified if it is desired to have the system perform with a maximum $M(\omega)$ of $1.3 (M_m = 1.3)$.

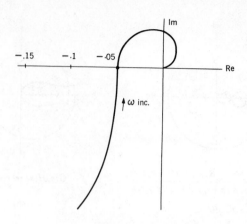

Prob. 10-5

10-6. A control system has the open-loop transfer function

$$G(s)H(s) = \frac{10K(5s + 1)}{s^2(0.2s + 1)}$$

where $H(s) = 10$ and K represents a variable gain setting.

a) Calculate the value of K that will result in an $M_m = 1.2$ for the closed-loop system. (HINT: Rearrange the block diagram and apply the techniques used for unity feedback systems.)

b) For the gain determined in part a, calculate the gain and phase margins.

c) Make a polar plot of the closed-loop frequency response

$$\frac{\Theta_o(j\omega)}{\Theta_i(j\omega)} = \frac{G(j\omega)}{1 + G(j\omega)H(j\omega)}$$

and verify the performance designed for in part a.

10-7. The term e^{-sD} arises in the transformation of the translated function, or physically, from dead time (see Eq. 6-25). When s is replaced by $j\omega$, this term is represented by a unit vector at an angle of $-\omega D$. It may be recalled that

$$e^{-j\omega D} = \text{Cos } \omega D - j \sin \omega D$$

Thus dead time may be readily represented on the polar plot by a unit vector with accumulating phase angle. Make a polar plot for each of the systems shown in parts *a* and *b* of the accompanying illustration. What may be said about the stability of a system with dead time?

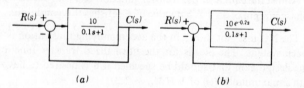

(a)   (b)

Prob. 10-7

10-8. For Prob. 10-7*b*, establish the value of dead time D that would make the system limitedly stable.

Chapter 11

System Analysis Using Logarithmic Plots (Bode Diagrams and Nichols Charts)

11-1. Introduction

The frequency-response approach to the analysis and prediction of control system dynamic behavior may be carried out using logarithmic plots as well as the polar plots discussed in Chapter 10. Again, the quantities under investigation are the magnitude and phase relationships of the open- and closed-loop frequency responses. A Bode attenuation diagram is a plot of the logarithm of the transfer function magnitude versus the logarithm of the frequency. The phase angle may also be plotted as a function of the logarithm of frequency.

These plots may be used to analyze system performance by first identifying, on the new coordinates, the nature of the various system merits discussed in Chapter 10. System stability, for example, was identified on the s plane by the absence of roots of the characteristic equation on the right-hand half of the plane. On the $G(s)H(s)$ plane, a system composed of stable elements is stable if the point $-1 + j0$ is not encircled by the closed contour of $G(s)H(s)$. It will be shown that there exists an interpretation of the simplified Nyquist criterion for ascertainment of system stability on the logarithmic plots. Gain and phase margins may again be specified, and the system design gain may be established from information on the Bode diagrams.

The main advantage in using the Bode plot rather than the polar plot is in the relative ease of plotting. Control systems are often composed of components approximated by blocks which have transfer relations made up of constants K, integrations $1/j\omega$, differentiations $j\omega$, time constants $1/(\tau j\omega + 1)$, and quadratic factors $1/[(j\omega\lambda)^2 + 2\zeta\lambda j\omega + 1]$. The Bode plot allows the systematic superposition of the effects of the various elements taken individually. In addition, Bode attenuation plots and phase angle plots may be obtained from experimental frequency analysis of components for which the actual transfer functions are not known. The plots may then be used to establish the actual transfer relations in s. With commercial hardware, an attenuation diagram may be the only source of dynamic information available.

This chapter will present the concepts of Bode diagrams (including their construction) and their use in stability analysis and in design and synthesis. The use of logarithmic coordinates will be extended to the use of Nichols charts with an example of their use in the design process. The many aspects of the application of logarithmic plots are discussed in the process of solving the illustrative examples.

11-2. Concepts of Bode Attenuation Diagrams

The open-loop transfer function of a control system, when s is replaced by $j\omega$, is, in general, a complex variable

$$G(j\omega)H(j\omega) = \alpha(\omega) + j\beta(\omega) \tag{11-1}$$

If one wishes to take the logarithm of $G(j\omega)H(j\omega)$, it is necessary to follow complex-variable theory which states that the logarithm of a complex number is another complex number,[1] defined by[2]

$$\ln G(j\omega)H(j\omega) = \ln R(\omega) + j[\phi(\omega) + 2\pi n] \tag{11-2}$$

for $n = 0, \pm 1, \pm 2, \ldots$ and where $R(\omega) = [\alpha^2(\omega) + \beta^2(\omega)]^{1/2}$ and $\phi(\omega) = \tan^{-1} \beta(\omega)/\alpha(\omega)$. The angle $2\pi n$ is present because of the multitude of possibilities for having the vector lie in the same place on the complex plane.

The principle value of the logarithm of Eq. 11-2 is

$$\ln G(j\omega)H(j\omega) = \ln R(\omega) + j\phi(\omega) \tag{11-3}$$

The logarithm of $G(j\omega)H(j\omega)$ is then composed of two separate functions of ω, the real part $\ln R(\omega)$ and the imaginary part $\phi(\omega)$. Thus, one way to completely represent the logarithm of the open-loop transfer function is to plot the real part $\ln R(\omega)$ and the imaginary part $\phi(\omega)$ versus the logarithm of ω.

Consider, as a general form of $G(j\omega)H(j\omega)$, the ratio of real and complex factors

$$G(j\omega)H(j\omega) = K \frac{N_1(j\omega)N_2(j\omega)}{D_1(j\omega)D_2(j\omega)} \tag{11-4}$$

---

[1] This may be verified by referring to any introductory text on complex variables.

Then, following the axioms of logarithms,

$\ln G(j\omega) H(j\omega)$

$$= \ln K + \ln N_1(j\omega) + \ln N_2(j\omega) - \ln D_1(j\omega) - \ln D_2(j\omega)$$

$\ln G(j\omega) H(j\omega)$

$$= \ln K + [\ln \mid N_1(j\omega) \mid + j\phi_{N_1}] + [\ln \mid N_2(j\omega) \mid + j\phi_{N_2}]$$
$$- [\ln \mid D_1(j\omega) \mid + j\phi_{D_1}] - [\ln \mid D_2(j\omega) \mid + j\phi_{D_2}]$$

It may be recalled from Chapter 6 that the algebraic sum of complex numbers is the algebraic sum of the real parts plus j times the algebraic sum of the imaginary parts. Thus,

$\ln G(j\omega) H(j\omega)$

$$= \ln K + \ln \mid N_1(j\omega) \mid + \ln \mid N_2(j\omega) \mid - \ln \mid D_1(j\omega) \mid$$
$$- \ln \mid D_2(j\omega) \mid + j(\phi_{N_1} + \phi_{N_2} - \phi_{D_1} - \phi_{D_2})$$

This relationship may now be used to obtain the two complete plots which represent the log $G(j\omega)H(j\omega)$. They are plots of

$\ln \mid G(j\omega) H(j\omega) \mid$

$$= \ln K + \ln \mid N_1(j\omega) \mid + \ln \mid N_2(j\omega) \mid$$
$$- \ln \mid D_1(j\omega) \mid - \ln \mid D_2(j\omega) \mid \qquad (11\text{-}5)$$

and
$$\phi(\omega) = \phi_{N_1} + \phi_{N_2} - \phi_{D_1} + \phi_{D_2} \qquad (11\text{-}6)$$

The amplitude relationship expressed by Eq. 11-5 may be expressed in decibels by taking $20 \log_{10} \mid G(j\omega) H(j\omega) \mid$ rather than the natural logarithm.[2] Thus,

$20 \log \mid G(j\omega) H(j\omega) \mid$

$$= 20 \log K + 20 \log \mid N_1(j\omega) \mid + 20 \log \mid N_2(j\omega) \mid$$
$$- 20 \log \mid D_1(j\omega) \mid - 20 \log \mid D_2(j\omega) \mid \qquad (11\text{-}7)$$

Plots of Eqs. 11-6 and 11-7 versus the logarithm of ω are Bode diagrams.

The simplicity of plotting the Bode attenuation diagram should now be apparent. The operations of multiplication and division have become addition and subtraction. The procedure is to compute (or look up in a table) the decibel values for each of the factors of $G(j\omega)H(j\omega)$ at specific values of ω, and then combine them algebraically.

It should be mentioned at this point that there are several alternatives in plotting the logarithm of the magnitude of $G(j\omega)H(j\omega)$. The method that will now be adopted is to plot the logarithm of $\mid G(j\omega)H(j\omega) \mid$ expressed in decibels labeled "Attenuation (decibels)" versus ω on semilog paper. The ordinate is then $20 \log \mid G(j\omega)H(j\omega) \mid$ plotted on equally spaced divisions, while the abscissa is ω plotted on a logarithmic scale.

---

[2]The symbol log as used in the following material is defined as $\log_{10}$.

The number of logarithmic cycles required on the abscissa is determined by the range of frequency over which the system is to be investigated. The advantages of using the decibel system and plotting the Bode diagrams on semilog paper will be illustrated in the course of solving the example problems.

EXAMPLE 11-1. Plot the Bode diagrams for the open-loop transfer function of a unity feedback system

$$G(j\omega) = \frac{100}{j\omega(j\omega + 10)}$$

Solution: The transfer function may first be put in normalized form by dividing numerator and denominator by 10, thus

$$G(j\omega) = 10 \frac{1}{j\omega(0.1j\omega + 1)}$$

Then,

$$20 \log |G(j\omega)| = 20 \log 10 - 20 \log \omega - 20 \log [(0.1\omega)^2 + 1]^{\frac{1}{2}}$$

and the angle is given by

$$\phi(\omega) = -\frac{\pi}{2} - \tan^{-1} 0.1\omega$$

A set of values are summarized in Table 11-1. The various terms may be obtained with the aid of the tables in Appendixes E and F.

TABLE 11-1

VALUES FOR EXAMPLE 11-1

ω	$20 \log |G(j\omega)|$ (decibels)				$\phi(\omega)$ (degrees)		
1	20 −	0	− 0	= 20	−90 − 5.7	=	−95.7
5	20 −	13.98	− 1.	= 5.02	−90 − 26.6	=	−116.6
8	20 −	18.06	− 2.16	= − 0.22	−90 − 38.7	=	−128.7
10	20 −	20	− 3	= − 3	−90 − 45.0	=	−135.0
15	20 −	23.5	− 5.1	= − 8.6	−90 − 56.3	=	−146.3
20	20 −	26.	− 7.	= − 13.	−90 − 63.4	=	−153.4
50	20 −	34.	− 14.	= − 28.	−90 − 78.7	=	−168.7

The attenuation and phase angle diagrams are given in Fig. 11-1. The left ordinate is scaled in decibels for the attenuation plot and the right ordinate is in degrees. The axis of 0 db coincides with the axis of − 180° phase for a reason that will become apparent in the following section. *Ans.*

11-3. Stability Analysis on Bode Diagrams

For control systems that satisfy the restriction for application of the simplified Nyquist criterion, stability analysis with the Bode diagrams is quite simple. The restriction is, of course, that the system be open-loop

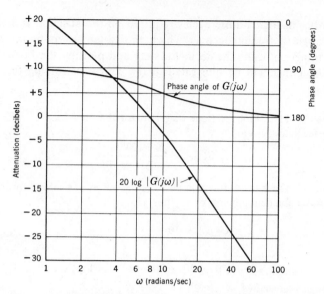

FIG. 11-1. Bode attenuation and phase angle plots for Example 11-1.

stable or, as illustrated in Sec. 10-4, that there be no poles of the open-loop transfer function on the right-hand side of the *s* plane.

Figure 11-2*a* illustrates the stability criterion on the Nyquist diagram. If the point $-1 + j0$ is not encircled by the closed plot of $G(j\omega)H(j\omega)$, the system is stable; if the point $-1 + j0$ is encircled, the system is unstable. At the threshold of instability the magnitude of $G(j\omega)H(j\omega)$ is unity and the phase angle is $-180°$. On the Bode diagram, this corresponds to the point of 0 db (20 log 1 = 0) and $-180°$ phase angle. Figure 11-2*b* is a Bode diagram with three possibilities for an attenuation plot and a corresponding phase angle curve. If the phase angle is less than $-180°$ when the attenuation is 0 db, the system is stable. The frequency at which the attenuation is 0 db is sometimes called the *crossover point.* The dashed attenuation curve illustrates the relative position of attenuation and phase angle plots for the threshold of instability. The combination of the phase angle curve and the highest attenuation curve shown would represent an unstable system.

In summary, a system which is open-loop stable will be closed-loop stable if the phase angle is less negative than $-180°$ at the frequency corresponding to 0 db attenuation. Or conversely, the system would be stable if the attenuation is less than 0 db (or negative) at the frequency corresponding to $-180°$ phase shift.

Gain Margin. The gain margin defined in Sec.10-5 may be expressed in decibels and identified on the Bode diagram as the amount of attenuation below 0 db where the phase angle is $-180°$. This is illustrated in Fig. 11-3.

A negative value of decibels at a phase angle of $-180°$ corresponds to a gain margin of greater than unity obtained from the Nyquist diagrams of Chap-

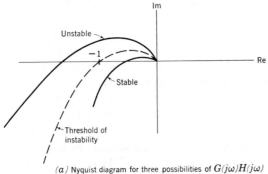

(a) Nyquist diagram for three possibilities of $G(j\omega)H(j\omega)$

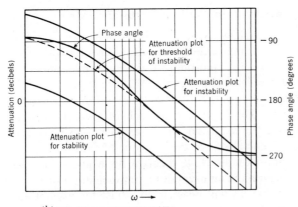

(b) Bode diagrams illustrating stability

FIG. 11-2. Interpretation of stability on Nyquist and Bode diagrams.

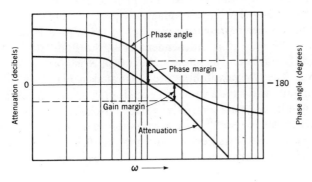

FIG. 11-3. Gain and phase margins.

ter 10. The relationship between decibel attenuation and gain margin obtained in Sec. 10-5 is

$$\text{Decibel attenuation} = 20 \log \frac{1}{m_G} \qquad (11\text{-}8)$$

For example, a gain margin of 2 corresponds to an attenuation of -6.02 db, while a gain margin of 10 corresponds to an attenuation of -20 db. The gain margin may also be specified in decibels. In the examples given, the gain margins would be 6.02 db and 20 db, respectively.

Phase Margin. The phase margin discussed in Sec. 10-5 may be obtained directly from the Bode diagrams and is the difference in angles, at the frequency corresponding to 0 db attenuation, indicated in Fig. 11-3.

As in Chapter 10, the gain and phase margin may be used in the design process as a measure of dynamic performance. In particular, they may be used as criteria for determining over-all system gain.

11-4. Simplifications in Plotting Bode Diagrams

Equation 11-7 suggests that the logarithm of $|G(j\omega)H(j\omega)|$ could be plotted by superimposing the contributions of each factor of $G(j\omega) \cdot H(j\omega)$. The process of plotting the Bode attenuation diagram may be simplified and speeded by recognizing that the factors of $G(j\omega)H(j\omega)$ are repeatedly found to be of the forms K, $(j\omega)^{\pm 1}$, $(\tau j\omega + 1)^{\pm 1}$, and $[(\lambda j\omega)^2 + 2\zeta\lambda(j\omega) + 1]^{\pm 1}$. The logarithm of each of these factors when plotted separately versus the logarithm of the frequency exhibits certain characteristics. With a knowledge of the behavior of each factor, it is possible to construct an approximate Bode diagram for a given $G(j\omega)H(j\omega)$. This approximate diagram is an asymptotic representation of the transfer function. It is possible, as will be seen, to adjust the asymptotic diagram to yield the actual plot by the use of simple correction factors. The first step in simplifying the plotting of Bode diagrams is then to analyze the behavior of each of the basic factors.

Bode Diagram of a Constant K. The gain constant K may be thought of as a complex number with zero imaginary part. The real part may be a plus or minus number representing a vector of magnitude K and lying on the real axis. Because K is independent of ω, the Bode attenuation diagram is a horizontal line. The phase angle is $0°$ if K is positive, or $-180°$ if K is negative. The two possibilities are illustrated in Fig. 11-4. A number greater than unity will have a value in decibels greater than 0, while a number less than unity will have a negative value in decibels.

Bode Diagrams of $(j\omega)^{\pm 1}$. The attenuation and phase angle plots for an integrating element $(1/j\omega)$ or a differentiating element $(j\omega)$ are straight lines. Consider the attenuation diagrams in decibels versus $\log \omega$. For an integration,

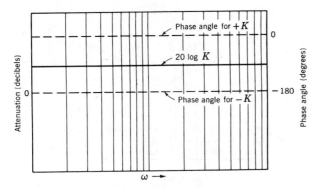

FIG. 11-4. Bode diagrams for a gain constant $\pm K$.

$$20 \log \left| \frac{1}{j\omega} \right| = -20 \log \omega \tag{11-9}$$

and for differentiation,

$$20 \log |j\omega| = 20 \log \omega \tag{11-10}$$

The slopes of each plot may be obtained by considering the change in each plot when the frequency is doubled (an octave). If $\omega_2 = 2\omega_1$, then, for an integration,

$$-20 \log \omega_2 - (-20 \log \omega_1) = -20 \log \omega_1 - 20 \log 2 + 20 \log \omega_1$$
$$= -20 \log 2 = -6.02 \text{ db}$$

Thus the slope of the Bode attenuation plot for an integration is approximately -6 db/octave.[3]

For a differentiation $(j\omega)$ the slope is approximately $+6$ db/octave or $+20$ db/decade. These plots are shown in Fig. 11-5a. Note that both lines cross 0 db at $\omega = 1$.

The phase angles are

$$\phi_{\text{int}} = -\tan^{-1} \frac{\omega}{0} = -\tan^{-1} \infty = -90°$$

$$\phi_{\text{diff}} = \tan^{-1} \infty = +90°$$

The phase angle plots for integration and differentiation are shown in Fig. 11-5b.

Bode Diagrams for $(\tau j\omega + 1)^{\pm 1}$. The third commonly encountered building block is the time constant commonly known as the simple lag $1/(\tau j\omega + 1)$ or the simple lead $(\tau j\omega + 1)$. First consider the simple lag.

---

[3]Some texts specify the slope on a decibel per decade basis. If the frequency is increased by a factor of 10, it may be shown that the drop in decibels is $-20 \log 10 = -20$, or -20 db/decade.

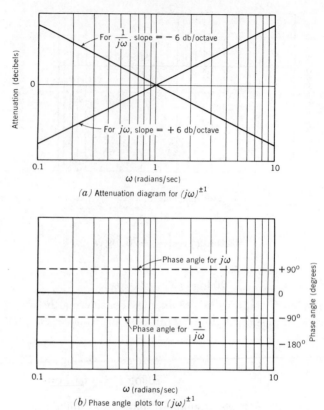

(a) Attenuation diagram for $(j\omega)^{\pm 1}$

(b) Phase angle plots for $(j\omega)^{\pm 1}$

FIG. 11-5. Bode diagrams for $(j\omega)^{\pm 1}$.

The attenuation plot is obtained from

$$20 \log \left| \frac{1}{\tau j\omega + 1} \right| = 20 \log \frac{1}{[(\tau\omega)^2 + 1]^{1/2}} = -20 \log [(\tau\omega)^2 + 1]^{1/2} \qquad (11\text{-}11)$$

and the phase angle is given by

$$\phi = -\tan^{-1} \tau\omega \qquad (11\text{-}12)$$

The Bode diagram for the transfer relationship may be easily plotted using Eqs. 11-11 and 11-12 or the tables of Appendixes E and F. It is possible, however, to obtain an asymptotic approximation of the attenuation curve by recognizing the behavior of Eq. 11-11 at various frequencies. At low frequencies,[4] $\tau\omega \ll 1$:

$$20 \log \left| \frac{1}{\tau j\omega + 1} \right| \cong 0$$

---

[4]Low and high frequencies as stated here are a relative matter. A low frequency is a value of ω such that $\omega\tau \ll 1$. Conversely, a value of ω such that $\omega\tau \gg 1$ is a high frequency.

and the curve approaches a horizontal asymptote at 0 db. The frequency $\omega = 1/\tau$ is called the *break point* and $\omega\tau = 1$. At high frequencies the plot is governed by the value of ω because $\omega\tau \gg 1$. Thus, the plot follows the slope of the attenuation asymptote of $1/j\omega$ and, as before, has a slope of -6 db/octave or -20 db/decade. At the break point, the actual curve is attenuated by

$$20 \log \frac{1}{\sqrt{2}} = 20 \log 0.707 = -3 \,\text{db}$$

Figure 11-6a illustrates the asymptotic approximation and the actual plot of $1/(\tau j\omega + 1)$ which is attenuated 3 db at $\omega = 1/\tau$.

Figure 11-6b illustrates the phase shift associated with a simple phase lag. At very low frequencies ($\omega\tau$ approaches 0) Eq. 11-12 shows that the phase angle approaches $0°$. At very high frequencies ($\omega\tau$ approaches ∞) the phase angle approaches $-90°$. At $\omega = 1/\tau$, $\phi = -\tan^{-1} 1 = -45°$.

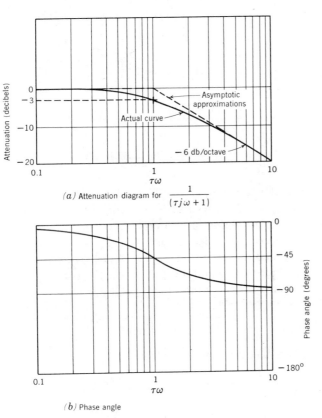

(a) Attenuation diagram for $\dfrac{1}{(\tau j\omega + 1)}$

(b) Phase angle

Fig. 11-6. Attenuation and phase angle for a simple lag, $1/(\tau j\omega + 1)$.

The same line of reasoning may be applied to the simple lead $(\tau j\omega + 1)$. The only difference in this case is that the imaginary quantity is in the numerator; thus, the attenuation diagram has a positive slope from the break point, and the phase shift is from $0°$ to $90°$. Figure 11-7 is a Bode plot for a simple lead. The term *lead* is derived from the fact that the output actually leads the input when $(\tau j\omega + 1)$ is the transfer function.

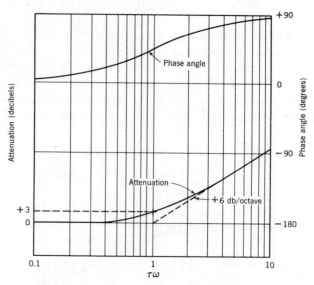

FIG. 11-7. Bode plots for simple lead $(\tau j\omega + 1)$.

Bode Diagram for $[(\lambda j\omega)^2 + 2\zeta\lambda(j\omega) + 1]^{\pm 1}$. The quadratic factor $(s^2 + 2\zeta\omega_n s + \omega_n^2)$ appears quite frequently in transfer functions, particularly in the denominator as a quadratic lag. If s is replaced by $(j\omega)$ and λ (a characteristic time defined as $\lambda = 1/\omega_n$) substituted, the quadratic lag may be defined in dimensionless terms as $1/[(\lambda j\omega)^2 + 2\zeta\lambda(j\omega) + 1]$, while the dimensionless quadratic lead is simply $[(\lambda j\omega)^2 + 2\zeta\lambda(j\omega) + 1]$.

First, consider the case of the quadratic lag

$$\frac{1}{(\lambda j\omega)^2 + 2\zeta\lambda(j\omega) + 1}$$

The attenuation diagram is obtained from

$$20 \log \left| \frac{1}{(\lambda j\omega)^2 + 2\zeta\lambda(j\omega) + 1} \right|$$
$$= -20 \log \{[1 - (\lambda\omega)^2]^2 + (2\zeta\lambda\omega)^2\}^{1/2} \quad \text{decibels} \quad (11\text{-}13)$$

and the phase angle diagram is obtained from

$$\phi = -\tan^{-1} \frac{2\zeta\omega\lambda}{1 - (\omega\lambda)^2} \qquad (11\text{-}14)$$

It should be noted that both the attenuation and the phase angle diagrams are functions of $(\omega\lambda)$ and the damping ratio ζ. Thus plotting Bode diagrams for the quadratic lag may be simplified by having a set of attenuation and phase diagrams for various values of the damping ratio ζ. It is also possible to consider the asymptotic relationships at low and high frequencies. At low frequencies $(\omega\lambda \ll 1)$ Eq. 11-13 becomes $-20 \log 1 = 0$ db, thus the plot approaches a horizontal asymptote of 0 db. At high frequencies $(\omega\lambda \gg 1)$ the plot is predominately $-20 \log |(\lambda j\omega)^2|$. Since $-20 \log |(\lambda j\omega)^2| = -2[20 \log |(\lambda j\omega)|]$, the asymptote will be a straight line with a negative slope of 12.04 db/octave or 40 db/decade. At $\omega\lambda \cong 1$ the plot is predominately governed by the amount of damping. If $\zeta < 0.707$, there will be a tendency for the curve to peak up above 0 db; if $\zeta > 0.707$, there will be attenuation of the curve. A set of attenuation and phase angle curves is given in Figs. 11-8a and b. A qualitative picture of the phase angle curve may be obtained from Eq. 11-14. At low frequency $(\omega\lambda \ll 1)$ ϕ approaches 0. At $\omega\lambda = 1$, $\phi = -90°$; at $\omega\lambda \gg 1$, ϕ approaches $-180°$.

The quadratic lead, which is less common in simple physical systems, may easily be plotted from information obtained from the quadratic lag plotted in Fig. 11-8. This may be logically verified by deriving equations similar to Eqs. 11-13 and 11-14 for the quadratic lead.

A table of values for evaluating the quadratic lag in terms of ζ and ω/ω_n or $\omega\lambda$ is given in Appendix H.

Summary of Simplified Rules for Plotting Bode Diagrams. The additive nature of transfer function plots on logarithmic coordinates allows a simplified building-block approach to plotting an approximate Bode attenuation diagram. The steps in this process are as follows:

1. Rearrange the transfer function $G(j\omega)H(j\omega)$ to a factored form, where the factors take on any of the four basic forms discussed.
2. Identify the break points associated with simple lead, lag, and quadratic factors.
3. Draw the approximate attenuation diagram with the proper asymptotic slopes between break points.
4. If an accurate plot is necessary for the analysis, draw in the actual attenuation diagram considering the correction at the break points. When more than a single time constant is present, the ± 3 db correction is within 10% only if the time constants differ by a factor of 4 or more. If several factors exist in the transfer function, it is advisable to calculate the actual attenuation at break-point frequencies and at any other frequencies for which an accurate plot is required.

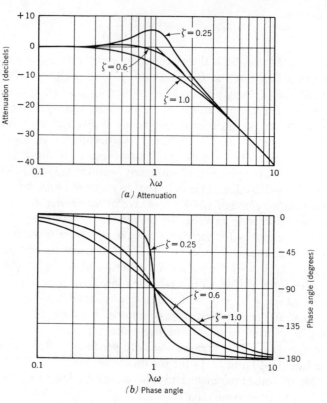

Fig. 11-8. Attenuation and phase angle curves for a quadratic lag $1/[(\lambda j\omega)^2 + 2\zeta\lambda j\omega + 1]$.

5. Phase angle plots may be first sketched by considering the general contribution of each factor at various frequencies. Asymptotic limits are readily determined. The final diagram is best plotted by calculation of the total angle at several points.

The real value of approximations and preliminary analyses which are possible when making Bode plots is not in getting a final accurate answer, but in establishing the general nature of the plots so that a minimum of actual calculation is necessary to obtain a precise plot for analysis.[5]

EXAMPLE 11-2. Plot the asymptotic attenuation and phase angle diagrams for a system with the open-loop transfer function

$$G(s)H(s) = \frac{25(s + 2)}{s^2 + 10.5s + 5}$$

and establish the nature of stability.

[5] Sample digital computer programs for these computations are given in Appendix A.

Solution: The first step is to look for real factors in the quadratic term. Factoring and letting $s = j\omega$,

$$G(j\omega)H(j\omega) = \frac{25(j\omega + 2)}{(j\omega + 10)(j\omega + 0.5)}$$

Normalizing each factor yields

$$G(j\omega)H(j\omega) = \frac{10(0.5j\omega + 1)}{(0.1j\omega + 1)(2j\omega + 1)}$$

This transfer function has break points at $\omega = 0.5$, 2, and 10 radians/sec. Below $\omega = 0.5$, the plot has zero slope at a decibel level of 20 db. At the first break point ($\omega = 0.5$) the slope changes to -20 db/decade and continues to $\omega = 2$, where the slope once again changes to zero. Above $\omega = 10$ the final time constant provides a -20 db/decade slope. Figure 11-9a shows the asymptotic attenuation plots for each of the terms in $G(j\omega)H(j\omega)$. The composite asymptotic plot is obtained by the algebraic addition of the component curves. The composite asymptotic plot and phase angle curve of $G(j\omega)H(j\omega)$ are given in Fig. 11-9b.

An approximate plot of the phase angle may be sketched by recognizing that for low values of frequency, the phase angle is essentially zero. At the first break

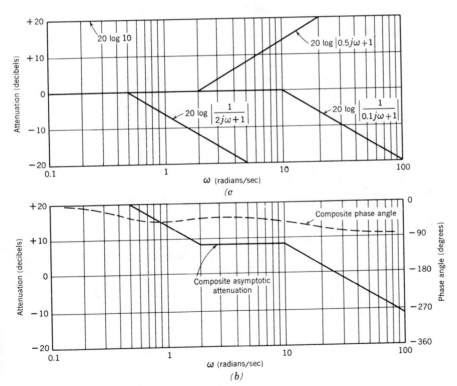

FIG. 11-9. Attenuation and phase angle plots for Example 11-2.

point there is a contribution of $-45°$ phase shift by the term $(2j\omega + 1)$ and a small contribution from the other two terms. At $\omega = 2$, the contribution from $(2j\omega + 1)$ is nearly $-90°$ while the contribution from $(0.5j\omega + 1)$ is $+45°$. The third term has only a small negative angle at this frequency, thus the net angle is approximately $-45°$. At $\omega = 10$, the nearly $\pm 90°$ contribution of the first two terms cancel and the third term contributes $-45°$. For large values of frequency, the phase angle is approximately $-90°$. The approximated phase angle plot is shown in Fig. 11-9b.

The system with this open-loop transfer function will be stable since there is less than $-180°$ phase shift at an attenuation of 0 db.   *Ans.*

EXAMPLE 11-3. Plot the Bode diagrams for the following transfer function and establish the nature of stability.

$$G(s)H(s) = \frac{10^3}{s(s + 2)(4s^2 + 12s + 10^2)}$$

Solution: The transfer function should first be rearranged to the previously analyzed form in $(j\omega)$. Thus

$$G(j\omega)H(j\omega) = \frac{5}{j\omega(0.5j\omega + 1)[0.04(j\omega)^2 + 0.12j\omega + 1]}$$

To simplify plotting the attenuation diagram, separate plots for each of the factors are shown in Figure 11-10a. The composite curve is then obtained by adding algebraically the decibel contribution of each of the four plots. A few points of the plot for the quadratic lag are obtained from Fig. 11-8a. When the individual terms are added algebraically at each frequency, the slope of the composite curve is cumulative. At the first break point the slope changes from -20 db/decade to -40 db/decade. At the break point for the quadratic lag factor the slope becomes -80 db/decade or approximately -24 db/octave (the latter slope is sometimes easier to use on a given piece of graph paper) due to the -40 db/decade slope added by the $(j\omega)^2$ term.

A similar procedure may be adopted for plotting the phase angle curve. In Fig. 11-10b, the phase angle curve of each factor is sketched. The algebraic sum of all the phase angles at each frequency provides the points for the composite phase angle curve.

If the phase angle curve is superimposed on the attenuation plot, it becomes apparent that the closed-loop system would be unstable. There is a positive value of decibels for the attenuation plot at a frequency corresponding to $-180°$ phase shift.   *Ans.*

An extension of the example would be to analyze what could be done to make the system stable. The most obvious approach is to shift the entire attenuation curve downward until some desirable gain margin is obtained. This is achieved by reducing the system gain K which, in this case, was 5. To obtain an accurate answer in a study of this kind, it would then be desirable to calculate a few specific values of attenuation and phase in the

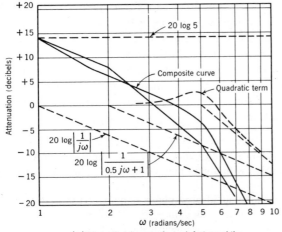

(a) Attenuation diagrams for each factor and the combined diagram for $G(j\omega)H(j\omega)$

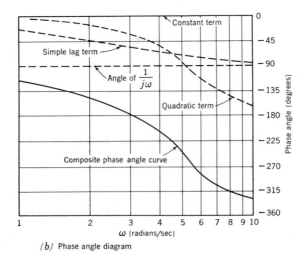

(b) Phase angle diagram

FIG. 11-10. Bode diagrams for Example 11-3.

region of $-180°$ phase shift. This type of problem is discussed more thoroughly in the following section.

11-5. Examples of System Analysis Using Bode Diagrams

The concepts and techniques that have been presented in Secs. 11-1 to 11-5 may be applied to engineering problems in control system analysis and design in several ways. The following examples will illustrate two such applications. The first example is a study in system design wherein it is desired to obtain the specific amplifier gain for a closed-loop electrohydraulic control system that will provide a specified gain margin. The second example illustrates the use of Bode plots for determination of the actual transfer function of a control system. In this case, the purpose of the investigation may be to establish a transfer function which is impractical to determine theoretically, to check a theoretically derived transfer function, or to facilitate redesign of some actual equipment which is not meeting the performance specifications.

EXAMPLE 11-4. A schematic drawing of an electrohydraulic positioning system is given in Fig. 11-11a. Determine the proper amplifier gain K_1 for the system to have stable operation with a gain margin of 10 db. The system parameters are given as follows:

K_1 = gain constant of an ideal amplifier = amp/volt (variable)
K_2 = servo valve and actuator gain = 76.7[(in.^{3}/sec)/amp]
τ_1 = servo valve and actuator time constant = 0.005 sec
K_3 = cylinder and load gain = 0.02(1/in.^{2})

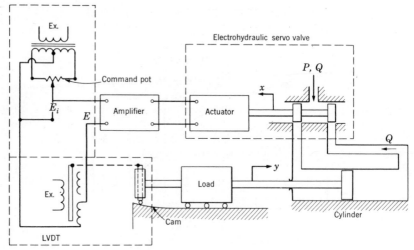

(a) Schematic drawing

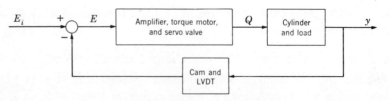

(b) Block-diagram configuration

FIG. 11-11. Electrohydraulic positioning system for Example 11-4.

λ_1 = cylinder and load characteristic time = 0.0083 sec (this corresponds to an undamped resonant frequency of 120.5 radians/sec or 19.2 cps)

ζ = total damping ratio of cylinder and load = 0.8

K_4 = LVDT and cam gain = 25 volts/in.

Solution: The first step in the solution is to obtain the transfer functions and block representation of the system. Figure 11-11b is a block diagram representation of the system illustrating one possible grouping of the elements into blocks. In proceeding, the transfer function of each block may be obtained by writing the differential equation that represents the behavior of each element in the system. It is possible, however, to draw upon the work of Chapter 7 to obtain the transfer relationships of the various components previously discussed. For example, the transfer function of a typical amplifier, torque motor or actuator, and servo valve was obtained in Eq. 7-40 as

$$\frac{Q(s)}{E(s)} = \frac{K}{\tau s + 1}$$

In this specific case,

$$\frac{Q(s)}{E(s)} = \frac{K_1 K_2}{\tau_1 s + 1}$$

Hydraulic cylinders and connected loads were also discussed in Sec. 7-9. The transfer function for this portion of the system is given by Eq. 7-32 as

$$\frac{Y(s)}{Q(s)} = \frac{A}{s\left[\frac{VM}{\beta} s^2 + \left(LM + \frac{Vc}{\beta}\right)s + (A^2 + Lc)\right]}$$

or in terms of a characteristic time $\lambda_1 = 1/\omega_n$, a total gain K_3, and the dimensionless damping ratio ζ,

$$\frac{Y(s)}{Q(s)} = \frac{K_3}{s(\lambda_1^2 s^2 + 2\zeta\lambda_1 s + 1)}$$

The block diagram with these transfer functions inserted is given in Figure 11-12a.

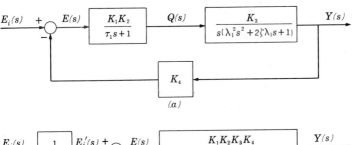

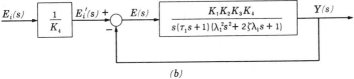

FIG. 11-12. Block diagrams and transfer relationships for Example 11-4.

It is convenient to put the system in the form of a unity feedback control system by rearranging the diagram to the form of Fig. 11-12b.

With the specific values for the variables inserted and s replaced by $j\omega$, the open-loop transfer function for the system becomes

$$G(j\omega)H(j\omega) = \frac{K_1(38.35)}{j\omega(0.005j\omega + 1)[68.9 \times 10^{-6}(j\omega)^2 + 0.0133j\omega + 1]}$$

To proceed with the determination of the proper K_1 for a specified gain margin, it is necessary to divide $G(j\omega)\,H(j\omega)$ by $K_1 K_2 K_3 K_4$ and plot a normalized transfer function similar to the procedure followed in setting system gains in Sec. 10-5. Thus,

$$G(j\omega)H(j\omega)_N = \frac{1}{j\omega(0.005j\omega + 1)[68.9 \times 10^{-6}(j\omega)^2 + 0.0133(j\omega) + 1]}$$

The Bode plots for this transfer function are given in Fig. 11-13. It is seen from these plots that with unity gain there is a gain margin of 40.8 db. It is desired to have

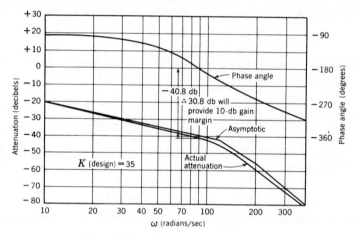

FIG. 11-13. Bode plots for Example 11-4.

a gain margin of only 10 db, therefore, the entire attenuation curve may be shifted vertically upward 30.8 db until the specified gain margin of 10 db is reached. A +30.8-db increase in all points of the attenuation curve corresponds, by reference to Appendix F, to a total system gain of 35. Therefore,

$$K_{\text{design}} = 35 = K_1 K_2 K_3 K_4$$

and thus

$$K_1 = \frac{35}{38.35} = 0.913\ \frac{\text{amp}}{\text{volt}} \qquad\qquad Ans.$$

EXAMPLE 11-5. It is desired to determine the actual transfer function for the synchro-positioning device shown in Fig. 11-14. In Sec. 7-7 the synchro transmitter and receiver configuration was considered. When wired as shown in Fig. 11-14, these devices may be used as a position transmitter or "electrical flexible shaft."

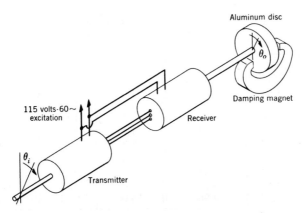

FIG. 11-14. Schematic drawing of a synchro transmitter
and receiver with load and damping.

This unit finds extensive use as a remote indicating device in instrument servo-mechanisms.

Figure 11-15 is a schematic drawing of a simple laboratory setup that may be used to obtain frequency response data. A d-c motor is used to drive a Scotch yoke which gives the rack simple harmonic motion. The flywheel is used to provide a constant speed rotation of the motor. A pinion on the transmitter is in contact with the rack, thus producing a sinusoidal rotation of the input shaft θ_i. The input and output positions may be recorded simultaneously on a dual-track recorder. By recording the input and output positions θ_i and θ_o at different motor speeds the following data were obtained.

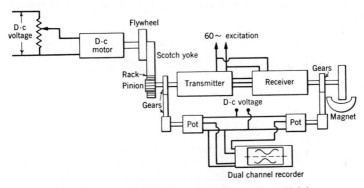

FIG. 11-15. Diagram of test setup for obtaining sinusoidal data.

Determine the over-all transfer function for the transmitter-receiver-load combination.

Solution: The first step in the analysis is to plot the attenuation versus frequency curve using the experimental data obtained from Table 11-2. Figure 11-16 is

TABLE 11-2

TEST DATA

f (cps)	$\dfrac{\theta_o}{\theta_i}$	$\dfrac{\theta_o}{\theta_i}$ (decibels)	f (cps)	$\dfrac{\theta_o}{\theta_i}$	$\dfrac{\theta_o}{\theta_i}$ (decibels)
0.15	1.0	0	1.3	1.84	5.3
0.23	1.01	0.09	1.4	1.82	5.2
0.48	1.14	1.14	1.5	1.67	4.45
0.79	1.26	2.01	1.6	1.25	1.94
0.86	1.34	2.54	1.7	1.09	0.75
0.97	1.42	3.05	2.0	0.75	−2.5
1.0	1.55	3.81	2.5	0.33	−9.63
1.1	1.68	4.51	3.0	0.23	−12.8
1.2	1.77	4.96	4.0	0.11	−19.2

a plot of the data. At low frequency the curve is asymptotic to a horizontal line at 0 db. At high frequency (above 2 cps) the curve is asymptotic to a line with a slope of −12 db/octave. The plot may be recognized as typical of a quadratic lag discussed in Sec. 11-4. The intersection of the asymptotes provides the break point. (However, it should be noted that the frequency scale is plotted in cycles per second.) Thus the transfer function is

$$\frac{\Theta_o(s)}{\Theta_i(s)} = \frac{1}{\lambda^2 s^2 + 2\zeta\lambda s + 1}$$

where $\lambda = 1/(1.39 \times 2\pi) = 0.115$ sec. The maximum amplitude is reached slightly before 1.39 cps and is 5.3 db or an M_m of 1.84. This corresponds to a damping ratio of $\zeta = 0.272$, which may be calculated using Eq. 11-13 for the maximum value of the curve and its corresponding frequency or estimated by the use of the curves in Fig. 11-8.

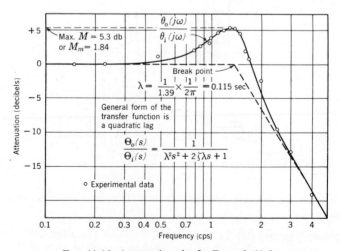

FIG. 11-16. Attenuation plot for Example 11-5.

The transfer function for this equipment is then

$$\frac{\Theta_o(s)}{\Theta_i(s)} = \frac{1}{0.013s^2 + 0.063s + 1} \qquad \qquad Ans.$$

11-6. Log-Magnitude Versus Phase Angle Representation of a Transfer Function

Equation 11-3 illustrated that the logarithm of a transfer function is completely represented by two quantities: the log-magnitude [represented by $20 \log |G(j\omega)H(j\omega)|$] and the phase angle $\phi(\omega)$. In the realm of frequency-response analysis, the independent variable is the frequency ω. Rather than plot two separate curves versus ω to represent the transfer function, it is possible to plot $20 \log |G(j\omega)H(j\omega)|$ versus $\phi(\omega)$ with ω as a parameter of the curve. It will be shown that this plot is particularly useful for design and synthesis of control systems when it is desired to set a system gain to meet a specific M_m criterion or when it is necessary to modify the shape of a transfer function plot. The closed-loop response of a system may be readily obtained when the loci of constant magnitude M and constant angle N are superimposed on the log-magnitude versus phase angle coordinates (commonly called a *Nichols chart*).

Equations for M and N contours for the Nichols chart may be developed from the geometry of the M and N circles on the Nyquist diagram.[6] Figure 11-17 shows the M and N contours on the log-magnitude phase angle coordinates. It should be noted that the plot is symmetrical about

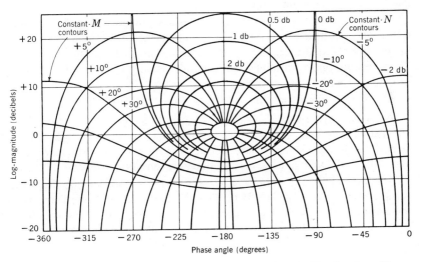

FIG. 11-17. Nichols chart illustrating the contours of constant M and constant N.

[6] For a complete derivation of these expressions the reader is referred to reference [3].

$-180°$. Most Nichols charts are drawn using the portion of the plot from $-180°$ to $0°$ of the abscissa because this is the most commonly encountered range of interest. If more of the plot is required, the plot may be turned over. Similarly, if it is desired to extend to the right from $0°$ or to the left from $-360°$, the plot is repeated for every period of $360°$.

It is perhaps valuable to summarize the meaning of M and N contours on the Nichols chart of Fig. 11-17. If a quantity $G(j\omega)$ is the forward transfer function of a unity feedback system, the closed-loop response is given by

$$\frac{C(j\omega)}{R(j\omega)} = \frac{G(j\omega)}{1 + G(j\omega)}$$

Then

$$\left|\frac{C(j\omega)}{R(j\omega)}\right| = M$$

at some ω, and

$$\text{Angle}\left[\frac{C(j\omega)}{R(j\omega)}\right] = N$$

at some ω. Thus, if $G(j\omega)$ is plotted on the Nichols chart, the values of M and N for the closed-loop response are ascertained by the intersection of $G(j\omega)$ and the M and N contours at various frequencies ω. Further, the M contour which is just tangent to the plot of $G(j\omega)$ is the maximum M for the closed-loop response, or M_m. The frequency at which M_m occurs is that frequency for $G(j\omega)$ at which the tangency takes place; the corresponding phase angle for the closed-loop response is given by the N contour which passes through the tangent point.

Gain margin and phase margin are quickly identified on the log-magnitude versus phase angle plot. The gain margin in decibels may be obtained directly as the negative of the ordinate at a phase angle of $-180°$. The phase margin is obtained from the abscissa at a log-magnitude value of 0 db and is given by

$$\text{Phase margin} = (180° + \text{phase angle at 0 db}) \qquad (11\text{-}15)$$

EXAMPLE 11-6. A log-magnitude versus phase plot for a unity feedback system is given on the Nichols chart of Fig. 11-18. It is desired to a) establish the nature of stability for the closed-loop system, b) determine the gain and phase margins, and c) plot the final closed-loop response for the system on a Bode diagram.

Solution: From the plot of Fig. 11-18, it is seen that the log-magnitude value is negative at a phase angle of $-180°$. Thus the system is stable. Furthermore, the log-magnitude is -6 db at this frequency, therefore the gain margin is 6 db. At 0 db the phase angle for $G(j\omega)$ is $-128°$, thus the phase margin is given by

$$180° - 128° = 52°$$

The closed-loop response of the system is given by

$$\frac{C(j\omega)}{R(j\omega)} = \frac{G(j\omega)}{1 + G(j\omega)}$$

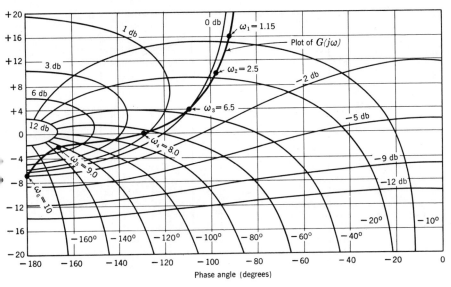

FIG. 11-18. Nichols chart for Example 11-6.

which has a magnitude $M(\omega)$ and phase angle $\phi(\omega)$ [obtained by the intersections of $G(j\omega)$ and the M and N contours] at each discrete frequency. At $\omega_1 = 1.15$ radians/sec, the plot of $G(j\omega)$ on Fig. 11-18 passes through the M and N contours at

$$M = -0.08 \text{ db} \quad \text{and} \quad \phi = -8.5°$$

Other values of magnitude and phase are listed in the following tabulation.

TABLE 11-3

VALUES FOR EXAMPLE 11-6

Frequency (radians/sec)	M (decibels)	ϕ (degrees)
$\omega_2 = 2.5$	-0.01	-18
$\omega_3 = 6.5$	0	-37
$\omega_4 = 8.0$	$+1.3$	-64
$\omega_5 = 9$	$+8.0$	-125
$\omega_6 = 10$	-2	-180

The closed-loop response may then be plotted from this information. The closed-loop frequency response for the system is given in Fig. 11-19. It may be noted that this closed-loop system is characterized by a rather "flat response," that is, there is little change in the magnitude of $C(j\omega)/R(j\omega)$ over the frequency range from $\omega = 1$ radian/sec to $\omega = 6.5$ radians/sec. It should be noted, however, that the response has considerable phase shift in the flat region (i.e., the output is lagging the input). Further, at 9 radians/sec the frequency response is characterized by a large resonant peak, specifically, $+8$ db. *Ans.*

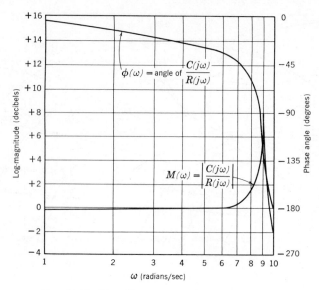

FIG. 11-19. Closed-loop response for Example 11-6.

Case of Nonunity Feedback. If a system has nonunity feedback, care must be exercised in manipulating quantities on the Bode and Nichols plots. Example 11-4 illustrated the approach to be taken when the feedback path involves a constant other than unity. The block diagram may be rearranged to put the system in the form of a unity feedback system.

If the feedback path of the system contains a frequency-sensitive element, that is, $H(j\omega)$ is not a constant, it is still possible to obtain the closed-loop response from the open-loop response. The concepts are similar to those expressed in Sec. 10-5. If $G(j\omega)H(j\omega)$ is plotted on the Nichols chart and values of $M(\omega)$ and $\phi(\omega)$ taken from the M and N contours, the resulting response if replotted on the Bode diagram is for a closed-loop response that is,

$$\text{Closed-loop response} = \frac{G(j\omega)H(j\omega)}{1 + G(j\omega)H(j\omega)} \qquad (11\text{-}16)$$

The true closed-loop response, however, is given by

$$\frac{C(j\omega)}{R(j\omega)} = \frac{G(j\omega)}{1 + G(j\omega)H(j\omega)} \qquad (11\text{-}17)$$

The relationship of Eq. 11-17 may be obtained from the expression of Eq. 11-16 by multiplying by $1/H(j\omega)$, thus

$$\frac{C(j\omega)}{R(j\omega)} = \frac{G(j\omega)H(j\omega)}{1 + G(j\omega)H(j\omega)} \times \frac{1}{H(j\omega)} \qquad (11\text{-}18)$$

On logarithmic coordinates, the relationships involved are

$$\log\left|\frac{C(j\omega)}{R(j\omega)}\right| = \log\left|\frac{G(j\omega)H(j\omega)}{1 + G(j\omega)H(j\omega)}\right| + \log\left|\frac{1}{H(j\omega)}\right| \qquad (11\text{-}19)$$

and

$$\text{Angle}\left[\frac{C(j\omega)}{R(j\omega)}\right] = \text{angle}\left[\frac{G(j\omega)H(j\omega)}{1 + G(j\omega)H(j\omega)}\right] + \text{angle}\left[\frac{1}{H(j\omega)}\right] \qquad (11\text{-}20)$$

The procedure for obtaining the closed-loop response from the open-loop response, for nonunity feedback, is as follows:

1. Plot $G(j\omega)H(j\omega)$ on the Nichols chart.
2. Pick values of $M(\omega)$ and $\phi(\omega)$ at particular frequencies.
3. Plot the response function

$$\frac{G(j\omega)H(j\omega)}{1 + G(j\omega)H(j\omega)}$$

   on the Bode plot.
4. Plot $1/H(j\omega)$ on the Bode plot.
5. Add the two attenuation and phase angle plots on the Bode diagram. The resultant plots for attenuation and phase angle are the closed-loop response $C(j\omega)/R(j\omega)$.

11-7. An Example of System Analysis Using Logarithmic Coordinates (Bode Diagrams and the Nichols Chart)

The application of Bode plots and Nichols charts to control system analysis and synthesis provides a rapid and complete, semigraphical means of analysis. Some of the basic techniques are illustrated in the following example.

EXAMPLE 11-7. Reconsider the simplified positioning system of Example 10-5, and apply the logarithmic plot techniques to the problem. The schematic drawing of the system is repeated in Fig. 11-20a, and the final block diagram is given in Fig. 11-20b. The open-loop transfer function is given as

$$G(j\omega)H(j\omega) = G(j\omega) = \frac{K_T}{j\omega(0.0824j\omega + 1)}$$

where $K_T = K_1 K_2 K_3/(K_4 + b)$ and $K_1 K_3/(K_4 + b) = 10.12$.

The following information is desired: a) Determine the nature of stability of the closed-loop system. b) Establish the amplifier gain required for the system to

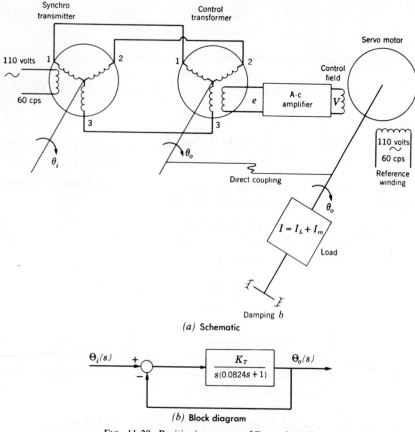

FIG. 11-20. Positioning system of Example 11-7.

have a maximum $M(\omega)$ of 1.2 ($M_m = 1.2$). c) For the gain established in part b, determine the phase margin of the system.

Solution: The normalized transfer function is

$$G_N(j\omega) = \frac{1}{j\omega(0.0824 j\omega + 1)}$$

which always leads to a stable closed-loop system. The Bode plot for this transfer function has a -6 db/octave slope up to the break point of $\omega = 1/0.0824 = 12.13$ radians/sec and then proceeds at a slope of -12 db/octave. At the break point, the actual curve is attenuated -3 db from the straight-line approximation and -1 db from the straight-line approximation at frequencies of 6.06 radians/sec and 24.3 radians/sec. The Bode diagram is plotted in Fig. 11-21.

The information of this plot may be transferred to the Nichols chart by picking points at several frequencies. Specifically, the log-magnitude and phase angle is read from Fig. 11-21 at 1, 1.5, 2.8, 4.2, 10, and 20 radians/sec. The log-magnitude plot of $G_N(j\omega)$ is given in Fig. 11-22. For the closed-loop system to have $M_m = 1.2$,

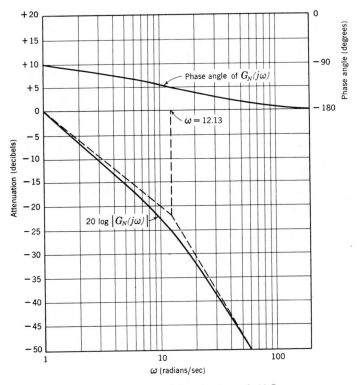

Fig. 11-21. Bode diagram for Example 11-7.

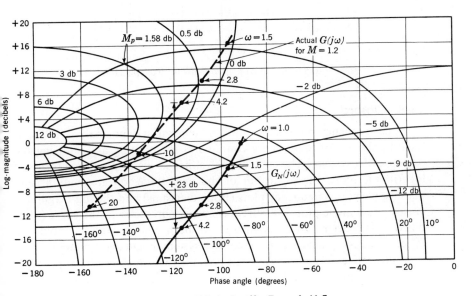

Fig. 11-22. Nichols chart for Example 11-7.

the plot of $G(j\omega)$ (the unnormalized open-loop transfer function) must be tangent to the M contour in decibels that corresponds to $M = 1.2$. Specifically, 20 log 1.2 = 1.58 db. The plot for $G_N(j\omega)$ may be translated vertically on Fig. 11-22 until some point is tangent to the $M = 1.58$ db contour. From the dashed curve it is seen that each point of $G_N(j\omega)$ must be translated vertically by $+23$ db to place the final curve tangent to $M = 1.58$ db. A rise on the logarithmic scale of 23 db is equivalent to a gain of approximately 14.1, thus the over-all design gain for the open-loop transfer function is $K_{\text{design}} = 14.1$.

The proper amplifier gain is then $K_2 = 14.1/10.12 = 1.39$, which is the same as that obtained in Example 10-5. The phase margin is again 48.5°. The Bode plot of the closed-loop response may be obtained by reading the values of the M and N contours passing through the plot of $G(j\omega)$ at the specific frequencies labeled in Fig. 11-22. Figure 11-23 is the closed-loop response obtained from the dashed curve of Fig. 11-22. The maximum magnitude ratio is 1.2 as required.   *Ans.*

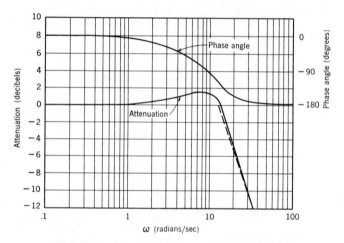

Fɪɢ. 11-23. Closed-loop response for Example 11-7.

11-8. Closure

The frequency response method of control system analysis is facilitated by the use of logarithmic coordinates for plotting and studying the response of a system. The control analyst has at his disposal two forms of logarithmic plots: the logarithm of the magnitude, or attenuation, and phase angle versus frequency plots (Bode diagrams) and the log magnitude versus phase angle plot where frequency is a parameter of the curve (plots on Nichols charts). The combined use of both plots provides a rapid and effective semigraphical means for control system design and synthesis.

Illustrative examples have been presented which show the application of the techniques for establishing system stability, designing a system gain to satisfy specified criteria, and formulating the closed-loop response from open-loop response. In this chapter, the sole means of adjusting a systems behavior was to modify the system gain. It was pointed out earlier that this is only one means of controlling the behavior of a closed-loop control system. It is quite often the case that the modification of system gain is not sufficient to provide the desired response. In this case, if it is not possible to alter the transfer function of the system components, it is necessary to alter the form of the open-loop transfer function by inserting compensating elements into the system. System compensation is another area of analysis, and it is the subject of Chapter 13. In this work the concepts presented for utilizing logarithmic coordinates for analysis will prove invaluable.

Problems

11-1. Plot Bode diagrams for the following open-loop transfer functions. For the attenuation curve in each case plot the asymptotic approximation first and then put in the actual curve by making corrections at the break points or calculating the actual value of the attenuation at specific frequencies.

a) $G(s)H(s) = \dfrac{15(0.001s + 1)}{0.01s + 1}$

b) $G(s)H(s) = \dfrac{150}{s(s^2 + 2s + 9)}$

c) $G(s)H(s) = \dfrac{25}{(0.1s + 1)(0.2s + 1)(0.01s + 1)}$

d) $G(s)H(s) = \dfrac{50}{s^2(s + 10)}$

e) $G(s)H(s) = \dfrac{100}{s^3 + 8s^2 + 25s + 26}$

11-2. From the Bode plots obtained in Prob. 11-1 establish whether or not the closed-loop systems with the stated transfer functions would be stable.

11-3. From the Bode plots obtained in Prob. 11-1 determine the gain and phase margins for the systems.

11-4. A speed regulator with unity feedback has the following open-loop transfer function

$$\frac{\Omega_o(s)}{E(s)} = \frac{K}{(0.015s + 1)(0.05s + 1)}$$

a) Using Bode diagrams establish the gain setting K that will result in a phase margin of 40°. b) Using a Nichols chart determine a new system gain to provide $M_m = 1.3$. For this value of M_m establish, directly from the Nichols chart, the value of the phase margin. c) Obtain the necessary information from the Nichols chart in part b and plot the closed-loop frequency response of the system.

11-5. An experimental study of a group of electrohydraulic elements gave the frequency response shown in the accompanying chart.

$\omega \left(\dfrac{\text{radians}}{\text{sec}} \right)$	$\left\| \dfrac{\text{Output}}{\text{Input}} \right\|$ (decibels)	ϕ (degrees)
0.2	+37	−108
0.5	+29	−155
0.6	+28	−170
0.7	+25.5	−189
0.8	+20.5	−204
1.0	+15	−226
2.0	− 2.5	−250
3.0	−13	−260

a) Plot attenuation and phase angle diagrams on 2-cycle, semilog paper. b) Using the best choice of asymptotic approximations to the data, establish a realistic transfer function for the system. Evaluate all the constants in the transfer function. c) If this group of electrohydraulic elements were used as the system elements of the unity feedback control system shown in the accompanying illustration, what

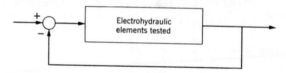

PROB. 11-5

would be the result? d) Redesign the control system and place a gain K_1 in series with the electrohydraulic elements. Calculate a value for K_1 such that the closed-loop system will have a stable operation with $M_m = 1.3$.

11-6. A control system with dead time is illustrated in the figure. Utilizing Bode diagrams determine the amount of dead time (seconds) which could be permitted before the system is in a condition of limited stability.

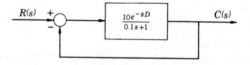

PROB. 11-6

11-7. Determine the maximum amount of dead time that could be tolerated in the system of Prob. 11-6 if the minimum desirable phase margin is 35°.

11-8. Plot the closed-loop frequency response for the system resulting in Prob. 11-7. What is M_m for the system? Compare the closed-loop response for a system with D which yields 35° phase margin with a system with $D = 0$.

11-9. The block diagram of a feedback control system is shown in the figure. a) Using Bode diagrams, determine the gain that will provide a phase margin

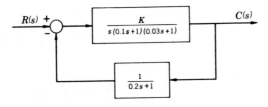

Prob. 11-9

Prob. 11-9

of 45°. b) Determine the maximum gain permissible for 45° phase margin if the system has unity feedback.

11-10. Plot the closed-loop response for cases a) and b) of Prob. 11-9.

Chapter 12

System Analysi

Using Root-Locus Plot

12-1. Introduction

The root-locus method for the analysis of control systems was deve
oped by W. R. Evans in 1950 [11, 12]. The method is related to the tra
sient-response approach to system analysis. The reader will recall th
with transient response techniques, the roots of the characteristic equ
tion must be obtained (a problem circumvented by frequency-respon
methods). The root-locus approach simplifies the root-solving proble
by providing an effective graphical procedure for determining all possib
roots of the characteristic equation for a control system.

An example will illustrate the potentialities of the root-locus metho
Consider the system block diagram of Fig. 12-1 in which

$$G(s) = \frac{K}{(\tau_1 s + 1)(\tau_2 s + 1)} \tag{12-}$$

where K is the system gain (typically a positive quantity) which can,
course, be varied in some manner. Equation 12-1 can be expressed as

$$G(s) = \frac{K/\tau_1 \tau_2}{(s + 1/\tau_1)(s + 1/\tau_2)}$$

or
$$G(s) = \frac{K'}{(s + a)(s + b)} \tag{12-}$$

where K' is proportional to K and the constants a and b are the inverse
the time constants of the system. In Eq. 12-2, $G(s)$ is expressed in the for
most convenient for root-locus work because a knowledge of the poles a
zeros of $G(s)$ is required.[1]

---

[1] Recall that, by definition, the zeros of $G(s)$ are values of s for which $G(s)$ is zero a
that the poles of $G(s)$ are values of s for which $G(s)$ is infinite. The particular $G(s)$ given
Eq. 12-2 has no finite zeros but has two poles which are $-a$ and $-b$.

FIG. 12-1. System block diagram.

The closed-loop transfer function for the system of Fig. 12-1 is

$$\frac{C(s)}{R(s)} = \frac{G(s)}{1 + G(s)} = \frac{K'}{s^2 + (a + b)s + (ab + K')}$$

from which the characteristic equation is

$$s^2 + (a + b)s + (ab + K') = 0 \qquad (12\text{-}3)$$

It can be seen that the roots of the characteristic equation depend on the system gain (as manifested by the value of K') and the system time constants (which establish the values of a and b). For a system of even moderate complexity, the task of obtaining the roots of the characteristic equation as a function of gain alone appears formidable. It is in this regard that the root-locus method is a powerful tool because it provides an effective means for determining all possible roots of a characteristic equation as K' is varied from zero to infinity. In addition, the root-locus method makes it fairly easy to accommodate changes in system constants (such as the time constants previously mentioned).

This chapter will first develop the concept underlying the root-locus method and will illustrate that concept by obtaining several root-locus plots. Next, guides for rapid plotting will be established. Following the presentation of graphical relationships on the s plane, procedures will be developed for setting system gains and obtaining system responses.

12-2. The Root-Locus Concept

The root-locus concept is important and should be understood before the method is used as a tool in control system analysis. In Sec. 10-2, it was established that, for a nonunity feedback control system (the general case), the roots of the characteristic equation also satisfy the expression

$$1 + G(s)H(s) = 0 \qquad (12\text{-}4)$$

In Eq. 12-4, the open-loop transfer function $G(s)H(s)$ includes the variable system gain constant and is expressed in factored form similar to that given in Eq. 12-2. Equation 12-4 may equally as well be expressed as

$$G(s)H(s) = -1 = 1\underline{/n\,180°} \qquad \text{for} \qquad n = \pm 1, \pm 3, \pm 5, \cdots \quad (12\text{-}5)$$

The key to the root-locus method is contained in Eq. 12-5 because any value of s which satisfies this relationship is a root of the characteristic equation.

Looked at from the graphical viewpoint on the s plane, the concept underlying the root-locus method is explained as follows. Any point on the s plane which satisfies the angular requirement of Eq. 12-5 is a possible root of the characteristic equation. A particular point will be a root of the characteristic equation when the value of K', Eq. 12-2, satisfies the magnitude requirement of Eq. 12-5. Thus, the procedure for obtaining a root-locus plot is first to seek out the loci of possible roots from the angular requirement. Then, the loci are scaled (i.e., values of K' are established for points along the loci) from the magnitude requirement. The completed root-locus plot will yield the roots of the characteristic equation for any particular value of system gain. The response for the system can then be determined.

12-3. Plots for Simple Transfer Functions

The concept underlying the root-locus method will be further illustrated by obtaining plots for three simple open-loop transfer functions.

Transfer Function with a Single Pole. A root-locus plot for the open-loop transfer function

$$G(s)H(s) = \frac{K'}{s + 2} \qquad (12\text{-}6)$$

will now be constructed. Because the root-locus method is a graphical one, the factor $(s + 2)$ must be represented as a vector. Thus, the pole -2 of $G(s)H(s)$ is plotted on the s plane of Fig. 12-2a and is designated by $\times$.

The relationship of Eq. 12-5 must be satisfied. Therefore,

$$\frac{K'}{s + 2} = 1\underline{/\pm 180°} \qquad (12\text{-}7)$$

The first step is to seek out points on the s plane which satisfy the angular requirement of Eq. 12-7. A general point s_1 (not on the real axis), Fig.

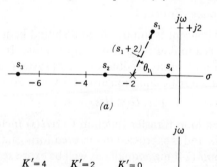

FIG. 12-2.   Plot for $G(s)H(s) = K'/(s + 2)$.

12-2a, is taken as a trial point; the vector representing $(s_1 + 2)$ is shown. From Eq. 12-7

$$\frac{K'}{s_1 + 2} = \frac{K'}{|s_1 + 2|\,/\theta_1^\circ} = \frac{K'}{|s_1 + 2|}\,/{-\theta_1^\circ}$$

Point s_1 is not a possible root of the system characteristic equation (or a point on the root locus) because the angular requirement is not met, that is, $\theta_1 \neq 180$. The same is true for all points not on the real axis.

Similar reasoning will show that the points s_2 and s_3 are on the root locus while point s_4 is not. Thus, the root locus lies along the real axis to the left of the pole -2 and nowhere else. The locus is shown as a heavy line in Fig. 12-2b.

The next step is to scale the locus, that is, to establish values for K' for various points along the locus. This is accomplished by satisfying the magnitude requirement of Eq. 12-7:

$$\left|\frac{K'}{s + 2}\right| = 1 \qquad \text{or} \qquad |K'| = |s + 2|$$

At $s = -2$,	$K' = |-2 + 2| = 0*$
At $s = -4$,	$K' = |-4 + 2| = 2$
At $s = -6$,	$K' = |-6 + 2| = 4$
As $s \to -\infty$,	$K' = |s + 2|$ and $K' \to \infty$

The completed root locus plot is given in Fig. 12-2b.

The use of the root-locus plot will now be illustrated. With a system gain such that $K' = 4$, the root of the characteristic equation (from the plot) is -6. But is it? From Eq. 10-3, for which $G(s)H(s) = N(s)/D(s)$, the characteristic equation is

$$N(s) + D(s) = s + 2 + K' = 0$$

If $K' = 4$, $s = -6$ and thus the root-locus plot provides the correct answer.

Transfer Function with Two Poles. Consider the open-loop transfer function

$$G(s)H(s) = \frac{K'}{(s + 1)(s + 3)} \tag{12-8}$$

To obtain the root-locus plot, the pole locations at -1 and -3 are plotted as shown in Fig. 12-3a. From angular considerations, it is rapidly established that the locus can exist on the real axis only between -1 and -3. [The total angle for the vectors $(s + 1)$ and $(s + 3)$ is $0°$ for points to the right of -1 and $360°$ for points to the left of -3. Neither combination satisfies the $\pm 180°$ angular requirement. However, for points between -1 and -3, the angle for $(s + 1)$ is $180°$ and that for $(s + 3)$ is $0°$ making the

---

* System gain is considered as a positive real quantity; therefore, $|K'| = K'$.

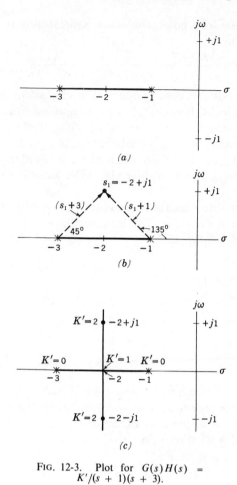

FIG. 12-3.   Plot for $G(s)H(s)$ =
$K'/(s + 1)(s + 3)$.

total angle 180°.] Thus, the locus along the real axis is as shown by the heavy line in Fig. 12-3*a*.

Can the locus leave the real axis? One should suspect that it can because Eq. 12-8 reveals that the system characteristic equation will be a second-degree equation from which complex roots are possible. A trial point, $s_1 = -2 + j1$, is selected as shown in Fig. 12-3*b*. This point satisfies the angular requirement because

$$\frac{K'}{(s_1 + 1)(s_1 + 3)} = \frac{K'}{(\,|\,s_1 + 1\,|\,\underline{/135°})(\,|\,s_1 + 3\,|\,\underline{/45°})}$$

$$= \frac{K'}{|\,(s_1 + 1)(s_1 + 3)\,|}\,\underline{/-180°}$$

In fact, all points for which the real part is -2 will meet the angular requirement. The locus, then, is as given in Fig. 12-3c.

The relationship to be used for scaling is

$$\left| \frac{K'}{(s + 1)(s + 3)} \right| = 1 \quad \text{or} \quad K' = |(s + 1)(s + 3)|$$

At $s = -1$, $\quad K' = |(-1 + 1)(-1 + 3)| = 0$

At $s = -3$, $\quad K' = |(-3 + 1)(-3 + 3)| = 0$

At $s = -2$, $\quad K' = |(-2 + 1)(-2 + 3)| = 1$

At $s = -2 + j1$, $\quad K' = |(s_1 + 1)(s_1 + 3)| = (1.414)(1.414) = 2$

As $s \rightarrow -2 \pm j\infty$, $\quad K' = |(s + 1)(s + 3)| \quad$ and $\quad K' \rightarrow \infty$

The completed plot is shown in Fig. 12-3c.

The plot can be checked by obtaining the system characteristic equation. The use of Eq. 12-8 yields

$$N(s) + D(s) = s^2 + 4s + (3 + K') = 0$$

With $K' = 1$, $\quad s^2 + 4s + 4 = 0 \quad$ and $\quad s_1 = s_2 = -2$

With $K' = 2$, $\quad s^2 + 4s + 5 = 0 \quad$ and $\quad s = -2 \pm j1$

Thus, the root-locus plot provides the correct roots.

Transfer Function with One Pole and One Zero. Consider the open-loop transfer function

$$G(s)H(s) = \frac{K'(s + 5)}{s + 1} \tag{12-9}$$

The pole at -1 and the zero at -5 are plotted as shown in Fig. 12-4. The pole location is designated by $\times$ and the zero location by $\circ$. From angular considerations, the locus exists only on the real axis between -1 and -5. The reader should verify this statement by trying other points.

The relationship to be used for scaling is

$$\left| \frac{K'(s + 5)}{s + 1} \right| = 1 \quad \text{or} \quad K' = \left| \frac{s + 1}{s + 5} \right|$$

At $s = -1$, $\quad K' = \left| \frac{-1 + 1}{-1 + 5} \right| = 0$

At $s = -3$, $\quad K' = \left| \frac{-3 + 1}{-3 + 5} \right| = 1$

At $s = -5$, $\quad K' = \left| \frac{-5 + 1}{-5 + 5} \right| = \infty$

The completed plot is given in Fig. 12-4. The reader should obtain the system characteristic equation and check the result of the plot with $K' = 1$.

At the close of this discussion, it is worth remarking that none of the systems considered could become unstable. The reason is that, in no case, did the root locus leave the left half of the *s* plane. Therefore, there can be no roots of the characteristic equation with positive real parts.

Root-locus plots can be obtained as illustrated. However, certain guides for plotting greatly facilitate the process. These guides will be developed in the following section.

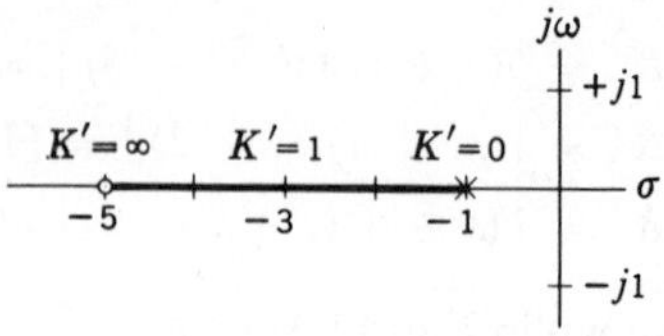

Fig. 12-4. Plot for $G(s)H(s) = K'(s + 5)/(s + 1)$.

12-4. Guides for Rapid Plotting

The discussion up to this point may lead the reader to believe that the construction of a root-locus plot proceeds mainly on a trial-and-error basis as the *s* plane is searched for possible roots of the characteristic equation. While a trial-and-error approach is in general necessary, the number of trials required is greatly reduced by following certain guides for plotting.

Guide 1. A root-locus plot will have as many loci, or branches, as there are roots of the system characteristic equation. The reason is that each root must move along a locus as K' is varied. However, because in control work the number of poles of $G(s)H(s)$ generally exceeds the number of zeros, the number of loci also equals the number of poles of $G(s)H(s)$. To illustrate, let

$$G(s)H(s) = \frac{K'N(s)}{D(s)} = \frac{K'(s + 1)}{s(s + 2)}$$

The characteristic equation is

$$K'N(s) + D(s) = s^2 + (2 + K')s + K' = 0$$

from which there will be two roots, the same number as the number of poles of $G(s)H(s)$.

Thus, in general, a root-locus plot will have as many loci as there are poles of $G(s)H(s)$.

Guide 2. The loci start at the poles of $G(s)H(s)$ with $K' = 0$ and end at the zeros of $G(s)H(s)$ with $K' = \infty$. The reason is as follows. Let

$$G(s)H(s) = \frac{K'N(s)}{D(s)}$$

The characteristic equation is

$$K'N(s) + D(s) = 0$$

With $K' = 0$, the roots of the characteristic equation are the zeros of $D(s)$ which are, of course, the poles of $G(s)H(s)$. For large values of K', $(K' \rightarrow \infty)$, $D(s)$ can be neglected. The roots of the characteristic equation are then the zeros of both $N(s)$ and $G(s)H(s)$.

Usually there are not as many finite zeros of $G(s)H(s)$ as there are poles in which case some of the loci are said to end at infinite zeros of $G(s)H(s)$. [Infinite zeros can be obtained by introducing additional factors of the form $(0s + 1)$ into $N(s)$; nothing, of course, is changed by so doing.] The guide becomes:

The loci start at the poles of $G(s)H(s)$ with $K' = 0$ and terminate either at the zeros of $G(s)H(s)$ or at infinity with $K' = \infty$.

Guide 3. The angular requirement of Eq. 12-5 limits the existence of the root locus on the real axis to certain portions determined by the location of the real poles and zeros of $G(s)H(s)$. Complex poles or zeros have no effect because, for a point on the real axis, the angles involved are equal and opposite. The guide is stated as follows:

The root locus exists on the real axis only to the left of an odd number of real poles and/or zeros.

The reader should verify this guide by sketching several pole-zero configurations and applying the angular requirement.

Guide 4. Complex roots must occur in conjugate pairs; therefore,

The root-locus plot is symmetrical about the real axis.[2]

Guide 5. The loci which terminate at infinity approach asymptotes in doing so. A knowledge of these asymptotes is helpful in defining the general appearance of a root-locus plot. The angles that the asymptotes make with the real axis can be obtained by applying the angular requirement.

Consider the pole-zero configuration of Fig. 12-5. For large values of

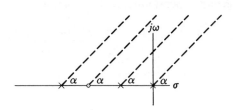

FIG. 12-5. Angle of asymptotes.

$s, (s \rightarrow \infty)$, the vectors shown are parallel, each making an angle α with the real axis. The angular requirement of Eq. 12-5 can be expressed as

$$\Sigma \theta_Z - \Sigma \theta_P = n 180° \quad \text{for} \quad n = \pm 1, \pm 3, \cdots \quad (12\text{-}10)$$

[2] At this point, the reader should review the plots of Figs. 12-2, 12-3, and 12-4 in the light of the first four guides.

where $\Sigma \theta_Z$ is the sum of the angles associated with the zeros, and $\Sigma \theta_P$ is the sum of the angles associated with the poles. Therefore,

$$Z\alpha - P\alpha = n180°$$

where Z is the number of zeros and P is the number of poles.

Thus, the angles that the asymptotes make with the real axis are

$$\alpha = \frac{n180°}{Z - P} \quad \text{for} \quad n = \pm 1, \pm 3, \cdots \quad (12\text{-}11)$$

To illustrate: For $G(s)H(s) = K'/(s + 1)(s + 3)$,

$$\alpha = \frac{\pm 180°}{0 - 2} = \pm 90°$$

The result is seen to agree with the plot given in Fig. 12-3c.

Guide 6. In addition to knowing the angles that the asymptotes make with the real axis, the point at which the asymptotes start must also be established. It is reasonable to expect that the asymptotes would begin at a single point representative of the entire pole-zero configuration. This point is sometimes called the *centroid* of the configuration and must lie on the real axis because poles and zeros off the real axis occur in conjugate pairs. To illustrate the idea, consider the two poles of Fig. 12-3c. The centroid is at -2 and this point is indeed where the $\pm 90°$ asymptotes start.

The discussion so far has offered no insight into determining the centroid for a configuration with both poles and zeros. One approach is to replace it with an equivalent one composed entirely of poles. Then the centroid can be established as previously illustrated. This plan of attack will be followed in obtaining an expression for the point at which the asymptotes begin.

Consider the general open-loop transfer function

$$G(s)H(s) = \frac{K'N(s)}{D(s)} = \frac{K'(s^i + a_{i-1}s^{i-1} + \cdots + a_1s + a_0)}{s^n + b_{n-1}s^{n-1} + \cdots + b_1s + b_0}$$

in which n is greater than i. Division of both numerator and denominator by $N(s)$ yields

$$G(s)H(s) = \frac{K'}{s^{n-i} + (b_{n-1} - a_{i-1})s^{n-i-1} + \cdots + \text{remainder}}$$

Now there are only poles (plus a remainder). For large values of s (far out along the asymptotes), the denominator may be approximated by the first two terms and $G(s)H(s)$ written as

$$G(s)H(s) = \frac{K'}{s^{n-i-1}[s + (b_{n-1} - a_{i-1})]}$$

Thus the original pole-zero configuration can be represented by $(n - i - 1)$

poles at zero and one pole at $-(b_{n-1} - a_{i-1})$. The centroid is then

$$\sigma_a = \frac{-(b_{n-1} - a_{i-1})}{n - i}$$

Now, recall that the sum of the roots of an equation equals the negative of the coefficient of the second-highest degree term. (See Appendix B.) Thus

$$-b_{n-1} = \Sigma P \qquad \text{and} \qquad +a_{i-1} = -\Sigma Z$$

where ΣP and ΣZ are the sums of the poles and zeros, respectively. Also

$$n = P \qquad \text{and} \qquad i = Z$$

Therefore, the point at which the asymptotes start is

$$\sigma_a = \frac{\Sigma P - \Sigma Z}{P - Z} \qquad (12\text{-}12)$$

Equation 12-12 can be used with $G(s)H(s) = K'/(s + 1)(s + 3)$; see Fig. 12-3.

$$\sigma_a = \frac{-4 - 0}{2 - 0} = -2^*$$

Guide 7. A *breakaway point* is a point at which two loci leave the real axis. Such a point always occurs between two adjacent poles when the root locus exists between them. The reason is that the loci start at each of the poles, move toward each other until they meet, and then leave the real axis in a symmetrical manner.

The determination of a breakaway point is based on angular considerations. Because each case is somewhat different, the general approach will be illustrated with a specific example. Consider the pole locations of Fig. 12-6 (which are the same as in Fig. 12-3). A breakaway point will

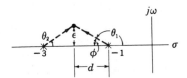

FIG. 12-6. Determination of breakaway point.

occur between -1 and -3. At a distance d to the left of -1, a trial point is selected which is above the real axis by some small distance ϵ. Thus

$$\tan \phi = \frac{\epsilon}{d} \qquad \text{and} \qquad \tan \theta_2 = \frac{\epsilon}{2 - d}$$

---

*It is worth remarking that, in general, loci do not leave the real axis at the starting point of the asymptotes (even though in Fig. 12-3 such is the case).

For small angles, the tangent of an angle can be replaced by the angle in radians. Therefore,

$$\phi = \frac{\epsilon}{d} \quad \text{and} \quad \theta_1 = \pi - \frac{\epsilon}{d} \quad \text{and} \quad \theta_2 = \frac{\epsilon}{2 - d}$$

The use of Eq. 12-10 yields

$$\Sigma\theta_Z - \Sigma\theta_P = 0 - (\theta_1 + \theta_2) = \pm\pi$$

$$0 - \left(\pi - \frac{\epsilon}{d} + \frac{\epsilon}{2 - d}\right) = \pm\pi$$

Therefore,

$$\frac{\epsilon}{d} - \frac{\epsilon}{2 - d} = 0 = \frac{1}{d} - \frac{1}{2 - d}$$

from which $d = 1$. Thus the breakaway point is at -2. Reference to Fig. 12-3 shows that -2 is indeed the point at which the loci leave the real axis.

The same approach can be used for more complicated cases. However, after some experience, the reader will find that breakaway points can be determined quickly without going through most of the preliminary steps indicated above. *Break-in points*, that is, points at which two loci join the real axis, are handled in a similar manner.

To summarize: A breakaway point is determined by selecting a trial point slightly off the real axis and applying the angular criterion.[3]

Guide 8. The angle at which the locus leaves a complex pole is called the *angle of departure* and can be obtained from angular considerations. The general approach will be illustrated with a specific example.

The poles for

$$G(s)H(s) = \frac{K'}{s(s + 1 + j1)(s + 1 - j1)}$$

are plotted in Fig. 12-7. For a point on the locus very close to the pole at $-1 + j1$, the vector angles are as shown. The angles for the vectors s and

---

[3]An alternate method is based on the fact that, at a breakaway point, the value of K' is at a maximum for the immediate portion of the real axis. Using the same illustrative example, Fig. 12-3, the characteristic equation is

$$s^2 + 4s + 3 + K' = 0$$

Solving for K',

$$K' = -s^2 - 4s - 3$$

Differentiation yields

$$\frac{dK'}{ds} = -2s - 4$$

With $dK'/ds = 0$, $s = -2$ (the breakaway point) and $K' = 1$.

The method can be used at a break-in point because there the value of K' is at a minimum for the immediate portion of the real axis.

This alternative approach is frequently of value when there are complex poles and/or zeros near the real axis.

$(s + 1 + j1)$ are 135° and 90°, respectively, while θ (which may be either positive or negative) is the angle associated with the vector $(s + 1 - j1)$.

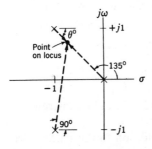

FIG. 12-7. Determination of angle of departure. (Point on locus is assumed to be very close to the pole at $-1 + j1$.)

Thus, θ is the angle of departure from the pole at $-1 + j1$. There are no zeros and so with Eq. 12-10

$$\Sigma\theta_Z - \Sigma\theta_P = 0 - (135 + 90 + \theta) = \pm 180°$$

$$\theta = -45°$$

The angle of departure is obtained by selecting a point on the locus very close to the pole in question and applying the angular criterion.

The reader should obtain the angle of departure from the other complex pole. The angle will be +45° as must be the case because of symmetry.

Guide 9. *The points at which loci cross the imaginary axis and the value of K' at crossing can be obtained from the system characteristic equation by letting $s = j\omega$.*

Consider the open-loop transfer function

$$G(s)H(s) = \frac{K'}{s(s + 3)(s + 5)}$$

The characteristic equation is

$$s^3 + 8s^2 + 15s + K' = 0$$

With $s = j\omega$,

$$j^3\omega^3 + 8j^2\omega^2 + 15j\omega + K' = 0$$

Both the real parts and the imaginary parts must equal zero:

$$-j\omega(\omega^2 - 15) = 0 \quad \text{and} \quad -8\omega^2 + K' = 0$$

from which

$$\omega = \pm(15)^{\frac{1}{2}} \quad \text{and} \quad K' = 8\omega^2 = 120$$

Thus, loci cross the imaginary axis at $\pm j(15)^{\frac{1}{2}}$ and the value of K' at these points is 120. Thus, for this case, a value of $K' = 120$ would put the closed-loop system on the threshold of instability.

Guide 10. A relationship from the study of polynomial equations is sometimes useful in scaling a root-locus plot.

With a characteristic equation in the form

$$s^n + a_{n-1}s^{n-1} + \cdots + a_1 s + a_0 = 0$$

the sum of the roots is

$$r_1 + r_2 + \cdots + r_n = -a_{n-1} \tag{12-13}$$

The application of Eq. 12-13 will be illustrated in several of the following examples.

Other Guides. In addition to the guides already developed, a tabulation of typical root-locus plots for various open-loop transfer functions is helpful in sketching the general shape of the plot for a given $G(s)H(s)$. Such tabulations can be found in a number of references [4, 14].

An aid for constructing an accurate root-locus plot is the Spirule.[4] This device is made of a transparent plastic material and consists of a protractor and a pivoted moveable arm. It provides a convenient means for angle measurement (locating points on the loci) and can also be used for scaling the plot.

It is worth remarking at this point that digital-computer programs are readily available which permit the automatic plotting of root locus plots.

Illustrative Examples. Three examples will now be presented to illustrate the procedure for obtaining root-locus plots.

EXAMPLE 12-1. Construct a root-locus plot for the open-loop transfer function

$$G(s)H(s) = \frac{K'}{s(s^2 + 2s + 2)} = \frac{K'}{s(s + 1 + j1)(s + 1 - j1)}$$

Scale several points along the loci. For what values of K' will the system be unstable?

Solution: The following information is obtained by using the guides presented.

There will be three loci. They start at 0, $-1 - j1$, $-1 + j1$, and end at infinity. A locus exists along the entire negative real axis. The use of Eq. 12-11 yields the angles of the asymptotes:

$$\alpha = \frac{n180°}{Z - P} = \frac{\pm 180°}{0 - 3} = \pm 60°$$

With Eq. 12-12, the asymptotes start at

$$\sigma_a = \frac{\Sigma P - \Sigma Z}{P - Z} = \frac{-2 - 0}{3 - 0} = -\frac{2}{3}$$

[4] A "Spirule" can be obtained from The Spirule Company, 9728 El Venado, Whittier, California.

There are no breakaway points because the locus does not leave the real axis. The angles of departure are $-45°$ and $+45°$ as determined under Guide 8. The system characteristic equation is

$$s^3 + 2s^2 + 2s + K' = 0$$

Letting $s = j\omega$,

$$j^3\omega^3 + 2j^2\omega^2 + 2j\omega + K' = 0$$

$$-j\omega(\omega^2 - 2) = 0 \quad \text{and} \quad \omega = \pm 1.414$$

$$-2\omega^2 + K' = 0 \quad \text{and} \quad K' = 4$$

Thus loci cross the imaginary axis at $\pm j 1.414$; the corresponding value of K' is 4.

The information obtained is used to draw the preliminary plot given in Fig. 12-8. Note that the same scale is used on both axes. With this much information, the root-locus plot can be sketched in. The completed plot is shown in Fig. 12-9. A Spirule was used to get better accuracy.

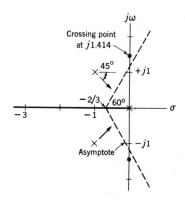

FIG. 12-8. Preliminary plot
for Example 12-1.

Scaling can now begin. From Guide 10 and the characteristic equation, the sum of the roots is -2. With $K' = 4$ (at the crossing points), two roots are $0 + j1.414$ and $0 - j1.414$. The third root is

$$0 + j1.414 + 0 - j1.414 + r_3 = -2$$

$$r_3 = -2$$

Thus, $K' = 4$ at -2 and is so marked. The value of K' at -1 can be determined from the magnitude requirement

$$K' = |s(s^2 + 2s + 2)|$$

With $s = -1$, $K' = 1$. The corresponding complex roots can be obtained since the sum of the real parts of all the roots is -2. Complex roots with real parts equal to $-\frac{1}{2}$ satisfy the requirement. A vertical line through $-\frac{1}{2}$, shown dashed in Fig. 12-9, determines the complex roots with $K' = 1$.

The system will be unstable with K' greater than 4 because roots will then lie in the right half plane. *Ans.*

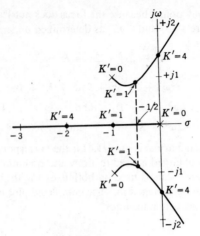

Fig. 12-9. Plot for $G(s)H(s) = K'/s(s^2 + 2s + 2)$.

Example 12-2. Construct a root-locus plot for the open-loop transfer function

$$G(s)H(s) = \frac{K'}{s(s + 1)(s + 8)}$$

For what values of K' will the system be unstable?

Solution: There will be three loci. They start at 0, -1, -8, and end at infinity. Loci exist on the real axis between 0 and -1 and between -8 and $-\infty$. Asymptote angles are:

$$\alpha = \frac{n\,180°}{Z - P} = \frac{\pm 180°}{0 - 3} = \pm 60°$$

Asymptotes start at

$$\sigma_a = \frac{\Sigma P - \Sigma Z}{P - Z} = \frac{-9 - 0}{3 - 0} = -3$$

A breakaway point will lie between the pole locations 0 and -1. A sketch showing

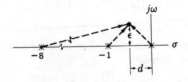

Fig. 12-10. Sketch for determining breakaway point.

a trial point just above the real axis is given in Fig. 12-10. With Eq. 12-10, the angular requirement can be expressed as

$$\Sigma \theta_Z - \Sigma \theta_P = \pm \pi$$

Therefore,

$$0 - \left(\pi - \frac{\epsilon}{d} + \frac{\epsilon}{1-d} + \frac{\epsilon}{8-d} \right) = \pm\pi$$

or

$$\frac{1}{d} - \frac{1}{1-d} - \frac{1}{8-d} = 0$$

from which a trial-and-error solution yields $d = 0.48$. There are no complex poles and thus no angles of departure. The system characteristic equation is

$$s^3 + 9s^2 + 8s + K' = 0$$

With $s = j\omega$,

$$j^3\omega^3 + 9j^2\omega^2 + 8j\omega + K' = 0$$

$$-j\omega(\omega^2 - 8) = 0 \quad \text{and} \quad \omega = \pm 2.83$$

$$-9\omega^2 + K' = 0 \quad \text{and} \quad K' = 72$$

Thus, loci cross the imaginary axis at $\pm j2.83$; the corresponding value of K' is 72. Because the sum of the roots is -9, $K' = 72$ at -9.

The completed plot is shown in Fig. 12-11. The system will be unstable with K' greater than 72.   *Ans.*

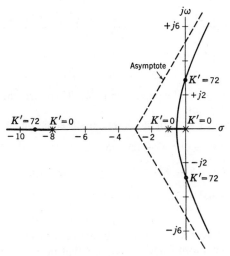

FIG. 12-11. Plot for $G(s)H(s) = K'/s(s + 1)(s + 8)$.

EXAMPLE 12-3. Construct a root-locus plot for the open-loop transfer function

$$G(s)H(s) = \frac{K'(s + 2)}{s(s + 1)(s + 8)}$$

Scale an arbitrary point on a locus using a graphical approach.

Solution: There will be three loci. They start at 0, -1, -8, and end at -2,

∞, ∞. Loci exist on the real axis between 0 and -1 and between -2 and -8. Asymptote angles are

$$\alpha = \frac{n\,180°}{Z - P} = \frac{\pm 180°}{1 - 3} = \pm 90°$$

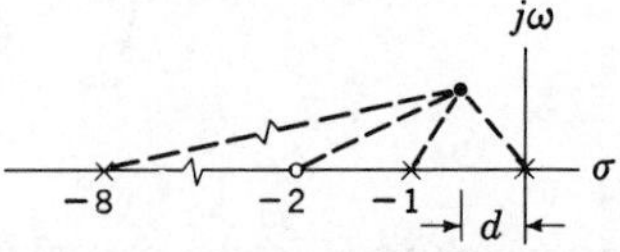

Asymptotes start at

$$\sigma_a = \frac{\Sigma P - \Sigma Z}{P - Z} = \frac{-9 - (-2)}{3 - 1} = -3\tfrac{1}{2}$$

Fig. 12-12. Sketch for determining breakaway point.

The breakaway point can be determined with the aid of the sketch given in Fig. 12-12. However, it is instructive to use the technique described in Footnote 3. The system characteristic equation is

$$s^3 + 9s^2 + (8 + K')s + 2K' = 0$$

Solution for K' yields

$$K' = \frac{-s^3 - 9s^2 - 8s}{s + 2}$$

Upon differentiation,

$$\frac{dK'}{ds} = \frac{(s + 2)(-3s^2 - 18s - 8) - (-s^3 - 9s^2 - 8s)(1)}{(s + 2)^2}$$

With $dK'/ds = 0$,

$$2s^3 + 15s^2 + 36s + 16 = 0$$

The breakaway point must lie between 0 and -1. Therefore, the root $s = -0.57$ defines the breakaway point.

The substitution of $s = j\omega$ in the characteristic equation yields

$$j^3\omega^3 + 9j^2\omega^2 + (8 + K')j\omega + 2K' = 0$$

$$-j\omega(\omega^2 - 8 - K') = 0 \quad \text{and} \quad \omega^2 = 8 + K'$$

$$-9\omega^2 + 2K' = 0 = -9(8 + K') + 2K'$$

$$-72 - 7K' = 0$$

There is no solution because K' has been assumed to be a positive real number. The conclusion that may be drawn from this result is that no loci cross the imaginary axis.

The completed plot is given in Fig. 12-13. A Spirule was used to obtain an accurate plot. It is noted that the system is stable for all values of K'.

An arbitrary point s_1 is chosen as shown in Fig. 12-13; the real part is -2. A value for K' will now be determined graphically. From the magnitude requirement

$$\left| \frac{K'(s + 2)}{s(s + 1)(s + 8)} \right| = 1$$

Therefore,

$$K' = \left| \frac{s_1(s_1 + 1)(s_1 + 8)}{s_1 + 2} \right|$$

The magnitudes for the vectors, shown as dashed lines in the figure, are scaled.

$$K' = \frac{2.82(2.22)(6.30)}{2.00} = 19.7$$

Thus, $K' = 19.7$ at point s_1.

A check can be made because, from the characteristic equation, the sum of the roots is seen to be -9. Therefore, the corresponding root on the real axis is

$$-2 + j\omega_1 - 2 - j\omega_1 + r_3 = -9$$
$$r_3 = -5$$

The value of K' at -5 is then 19.7. But is it?

$$K' = \left| \frac{s(s + 1)(s + 8)}{s + 2} \right|$$

With $s = -5$,

$$K' = \frac{5(4)(3)}{3} = 20$$

The true value for K' is 20. However, the graphical solution provided reasonable accuracy.   *Ans.*

It should be noted that the system of Example 12-3 was the same as that of Example 12-2 except that a zero at -2 was introduced to improve the performance of the system. A comparison of Figs. 12-11 and 12-13

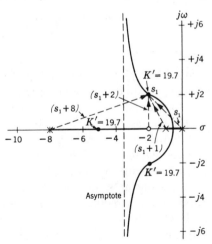

FIG. 12-13. Plot for $G(s)H(s) = K'(s + 2)/s(s + 1)(s + 8)$.

indicates the benefit obtained; the zero has the effect of drawing the loci away from the imaginary axis and instability. Further comment on system compensation will be reserved for Chapter 13.

Figure 12-14 provides sketches of four additional root-locus plots.

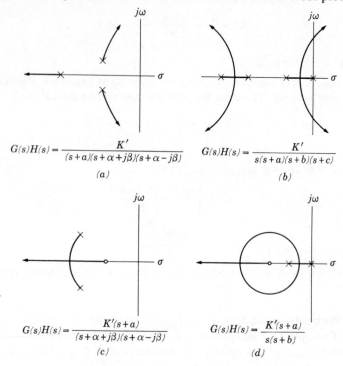

$$G(s)H(s) = \frac{K'}{(s+a)(s+\alpha+j\beta)(s+\alpha-j\beta)}$$

(a)

$$G(s)H(s) = \frac{K'}{s(s+a)(s+b)(s+c)}$$

(b)

$$G(s)H(s) = \frac{K'(s+a)}{(s+\alpha+j\beta)(s+\alpha-j\beta)}$$

(c)

$$G(s)H(s) = \frac{K'(s+a)}{s(s+b)}$$

(d)

FIG. 12-14. Sketches of root-locus plots.

12-5. Graphical Relationships

The completed root-locus plot can be used in the design process to establish the system gain for desired response characteristics. For example, the designer may wish to keep all time constants smaller than some particular value or to keep damping ratios greater than some assigned value. To do this, he must locate the roots of the characteristic equation at points on the s plane corresponding to the desired values of time constant or damping ratio. Thus, a knowledge of the locations on the s plane corresponding to various time constants, damping ratios, undamped natural frequencies, and damped natural frequencies is required. These graphical relationships will now be developed.

The closed-loop transfer function for a system may be written as

$$\frac{C(s)}{R(s)} = \frac{N(s)}{D(s)} = \frac{N(s)}{(s + r_1)(s + r_2) \cdots (s + r_n)}$$

in which $D(s)$ is the characteristic function in factored form and $-r_1$, $-r_2, \ldots, -r_n$ are the roots of the system characteristic equation. The roots may be real or complex numbers. Therefore,

$$\frac{C(s)}{R(s)} = \frac{N(s)}{(s + a)(s + b + jg)(s + b - jg) \cdots} \tag{12-14}$$

The time response is

$$c(t) = C_1 e^{-at} + C_2 e^{-bt} \sin(gt + \phi) + \cdots^5$$

or

$$c(t) = C_1 e^{-t/\tau_1} + C_2 e^{-t/\tau_2} \sin(\omega_d t + \phi) + \cdots$$

or

$$c(t) = C_1 e^{-t/\tau_1} + C_2 e^{-\zeta\omega_n t} \sin[\omega_n(1 - \zeta^2)^{1/2} t + \phi] + \cdots$$

where τ_1 and τ_2 are time constants, ω_d is damped natural frequency, ζ is the damping ratio, and ω_n is the natural frequency. Thus, it is seen that Eq. 12-14 can be expressed as either

$$\frac{C(s)}{R(s)} = \frac{N(s)}{(s + 1/\tau_1)(s + 1/\tau_2 + j\omega_d)(s + 1/\tau_2 - j\omega_d) \cdots} \tag{12-15}$$

or

$$\frac{C(s)}{R(s)} = \frac{N(s)}{[s + 1/\tau_1][s + \zeta\omega_n + j\omega_n(1 - \zeta^2)^{1/2}][s + \zeta\omega_n - j\omega_n(1 - \zeta^2)^{1/2}] \cdots} \tag{12-16}$$

A comparison of Eqs. 12-15 and 12-16 with Eq. 12-14 reveals that lines parallel to the imaginary axis can be considered as lines of constant $1/\tau$ (and consequently lines of constant τ) or as lines of constant $\zeta\omega_n$. Such a line is shown in Fig. 12-15a. Further, lines parallel to the real axis are lines of constant ω_d (Fig. 12-15a).

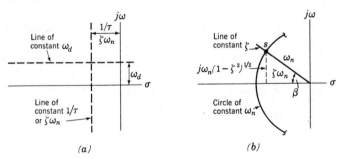

FIG. 12-15. Graphical relationships on the s plane.

Lines of constant ζ are radial lines from the origin. In Fig. 12-15b, an arbitrary point $s = -\zeta\omega_n + j\omega_n(1 - \zeta^2)^{1/2}$ is plotted. From the figure,

$$\cos \beta = \zeta \tag{12-17}$$

Equation 12-17 can be used to determine the inclination of lines of constant ζ. Figure 12-15b also reveals that circles about the origin are circles of constant ω_n.

---

[5]The additional terms indicated include other similar transient terms and also any related to the type of input.

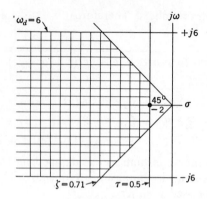

FIG. 12-16. Use of graphical
relationships.

The use of the graphical relationships is illustrated in Fig. 12-16. The roots of the characteristic equation must lie within the shaded area if the requirements are

$$\tau \leq 0.5 \text{ sec} \qquad (\sigma = -1/0.5 = -2)$$
$$\omega_d \leq 6 \text{ radians/sec}$$
$$\zeta \geq 0.71 \qquad (\beta = \cos^{-1} 0.71 = 45°)$$

12-6. Setting the System Gain

The procedure for setting system gains will be illustrated with the control system of Fig. 12-17. A value for the controller gain K_i is to be obtained such that (as required by system specifications) either the maximum time constant is 1 sec or the minimum damping ratio is 0.5.

The open-loop transfer function for the system is[6]

$$G(s)H(s) = \frac{5K_i}{s(0.25s + 1)(0.1s + 1)}$$
$$= \frac{200K_i}{s(s + 4)(s + 10)} = \frac{K'}{s(s + 4)(s + 10)} \qquad (12\text{-}18)$$

The root-locus plot, which is similar to that of Example 12-2, is given in Fig. 12-18. A line is drawn passing through -1 and parallel to the imaginary axis; it is the line of $1/\tau = 1$ or $\tau = 1$. With Eq. 12-17, $\cos \beta = \zeta = 0.5$ and $\beta = 60°$. A radial line at an angle of $60°$ with the negative real axis provides the line of $\zeta = 0.5$.

It is seen that the maximum allowable system gain is limited by the damping ratio requirement. Therefore, the point s_1 on the locus at which

---

[6]Note that $G(s)H(s)$ is the same for either a reference input or a disturbance input.

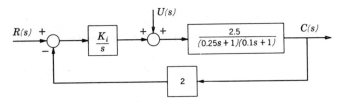

FIG. 12-17. System block diagram.

the line of $\zeta = 0.5$ crosses is scaled to determine the value of K'. From Eq. 12-18

$$K' = |s(s + 4)(s + 10)| = |s_1(s_1 + 4)(s_1 + 10)|$$

The vector lengths are measured on the plot with the result

$$K' = 2.85(3.55)(8.90) = 90$$

The controller gain K_i is now found to be

$$200K_i = K' = 90$$

$$K_i = 0.45 \qquad\qquad (12\text{-}19)$$

With $K_i = 0.45\ (K' = 90)$ the roots of the characteristic equation (from the

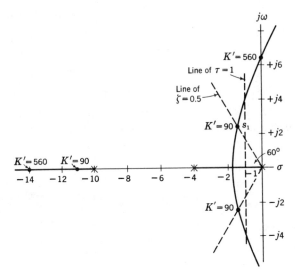

FIG. 12-18. Plot for $G(s)H(s) = K'/s(s + 4)(s + 10)$.

plot) are -11.1 and $-1.45 \pm j2.47$. Thus, the maximum time constant is $1/1.45 = 0.69$ sec and the damped natural frequency is 2.47 radians/sec.

12-7. System Transient Response

The root-locus plot can also be used to determine the transient response of a system. The approach will now be illustrated.

Consider the response of the system of Fig. 12-17 (with K_i = 0.45 as

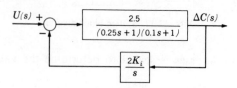

FIG. 12-19. Block diagram with disturbance input.

previously determined) to a unit step disturbance input. The appropriate block diagram is given in Fig. 12-19.

$$G(s) = \frac{2.5}{(0.25s + 1)(0.1s + 1)} = \frac{100}{(s + 4)(s + 10)}$$

$$H(s) = \frac{2K_i}{s}$$

$$G(s)H(s) = \frac{200K_i}{s(s + 4)(s + 10)} = \frac{K'}{s(s + 4)(s + 10)}$$

The root-locus plot is that of Fig. 12-18. With K_i = 0.45 (K' = 90), the closed-loop transfer function is

$$\frac{\Delta C(s)}{U(s)} = \frac{G(s)}{1 + G(s)H(s)} = \frac{100s}{s(s + 4)(s + 10) + 90} \qquad (12\text{-}20)$$

With a unit step disturbance, all initial conditions zero, and the known roots of the characteristic equation (with K' = 90 from Fig. 12-18)

$$\Delta C(s) = \frac{1}{s}\left[\frac{100s}{(s + 11.1)(s + 1.45 + j2.47)(s + 1.45 - j2.47)}\right]$$

$$\Delta C(s) = \frac{C_1}{s + 11.1} + \frac{C_2}{s + 1.45 + j2.47} + \frac{C_3}{s + 1.45 - j2.47}$$

From Chapter 6,

$$C_1 = \frac{100}{(s + 1.45 + j2.47)(s + 1.45 - j2.47)}\bigg|_{s = -11.1}$$

$$C_2 = \frac{100}{(s + 11.1)(s + 1.45 - j2.47)}\bigg|_{s = -1.45 - j2.47}$$

$$C_3 = \frac{100}{(s + 11.1)(s + 1.45 + j2.47)}\bigg|_{s = -1.45 + j2.47}$$

The constants can be evaluated by obtaining the vector quantities graphically from the root-locus plot (Fig. 12-18). Figure 12-20 shows how the

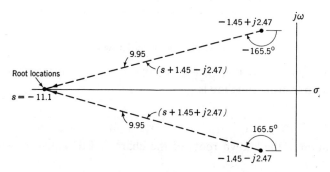

FIG. 12-20. Obtaining vector quantities graphically.

necessary information for the evaluation of C_1 is determined from the root-locus plot. Therefore,

$$C_1 = \frac{100}{(9.95\,\underline{/165.5°})(9.95\,\underline{/-165.5°})} = 1.01$$

In a similar manner,

$$C_2 = \frac{100}{(9.95\,\underline{/-14.5°})(4.95\,\underline{/-90°})} = 2.03\,\underline{/104.5°}$$

$$C_3 = \frac{100}{(9.95\,\underline{/14.5°})(4.95\,\underline{/90°})} = 2.03\,\underline{/-104.5°}$$

The transient response, by the method of Chapter 6, is then found to be

$$\Delta c(t) = 1.01e^{-11.1t} + 4.06e^{-1.45t}\sin(2.47t - 14.5°) \qquad (12\text{-}21)$$

12-8. System Frequency Response

Although the root-locus plot is generally thought of as being related to the transient response approach to system dynamics, it can also be used to obtain frequency response information. The procedure involved will now be illustrated.

The system block diagram of Fig. 12-21 is that of Fig. 12-17 except that $U(s)$ is zero. The controller gain K_i is 0.45 as previously determined in Sec. 12-6. From Fig. 12-21,

$$G(s) = \frac{2.5K_i}{s(0.25s + 1)(0.1s + 1)} = \frac{100K_i}{s(s + 4)(s + 10)}$$

$$H(s) = 2$$

$$G(s)H(s) = \frac{200K_i}{s(s + 4)(s + 10)} = \frac{K'}{s(s + 4)(s + 10)} \qquad (12\text{-}22)$$

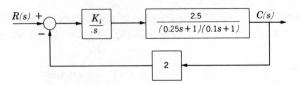

FIG. 12-21. System block diagram.

The root-locus plot is that of Fig. 12-18. With $K_i = 0.45$ ($K' = 90$), the closed-loop transfer function is

$$\frac{C(s)}{R(s)} = \frac{G(s)}{1 + G(s)H(s)} = \frac{45}{s(s + 4)(s + 10) + 90} \tag{12-23}$$

Substitution of the known roots of the characteristic equation with $K' = 90$ yields

$$\frac{C(s)}{R(s)} = \frac{45}{(s + 11.1)(s + 1.45 + j2.47)(s + 1.45 - j2.47)} \tag{12-24}$$

Frequency-response information can be obtained from Eq. 12-24 by letting $s = j\omega$. However, because the roots of the characteristic equation are already plotted on the s plane of Fig. 12-18, the graphical approach of Sec. 9-5 can be used. The vector quantities are scaled on the root-locus plot

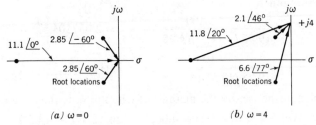

(a) $\omega = 0$ (b) $\omega = 4$

FIG. 12-22. Graphical determination of system frequency response.

as illustrated by the sketches of Fig. 12-22. Therefore, at $\omega = 0$,

$$\frac{C(j\omega)}{R(j\omega)} = \frac{45}{(11.1)(2.85)(2.85)\underline{/0° + 60° - 60°}} = 0.499\underline{/0°} \tag{12-25}$$

At $\omega = 4$,

$$\frac{C(j\omega)}{R(j\omega)} = \frac{45}{(11.8)(6.6)(2.1)\underline{/20° + 77° + 46°}} = 0.275\underline{/-143°} \tag{12-26}$$

Other values can be obtained in a similar manner and the results plotted.

Note that the magnitude of $C(j\omega)/R(j\omega)$ with $\omega = 0$ should be 0.5 (instead of the usual 1.0) because $H(s) = 2$ in the block diagram of Fig. 12-21. The value obtained graphically, Eq. 12-25, is 0.499—better accuracy than is usually expected.

12-9. Closure

The root-locus approach to control system analysis and design is quite versatile as has been demonstrated in this chapter. Not only can a system gain be established for the desired response characteristics, but both transient and frequency response information can be obtained. After some experience with the method, the control engineer can quickly evaluate the effects of possible changes in system parameters, including the addition of other elements to the system, by visualizing the plots resulting from various pole-zero configurations. Thus, the root-locus method is a tool of considerable value to the control engineer.

Problems

NOTE: In making root-locus plots, be sure to use the same scale on both the real and imaginary axes.

12-1. Construct root-locus plots for the following open-loop transfer functions. Scale several points; check the results by solving the system characteristic equations.

a) $G(s)H(s) = \dfrac{K'}{s}$

d) $G(s)H(s) = \dfrac{K'}{s^2 + 2s + 5}$

b) $G(s)H(s) = \dfrac{K'}{s^2}$

e) $G(s)H(s) = \dfrac{K'}{s(s + 4)}$

c) $G(s)H(s) = \dfrac{K'(s + 1)}{s + 5}$

f) $G(s)H(s) = \dfrac{K'(s + 3)}{(s + 1)(s + 6)}$

12-2. Using the guides developed in Sec. 12-4, sketch root-locus plots for the following transfer functions. Scale several points. For what values of K' will the systems be unstable?

a) $G(s)H(s) = \dfrac{K'}{s(s^2 + 4s + 8)}$

b) $G(s)H(s) = \dfrac{K'}{(s + 3)(s^2 + 2s + 2)}$

c) $G(s)H(s) = \dfrac{K'}{(s + 1)(s + 3)(s + 5)}$

d) $G(s)H(s) = \dfrac{K'(s + 1)}{s^2 + 4s + 8}$

12-3. Sketch root-locus plots for the following transfer functions. Scale several points. For what values of K' will the systems be unstable?

a) $G(s)H(s) = \dfrac{K'(s + 2)}{s^2}$

b) $G(s)H(s) = \dfrac{K'}{s(s + 2)(s + 4)(s + 6)}$

12-4. For each of the systems shown in the accompanying illustration, sketch the root-locus plot. Determine the value of K' required for a damping ratio of 0.5. With K' set, what will be the maximum time constant and the damped natural frequency?

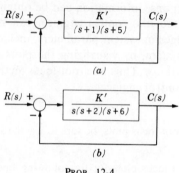

(a)

(b)

PROB. 12-4

12-5. Obtain the transient response, for a unit step change of reference input, for the systems of a) Prob. 12-4a and b) Prob. 12-4b. The values of K' are to be those determined in Prob. 12-4.

12-6. Obtain frequency response data, for several values of ω, for the systems of a) Prob. 12-4a and b) Prob. 12-4b. Plot the results. The values of K' are to be those determined in Prob. 12-4.

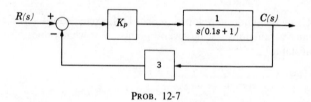

PROB. 12-7

12-7. For the system shown in the accompanying illustration, determine the controller gain K_p required for a damping ratio of 0.71.

Chapter 13

System Compensation

13-1. Introduction

Automatic control systems are designed and constructed to perform specific functional tasks. As illustrated in previous chapters, this may mean the control of position, velocity, temperature, or some other physical quantity. The conception of a control system design starts by defining the output variable (e.g. position, speed, temperature, pressure, and so forth) and then determining the required specifications for the control of that variable. As pointed out in Chapter 1, these specifications relate to stability, accuracy, and speed of response. In the design process the control engineer might first choose the control media (e.g. mechanical, electrical, hydraulic, or pneumatic) and then the basic control system elements that can be assembled to meet the desired end. In actual practice, several alternatives may be analyzed, and the final judgment made on the basis of projected over-all performance and economy. Knowing the transfer functions for the elements, a "paper" system may be set up in block diagram form, the performance analyzed, and the system design gain established by the techniques introduced in Chapters 10, 11, and 12.

The problem that often arises at this point, however, is that the adjustment of system gain alone does not provide sufficient alteration of the control system's behavior to meet the desired set of specifications. Or, for example, the combination of control elements may be unstable for all settings of gain. Or, perhaps, an over-all gain constant sufficiently high to reduce the steady-state error of the controlled variable to an acceptable value also results in an unstable system. It has been pointed out that increasing the gain generally reduces system error and improves the speed of response, but this is usually at the expense of the degree of stability.

When gain adjustment alone is found not to suffice in the design of a control system, it is necessary (when using frequency-response techniques as will be the case in this chapter) to modify the actual shape of the plot of $G(j\omega)H(j\omega)$. This modification may be accomplished by redesign of the system and control elements when it is physically and economically possible to do so. When modification of the basic elements is not possible or desirable, reshaping of the $G(j\omega)H(j\omega)$ plots may be accomplished by placing a *compensating element* in the system. A compensating element is a device having a particular transfer function which favorably alters the over-all transfer relationship of the system. Compensating elements may be placed in series with the forward transfer function, *series compensation*; or placed in an inner feedback path around some system element or elements, *feedback compensation* (also commonly called parallel compensation).

Compensating devices that may be used for either series or parallel compensation have been categorized in control literature in three broad groups. They include: phase-lag controllers, phase-lead controllers,[1] and controllers which provide a combination of lead and lag (or vice versa) called compound controllers. The basic nature of different compensating elements and their effect on system behavior when located at various places in the system is the subject of this chapter. The application of system compensation concepts is exemplified by the illustrative examples presented.

Control engineers have developed a number of techniques to improve the performance of a system or to force a system to behave in a desired manner. Cascade control, which is a form of feedback compensation, is an example. Feedforward compensation is another example. Further, it is possible to establish a procedure for the design of a controller such that the overall closed-loop response will have a particular form. These techniques will be discussed in the material to follow.

13-2. General Considerations for Choosing Compensating Characteristics

There are many specific criteria which can be used to determine the form of a compensating element and the particular numerical values for its parameters. However, it is possible to make some generalizations about the characteristics that nearly all control systems should possess. These generalized characteristics may then serve as guides for choosing specific compensating elements.

This section will discuss the general considerations for choosing the character of a compensating element based on the frequency response of the control system in question. Typically, the discussion here refers to frequency-response information presented on Bode diagrams.

First, in the lower range of frequency (frequencies below the crossover point) it is desirable that the output of the closed-loop system follow the reference input with a minimum of error. In particular, at zero frequency, the steady-state error should be reduced to an absolute minimum. For the open-loop system on the Bode diagram, this corresponds to as high a db level at low frequencies as possible. This concept will be illustrated in a later example.

---

[1] The behavior of these elements and their Bode plots was discussed in Sec. 11-4 in the discussion of simple lag and simple lead, respectively.

Second, in the middle range of frequencies (frequencies near the cross-over point), the slope of the attenuation curve should be limited to -20 db/decade or -6 db/octave. This generalization refers primarily to systems which are potentially unstable systems (e.g., systems which may be unstable for certain gains or locations of the break points). Where system stability is a function of the gain setting and the location of the break points, this criterion will almost certainly ensure stability if the -20 db/decade slope extends for a decade above and below the crossover point.

Third, at high frequencies (frequencies above the crossover point), it is generally desirable to attenuate as rapidly as possible, that is, the greater the negative slope of the attenuation curve the better. The purpose here is to reduce the effect of any high-frequency noise or high-frequency disturbances which may find their way into the system. Electronically, noise may be generated within circuits, or disturbances may be picked up by sensitive electronic equipment. Mechanically, gear teeth and external torque fluctuations give rise to undesirable signals that may be eliminated if there is sufficient attenuation at the proper frequency.

It is perhaps appropriate at this time to point out that system compensation is one area of study of automatic controls that should not, in general, be attacked by establishing a rigid set of rules. Rather, the compensation of each new system requires the creative application of the basic principles to obtain a system that meets the stated set of specifications. The approach taken in the remainder of this chapter is to present some of the concepts and then to illustrate some types of problems that arise by carrying through illustrative examples.

13-3. Series Compensation

The general class of series compensation devices is illustrated in Fig. 13-1 where the compensating element $G_c(s)$ is placed in series with the for-

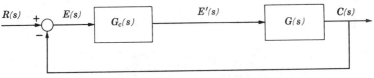

FIG. 13-1. Series compensation in a control system.

ward transfer function $G(s)$. The compensating controller may be either a simple lag, simple lead, or a compound controller depending on the nature of the modification needed to make the system perform as desired.

Figure 13-2 illustrates some of the possibilities for transfer functions which may be used in series compensating blocks. Beside each block diagram is the asymptotic Bode attenuation diagram for that specific compensating element. Figure 13-2*a* illustrates the use of a lead network which might typically improve the response at frequencies near or above crossover or might be utilized to allow a higher system gain setting and still maintain a desired degree of stability. Figure 13-2c illustrates a simple lag which is generally of lesser utility in compensation.

The transfer functions given in Figs. 13-2e and g are examples of compound compensating elements which are commonly found in control systems. The "proportional plus integral" device in Fig. 13-2e acquires its name by virtue of the fact that the output of the element is equal to a

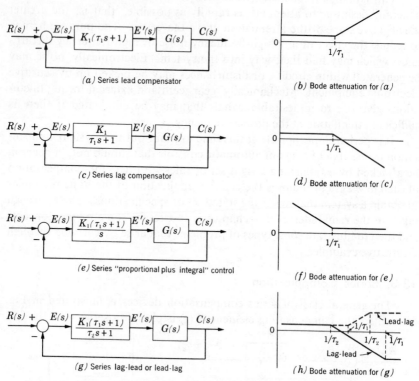

(a) Series lead compensator

(b) Bode attenuation for (a)

(c) Series lag compensator

(d) Bode attenuation for (c)

(e) Series "proportional plus integral" control

(f) Bode attenuation for (e)

(g) Series lag-lead or lead-lag

(h) Bode attenuation for (g)

Fig. 13-2. Typical possibilities for series compensation blocks and their asymptotic Bode attenuation diagrams.

constant times the input plus a constant times the integral of the input. Thus, if $E(s)$ is the input and $E'(s)$ is the output,

$$E'(s) = K_p E(s) + K_i \frac{E(s)}{s} = \frac{K_p s + K_i}{s} E(s)$$

or

$$\frac{E'(s)}{E(s)} = K_1 \frac{(\tau_1 s + 1)}{s} \tag{13-1}$$

It will be shown in a later example that this series compensator is extremely useful for improving the low-frequency response of a type 1 system.

Figure 13-2g illustrates the use of a commonly encountered compound compensating element having the property of a lag-lead or a lead-lag depending on the values of the time constants τ_1 and τ_2. If $\tau_1 > \tau_2$, the compensation is lead-lag; whereas if $\tau_1 < \tau_2$, the compensation is lag-lead. Each of these cases is shown in Fig. 13-2h.

An example will illustrate the application of series compensation.

EXAMPLE 13-1. Consider the basic control system of Fig. 13-3a. The control-system element has the characteristic of a double integration or, as it is sometimes expressed, a double pole at the origin. The problem is to design a compensating element that may be placed in series with the system element to provide stable operation and a maximum magnitude ratio of $M_m = 1.26$ (2.0 db) at about 100 radians/sec.

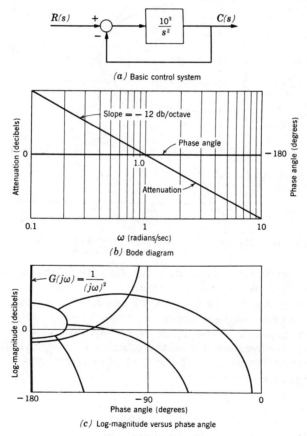

(a) Basic control system

(b) Bode diagram

(c) Log-magnitude versus phase angle

FIG. 13-3. Simple control system and plots for Example 13-1.

Solution: Figure 13-3*b* is a Bode diagram for the normalized system element $G_N(j\omega) = 1/(j\omega)^2$. It is obvious that there is no value of gain for which the system can be stable because the phase angle is always $-180°$ and thus the phase margin is always $0°$. A log-magnitude versus phase angle plot is given in Fig. 13-3*c*. To meet the specifications established, it is necessary to give the system some positive phase shift at lower frequencies so that there will be some positive phase margin at the 0 db crossover point. In particular, the phase shift must be sufficient to make the log-magnitude versus phase angle plot loop to the right and become tangent to the M contour of 2 db as shown in Fig. 13-4. This would ensure a maximum magnitude ratio of $M_m = 2.0$ db or $M_m = 1.26$. It was specified that this

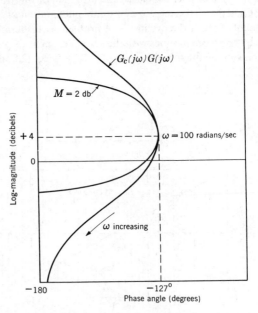

FIG. 13-4. Desirable log-magnitude plot
for the given specifications.

tangency should take place at a frequency of $\omega = 100$ radians/sec. Further, it is advantageous to have the phase angle return to $-180°$ at high frequencies. This is often desirable because it means that there will be greater attenuation of any higher frequency noise or disturbance signals.

The question now before the designer is, What type of series compensator will provide a positive phase shift at low frequencies and a negative phase shift at high frequencies? The answer is, of course, a lead-lag element as pictured in Fig. 13-2*g* with the transfer relation

$$\frac{E'(s)}{E(s)} = \frac{K_1(\tau_1 s + 1)}{\tau_2 s + 1} = G_c(s)$$

The problem now is to determine the values of K_1, τ_1, and τ_2 such that the desired specifications are met. It is seen in Fig. 13-4 that if a symmetrical phase shift is obtained about ω = 100 radians/sec, the plot of $G_c(j\omega)G(j\omega)$ will be tangent to M = 2.0 db at the right extreme of the contour. This corresponds to

$$20 \log |G_c(j\omega)G(j\omega)| = 4 \text{ db}$$

at an angle of $-127°$ obtained from Fig. 13-4.

If the tangency is to occur at a frequency of ω = 100 radians/sec, then the lead-lag compensating element must provide a net $+53°$ ($180° - 127°$) phase shift at 100 radians/sec. For symmetry of the phase shift about 100 radians/sec, it is necessary to choose τ_1, n times greater than $1/100$ sec, and τ_2, $1/n$ times $1/100$ sec where n is a number greater than one. By plotting the maximum phase shift obtained by the lead-lag element versus n, the correct value of n may be chosen to give the required $+53°$ phase lead at 100 radians/sec. Such a curve is given in Fig. 13-5.

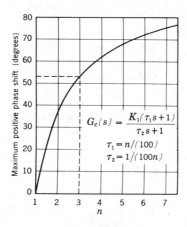

FIG. 13-5. Plot of maximum positive phase shift as a function of n for a lead-lag element.

For a value of n = 3, it is seen that the phase angle is $+53.1°$ at ω = 100 radians/sec and thus is the correct value for this case. The values for the time constants are therefore τ_1 = 0.03 sec and τ_2 = 0.00333 sec, and the break frequencies become $\omega_1 = 1/\tau_1$ = 33.3 radians/sec and $\omega_2 = 1/\tau_2$ = 300 radians/sec. The normalized transfer function for the compensating element is then

$$G_{cN}(j\omega) = \frac{0.03 j\omega + 1}{0.00333j\omega + 1}$$

The Bode diagram for this transfer function is given in Fig. 13-6. With the transfer function normalized, the attenuation is 0 db below the first break frequency. At ω = 33.3 radians/sec, the slope changes to $+$ 6 db/octave and continues until the second break frequency is reached at ω = 300 radians/sec.

The over-all open-loop transfer function of the compensated system is

$$G_c(j\omega)G(j\omega) = \frac{K_1 10^3(0.03j\omega + 1)}{(j\omega)^2(0.00333j\omega + 1)}$$

The normalized transfer function is

$$G_{cN}(j\omega)G_N(j\omega) = \frac{0.03j\omega + 1}{(j\omega)^2(0.00333j\omega + 1)}$$

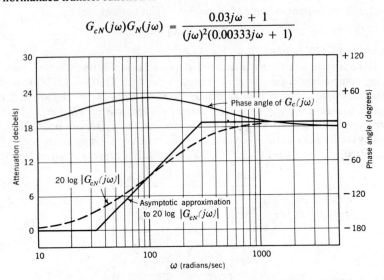

FIG. 13-6. Bode diagram for the compensating element of Example 13-1.

The composite attenuation plot is that given by curve (*a*) of Fig. 13-7. As pointed out previously, the plot must pass through the point of +4 db and −127° phase shift at $\omega = 100$ radians/sec. To do this, the attenuation plot must be shifted upward

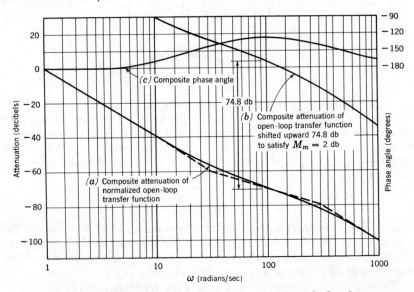

FIG. 13-7. Bode diagram for open-loop compensated transfer functions.

74.8 db in Fig. 13-7. This corresponds to an over-all system design gain of

$$K_{\text{design}} = 5.5 \times 10^3$$

Thus, $K_1 = 5.5$. The actual open-loop transfer function attenuation plot is given in curve (*b*) of Fig. 13-7. The final system block diagram is given in Fig. 13-8.   Ans.

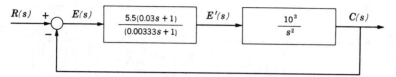

Fig. 13-8. System block diagram for compensated control system of Example 13-1.

It is valuable at this point to extend the analysis to look at the actual log magnitude versus phase angle curve that has been obtained and at the resulting closed-loop response on the Bode diagram.

The log-magnitude versus phase angle curve is plotted on the Nichols chart of Fig. 13-9 by choosing several frequencies and transferring the attenuation and phase angle data from curves (*b*) and (*c*) of Fig. 13-7. The closed-loop response is then constructed from the intersection of this plot with the *M* and *N* contours shown on Fig. 13-9. The final result is the closed-loop frequency response of the Bode diagram of Fig. 13-10. It may be observed that the closed-loop response has a maximum value of 2 db at $\omega = 100$ radians/sec. It may be further noticed that the frequency response of this system is "flat" and there is very little phase shift up to 20 radians/sec.

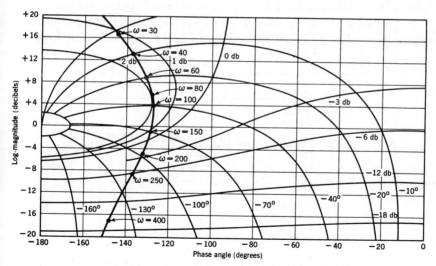

Fig. 13-9. Log-magnitude plot for open-loop transfer function of Example 13-1.

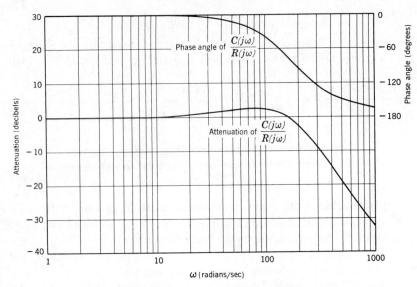

FIG. 13-10. Closed-loop response for Example 13-1.

13-4. Feedback Compensation and Cascade Control

It was pointed out in the introductory remarks of this chapter that an alternative to placing a compensating element in series with the system element is to feed back around the system element and place a compensating device in this feedback path (often termed *parallel compensation*). Feedback compensation may be used to improve system stability, reduce steady-state error, and improve the speed of response and frequency response.

Feedback Compensation. The concept of feedback compensation is illustrated in Fig. 13-11, wherein $G(s)$ in Fig. 13-11a is the transfer func-

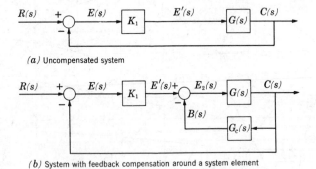

(a) Uncompensated system

(b) System with feedback compensation around a system element

FIG. 13-11. Feedback compensation.

tion of one or more of the system elements. An inner feedback path may be constructed around $G(s)$ with the compensating element $G_c(s)$ inserted, as shown in Fig. 13-11*b*. The use of feedback compensation as illustrated here results in the creation of a multiple-loop system.

The form of the compensating element $G_c(s)$ transfer function may be similar to any of those illustrated by Fig. 13-2 in the discussion of series compensation, depending on the nature of the compensation required. Consider the transfer relation for the inner loop of Fig. 13-11*b*:

$$\frac{C(s)}{E'(s)} = \frac{G(s)}{1 + G(s)G_c(s)} \tag{13-2}$$

The block diagram of Fig. 13-11*b* may now be replaced by the block diagram of Fig. 13-12 where the transfer functions of the various elements are lumped into a simple block.

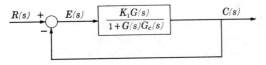

FIG. 13-12. Lumped compensated system.

Tachometer Feedback as a Compensating Device. Tachometer feedback is so commonly found in practical positioning control systems that it is worthy of special attention in this discussion of feedback compensation. If the output of a certain element is rotary position θ_o, a tachometer connected to the output provides a signal proportional to the velocity. Thus, the transfer function for a tachometer on a positioning device is given by

$$\frac{B(s)}{\Theta_o(s)} = K_t s \tag{13-3}$$

A typical simplified block diagram of a positioning system is given in Fig. 13-13. Consider for purposes of illustration that the transfer function $G(s) = K_2/s^2$. This was the transfer function chosen for Example 13-1 and it was established that closing the loop by an outer unity feedback path results in a system that is always unstable. Using the form of Eq. 13-2,

$$\frac{\Theta_o(s)}{E'(s)} = \frac{K_2}{s(s + K_2 K_t)} = \frac{1/K_t}{s[(1/K_2 K_t)s + 1]} \tag{13-4}$$

and the entire forward transfer function may be lumped into one block giving

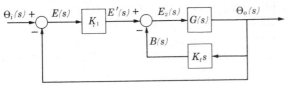

FIG. 13-13. Typical positioning system with internal tachometer feedback.

$$\frac{\Theta_o(s)}{E(s)} = \frac{K_1/K_t}{s(\tau_1 s + 1)} \tag{13-5}$$

where $\tau_1 = 1/K_2K_t$.

The lumped system with tachometer feedback is represented in the block diagram of Fig. 13-14. It should be readily apparent at this point that the open-loop transfer function of Eq. 13-5 is that of a stable system. Further, it may be verified, by expansion of $\Theta_o(s)/\Theta_i(s)$, that when a unity feedback loop is placed around the forward transfer function, Eq. 13-5, the closed-loop response is that of a second-order damped system and that

$$\zeta = \frac{K_t}{2} (K_2/K_1)^{1/2} \tag{13-6}$$

It is apparent that tachometer feedback is a source of damping in the system. Further, the time constant of the system with tachometer feedback is $1/(K_2K_t)$. Thus increasing the tachometer feedback reduces the time constant and improves the speed of response for a particular allowable gain setting.

It should be emphasized that tachometer feedback is a specific type of feedback belonging to a wide class of *rate feedback* devices. The tachometer output is a signal proportional to the rate of change of a position. Similarly, a rate gyro in aircraft auto-pilots serves the same purpose.

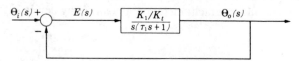

FIG. 13-14. Reduced system block diagram.

Cascade Control. Cascade control is a form of feedback compensation that arose in the process industry. In principle, cascade control involves manipulating the set-point of one controller by an additional feedback loop. Thus, cascade control involves two feedback loops, an outer loop and an inner loop which modifies the outer loop. The result of cascade control is, of course, improved response, particularly in regard to disturbances within the inner loop.

As an example, consider the two-tank system illustrated in Fig. 13-15a. The block diagram of the tandem tank setup is shown in Fig. 13-15b. The system is subject to a disturbance in the form of pressure fluctuations upstream of the control valve. Figure 13-16 illustrates the tank system with a single feedback loop and a proportional controller. For disturbance inputs $Q_u(s)$ resulting from pressure fluctuations, the closed-loop transfer function is

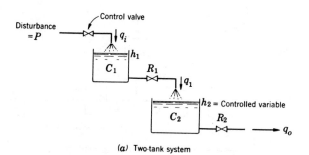

(a) Two-tank system

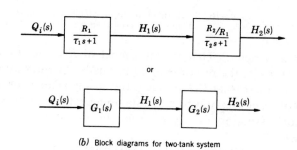

(b) Block diagrams for two-tank system

FIG. 13-15. A two-tank system with a supply
pressure disturbance.

$$\frac{\Delta H_{2c}(s)}{Q_u(s)} = \frac{G_1(s)\,G_2(s)}{1 + G_1(s)\,G_2(s)\,K_1 K_v K_f} \qquad (13\text{-}7)$$

Substituting the transfer functions of the tanks,

$$\frac{\Delta H_{2c}(s)}{Q_u(s)} = \frac{R_2}{\tau_1 \tau_2 s^2 + (\tau_1 + \tau_2)s + (1 + R_2 K_1 K_v K_f)} \qquad (13\text{-}8)$$

The system undamped natural frequency is then given by

$$\omega_n = \sqrt{\frac{1 + R_2 K_1 K_v K_f}{\tau_1 \tau_2}} \qquad (13\text{-}9)$$

and the damping ratio is

$$\zeta = \frac{\tau_1 + \tau_2}{2\sqrt{\tau_1 \tau_2 (1 + R_2 K_1 K_v K_f)}} \qquad (13\text{-}10)$$

Steady-state deviations in the second tank liquid level due to a unit step
change of disturbance input are obtained using the final value theorem

$$\Delta H_{2css} = \lim_{s \to 0} s\,\Delta H_{2c}(s) = \frac{R_2}{1 + R_2 K_1 K_v K_f} \qquad (13\text{-}11)$$

The parameters R_2, τ_1, τ_2, and K_v are fixed for a given tank system. The
product of the controller gain K_1 and feedback transducer K_f is the only

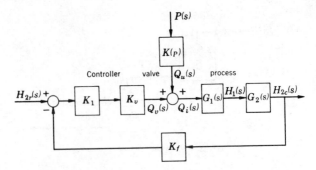

FIG. 13-16. Block diagram using single loop and
proportional controller.

controller parameter. That is, adjustment of $K_1 K_f$ determines both the
transient nature of the response and the steady-state deviation due to the
disturbance. Increasing $K_1 K_f$ reduces the steady-state error and increases
the speed of response; however, the damping ratio is reduced making the
system more oscillatory for a given tolerable deviation or controlled variable off-set.

Figure 13-17 illustrates a block diagram for the same tank system

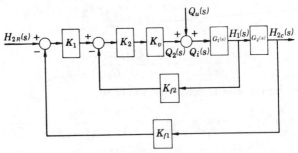

FIG. 13-17. Block diagram with cascade control.

with cascade control. Reducing the block diagram for the case where
$H_{2r} = 0$, yields

$$\frac{\Delta H_{2c}(s)}{Q_u(s)} = \frac{R_2}{\tau_1 \tau_2 s^2 + (\tau_1 + \tau_2 + K_2 K_v K_{f2} R_2 \tau_2) s + (1 + K_2 K_v K_{f2} R_2 + K_1 K_2 K_v K_{f1} R_2)} \qquad (13\text{-}12)$$

from which

$$\omega_n = \sqrt{\frac{1 + K_2 K_v K_{f2} R_2 + K_1 K_2 K_v K_{f1} R_2}{\tau_1 \tau_2}} \qquad (13\text{-}13)$$

and

$$\zeta = \frac{\tau_1 + \tau_2 + K_2 K_v K_{f2} R_2 \tau_2}{2 \sqrt{\tau_1 \tau_2 (1 + K_2 K_v K_{f2} R_2 + K_1 K_2 K_v K_{f1} R_2)}} \qquad (13\text{-}14)$$

The steady-state deviation to a unit step disturbance of flow input is given by

$$\Delta H_{2css} = \lim_{s \to 0} s\,\Delta H_{2c}(s) = \frac{R_2}{1 + K_2 K_v K_{f2} R_2 + K_1 K_2 K_{f2} K_v R_2} \quad (13\text{-}15)$$

The result of incorporating the inner loop may be seen by a careful comparison of Eqs. 13-9 through 13-11 for the simple single feedback system, with Eqs. 13-13 through 13-15 for the cascade control. Increasing the inner loop gain will increase the speed of response and decrease the steady-state offset. At the same time, the inner loop gain adds damping. A greater degree of control for the overall system has been achieved. In practice the designer must be cognizant of other constraints such as maximum available gains on a controller and maximum permissible gains for tolerable signal to noise ratios.

There is also a degree of intuitive insight that may be gained from Fig. 13-17 regarding cascade control. Suppose, for example, that the upstream pressure on the valve is reduced. This results in a reduced head in the first tank. The inner feedback loop will sense a head decrease and open the valve before the second tank head actually decreases. Thus a compensating flow is obtained sooner. With corrective action coming from both feedback signals, the steady-state deviation of the second head is also reduced.

13-5. Feedforward Compensation

Feedforward control is a useful technique for reducing or eliminating system response to disturbances. The principle involved is to feedforward the disturbance signal and subtract it from the reference, thereby reducing the effect of the disturbance. Figure 13-18 illustrates the concept.

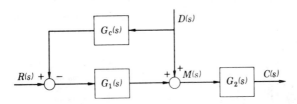

FIG. 13-18. Concept of feedforward control.

A system is composed of elements with transfer functions $G_1(s)$ and $G_2(s)$. In the case of no reference change $r(t) = 0$, but a disturbance $d(t)$, the manipulated variable is obtained as

$$M(s) = D(s) - G_c(s)\,G_1(s)\,D(s)$$
$$M(s) = D(s)\,[1 - G_c(s)\,G_1(s)] \quad (13\text{-}16)$$

If the disturbance is such that it may be measured and $G_1(s)$ is known, and a feedforward controller may be designed such that $G_c(s) = 1/G_1(s)$ then the effect of the disturbance is totally eliminated.

Unfortunately, complete elimination of deviations due to disturbances is seldom realizable. This becomes readily apparent, when it is recognized that $G_1(s)$ is likely to possess lags and even dead times. In the

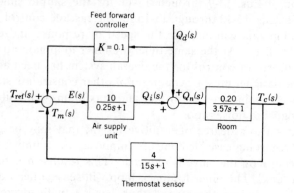

Fig. 13-19. Block diagram of a cloud-loop heating system with feedforward control.

case of a simple lag it may be possible to do a fair compensation job with a lead network; however, dead-time compensation is physically unrealizable. Because perfect feedforward control is difficult, if not impossible, it is generally combined with feedback. A simple feedforward path may substantially improve a feedback system performance in regard to disturbances.

As an example of a practical case, consider the heating system shown

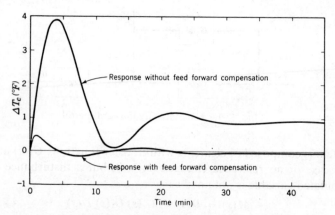

Fig. 13-20. Computer results showing effect of feedforward compensation on the system of Figure 13-19.

in block diagram form in Fig. 13-19. A heat disturbance Q_d to the room is to be compensated by a simple proportional feedforward controller. A study of the performance of the system will reveal the substantial improvement shown by the curves of Fig. 13-20. A careful study of Fig. 13-19 will verify that the steady-state deviation for the uncompensated system is eliminated by feedforward compensation. The gross improvement in the dynamics is due to the relationship of the time constants in $G_1(s)$ and $G_2(s)$. In this case the rapidly responding air supply unit is correcting for the disturbance before the slower room model can respond thus providing effective compensation.

13-6. Compensation Design for Desired Overall Closed-Loop Response

When the control designer is faced with the question of selecting a controller or series compensating element for a system, it is necessary to decide on what basis the device is to be chosen. If the controlled system elements (or plant as it is often called in process control) is well defined, there is a relatively direct technique for establishing the mathematical form of the controller based on a desired closed-loop response. The closed-loop response to commands, disturbances, or noise inputs may be taken into consideration.

Figure 13-21 shows the block diagram of a general closed-loop sys-

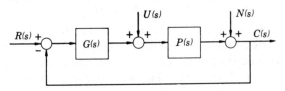

FIG. 13-21. The generalized feedback control system.

tem. The parameters indicated represent the following; input or command $R(s)$, disturbance $U(s)$, noise $N(s)$, response or output $C(s)$, plant $P(s)$, and the controller to be selected $G(s)$. Noise $N(s)$ may be thought of as periodic or stochastic fluctuations of the output variable resulting from a multitude of unpredictable disturbances within the plant originating at a variety of points.

There are three possible closed-loop response functions to consider. They are determined from the transfer functions of $C(s)/R(s)$, $C(s)/U(s)$ and $C(s)/N(s)$. For commands only,

$$\frac{C(s)}{R(s)} = \frac{G(s)\,P(s)}{1 + G(s)\,P(s)} \tag{13-17}$$

for disturbance only,

$$\frac{C(s)}{U(s)} = \frac{P(s)}{1 + G(s)\,P(s)} \tag{13-18}$$

and for noise only,

$$\frac{C(s)}{N(s)} = \frac{1}{1 + G(s)\,P(s)} \tag{13-19}$$

The design of $G(s)$ follows by establishing a specification for the relationships of Eqs. 13-17 through 13-19. Let $K_R(s)$ be the desired closed-loop transfer function for command inputs. Then

$$\frac{C(s)}{R(s)} = \frac{G(s)\,P(s)}{1 + G(s)\,P(s)} = K_R(s)$$

and

$$G(s) = \frac{1}{P(s)}\frac{K_R(s)}{1 - K_R(s)} \tag{13-20}$$

Consider $K_u(s)$ as a desired transfer function for disturbances

$$\frac{C(s)}{U(s)} = K_u(s)$$

or

$$G(s) = \frac{P(s) - K_u(s)}{P(s)\,K_u(s)} \tag{13-21}$$

For noise, let $K_N(s)$ be the desired transfer function. Then

$$\frac{C(s)}{N(s)} = K_N(s)$$

and

$$G(s) = \frac{1 - K_N(s)}{P(s)\,K_N(s)} \tag{13-22}$$

Equations 13-20 through 13-22 provide three alternatives for selecting $G(s)$ on the basis of some desired performance. Clearly, only one choice may be made independently. Once $G(s)$ is established from either Eq. 13-20, 13-21, or 13-22, the remaining two transfer functions may be calculated.

13-7. Physical Devices for System Compensation

The necessity to compensate a control system ultimately leads the designer to the realization that some physical device must be designed to provide the theoretically desirable input-output relationships. In other words, a combination of components must be synthesized to obtain the prescribed compensating element transfer function form with the necessary values of time constants and gains. The problem may be complicated by demands that all of the control elements be of a particular physical nature such as completely hydraulic, pneumatic, mechanical, or electrical. Hopefully, the system is such that the designer is free to choose from a wide variety of elements and combinations of elements.

The design of clever, economical, and dependable control system compensating devices provides an opportunity for the designer to exercise a high degree of ingenuity and creative imagination. Dependability is a prime consideration in the design of a compensating element, particularly when it is providing the necessary stability for satisfactory operation. If a compensating device should fail or alter its characteristics in this latter regard, a system may become unstable and cause destruction.

The following is a summary of some simple devices that may be utilized to obtain desired transfer functions for some of the basic compensating forms discussed.

Simple Lag. A simple lag element is easily obtainable with mechanical or electrical elements. The simple lag arises from transformation of a first-order differential equation. Such an equation arises, for example, from the series connection of a spring and dashpot as shown in Fig. 13-22a. As the spring force must equal the damper force

$$k(x_i - x_o) = c \, \frac{dx_o}{dt}$$

or in transformed form with all initial conditions zero

$$\frac{X_o(s)}{X_i(s)} = \frac{1}{(c/k)s + 1} = \frac{1}{\tau s + 1} \tag{13-23}$$

If the signal to be sensed is a displacement (translation or rotation) and a displacement signal output may be compared with a reference input in some fashion, this device would be quite satisfactory.

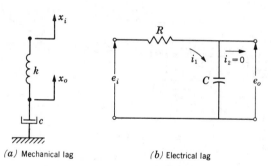

(a) Mechanical lag *(b)* Electrical lag

FIG. 13-22. Devices for the simple lag.

Figure 13-22b illustrates an electrical resistance-capacitance network where the input and output signals are voltages. If $i_2 = 0$, the output voltage is obtained from

$$e_i - RC \, \frac{de_o}{dt} = e_o$$

or in transformed form,

$$\frac{E_o(s)}{E_i(s)} = \frac{1}{RC\,s + 1} \tag{13-24}$$

One point which is too infrequently emphasized in the application of devices such as those shown in Fig.13-22 is that certain impedance properties must exist for the proper operation of these devices. For example, in the mechanical system in Fig. 13-22a the device that is used to sense the position x_o must not have appreciable mass nor can it require much force relative to the spring and dashpot forces. If these conditions are not satisfied, then the dashpot force and the spring force will not be equal and the relationship of Eq. 13-23 will not hold. If, in the electrical system, the impedance of the voltage-measuring device is not sufficiently high, current i_2 will not be close to 0, and Eq. 13-24 will not be the proper transfer relationship for the actual network. The problem of impedance matching is often solved in electrical systems by the use of amplifiers and preamplifiers in cascade with control system components. The electronic amplifier can be used in automatic controls to provide desired impedance matching as well as a voltage, current, or power gain. Further, an amplifier may have incorporated within itself a lag, lag-lead, or other transfer relationship particularly designed for system compensation.

A good example of a hydromechanical device that could be used to obtain a simple lag was discussed in Sec. 5-5. This device is pictured in Fig. 13-23, and, in addition, is one which provides a high input impedance (that

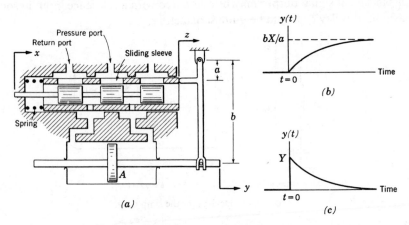

FIG. 13-23. A hydromechanical lag.

is, a relatively low force is required to move the valve spool) and power amplification (that is, the output force can be quite large for a relatively small input force). Here again, the input and output quantities are displacements. It should be remembered that mechanical motions may be translated into electrical signals through the use of potentiometers and linear variable differential transformers discussed in Sec. 7-7.

Lag-Lead. The lag-lead device is an extremely versatile and useful compensating device. Mechanically, a lag-lead transfer function may be obtained by the use of a spring and two dashpots as shown in Fig.13-24a. From the summation of forces at x_o

$$c_2 \frac{dx_o}{dt} = k(x_i - x_o) + c_1\left(\frac{dx_i}{dt} - \frac{dx_o}{dt}\right)$$

In transformed form this becomes

$$c_2 s X_o(s) + c_1 s X_o(s) + k X_o(s) = k X_i(s) + c_1 s X_i(s)$$

or
$$\frac{X_o(s)}{X_i(s)} = \frac{(c_1/k)\,s + 1}{[(c_1 + c_2)/k]\,s + 1} = \frac{\tau_2 s + 1}{\tau_1 s + 1} \qquad (13\text{-}25)$$

Again, it is important that the position x_o be sensed in such a way as not to apply a force at that point in the device nor add appreciable mass. In addition, the springs and dampers should be chosen so that the force required to produce the motion x_i does not effect the actual deflection x_i that is being sensed. To reiterate, this is an impedance problem and it must always be considered in the design of control devices.

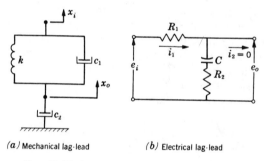

(a) Mechanical lag-lead   *(b)* Electrical lag-lead

FIG. 13-24. Devices for lag-lead compensation.

Figure 13-24b is the electrical counterpart of the mechanical lag-lead. With $i_2 = 0$, the transfer relationship for the input and output voltages is

$$\frac{E_o(s)}{E_i(s)} = \frac{R_2 C s + 1}{C(R_1 + R_2)\,s + 1} = \frac{\tau_2 s + 1}{\tau_1 s + 1} \qquad (13\text{-}26)$$

The assumption $i_2 = 0$ is again justified if the impedance of the measuring device is sufficiently high. It should be pointed out that it is possible to work with a finite impedance so that $i_2 \neq 0$ just as it is possible to consider an additional force at x_o in Fig. 13-24a. The derivation of the transfer relationship is, however, complicated and the relationship expressed by Eqs. 13-25 and 13-26 are no longer valid.

Devices which may be used for compensation are numerous and specific ideas for compensating components are found in the literature and brochures of component suppliers. References [6] and [14], for example, provide a good source of information on design and synthesis. It is intended here to introduce some of the concepts and techniques that are available and to alert the reader to the fact that there is a wealth of literature on clever, useful devices.

13-8. An Example of System Compensation

One widely employed type of series compensator for feedback control systems is the addition of an integrator to a series proportional controller. This device, commonly called a proportional plus integral controller, was discussed in Sec. 13-2, and its dynamic behavior is defined by Eq. 13-1 as

$$\frac{E'(s)}{E(s)} = \frac{K_1(\tau_1 s + 1)}{s}$$

It is of considerable value at this point to illustrate with a specific example how this compensating device, with properly chosen parameters, can improve the response of a typical positioning system.

EXAMPLE 13-2. Design a series compensator that will eliminate the steady-state error for a constant velocity input to the simplified two-phase servomotor positioning system of Example 10-5. The resulting system should have no less than 40° phase margin and a maximum magnitude ratio of less than 1.4. The block diagram for the system designed in Example 10-5 is given in Fig. 13-25. The Bode diagram for the system is shown in Fig. 13-26.

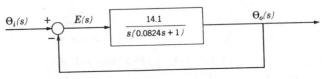

FIG. 13-25. Block diagram for the simplified positioning system of Example 13-2.

Solution: To decide on the form of the compensating device, it is first necessary to analyze the error obtained for the input specified. The forward transfer function for the positioning system is

$$G(s) = \frac{K_T}{s(\tau_m s + 1)} = \frac{\Theta_o(s)}{E(s)}$$

where, for this system, $K_T = 14.1$ and $\tau_m = 0.0824$.

The error $E(s)$ is given by $E(s) = \Theta_i(s) - \Theta_o(s)$ and

$$\frac{\Theta_o(s)}{\Theta_i(s)} = \frac{G(s)}{1 + G(s)}$$

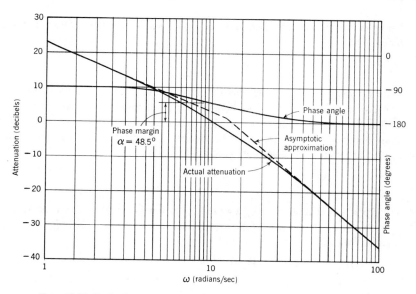

FIG. 13-26. Bode diagram for the uncompensated system of Example 13-2.

for the unity feedback shown in Fig. 13-25. Combining the latter two equations gives

$$E(s) = \Theta_i(s) \frac{1}{1 + G(s)}$$

or

$$E(s) = \Theta_i(s) \left[\frac{s(\tau_m s + 1)}{s(\tau_m s + 1) + K_T} \right]$$

If $\theta_i(t) = \omega_i t$, then $\Theta_i(s) = \omega_i/s^2$. Therefore,

$$E(s) = \frac{\omega_i}{s^2} \left[\frac{s(\tau_m s + 1)}{s(\tau_m s + 1) + K_T} \right] = \frac{\omega_i(\tau_m s + 1)}{s^2(\tau_m s + 1) + K_T s}$$

Applying the final value theorem (Eq. 6-28) gives

$$\lim_{t \to \infty} \epsilon(t) = \lim_{s \to 0} sE(s) = \frac{\omega_i}{K_T}$$

It is seen that there would be a steady-state difference between the input and output positions of ω_i/K_T if the system input were a steady angular velocity. Specifically, if the input shaft were rotating at 10 radians/sec, the output would differ from the input by $\omega_i/K_T = 10/14.1 = 0.71$ radians, or $40.7°$.

If another pole at the origin were introduced in $G(s)$, that is, $G(s) = K_T/s^2(\tau_m s + 1)$, the final value theorem for the previous input $\omega_i t$ would yield

$$\lim_{t \to \infty} \epsilon(t) = \lim_{s \to 0} sE(s) = 0$$

The steady-state error for the constant velocity input is then eliminated as desired.

However, the problem now is, that a unity feedback system with a forward transfer function of

$$G(s) = \frac{K_T}{s^2(\tau_m s + 1)}$$

is always unstable. (The reader should verify this by sketching either Nyquist, Bode, or root-locus plots.) On the Bode diagram there will always be at least $-180°$ phase shift at the crossover point. To provide stability, therefore, it is necessary to have some positive phase shift prior to the crossover point on the Bode diagram. Specifically, the compensating element may be of the form

$$\frac{K_1(\tau_1 s + 1)}{s}$$

and is precisely that obtained by a series proportional plus integral controller.

The block diagram for the compensated system is given in Fig. 13-27. The forward transfer function is

$$\frac{\Theta_o(s)}{E(s)} = \frac{14.1 K_1(\tau_1 s + 1)}{s^2(0.0824 s + 1)}$$

Solving for $E(s)$ in terms of $\Theta_i(s)$ by the previous method

$$E(s) = \Theta_i(s) \frac{s^2(0.0824 s + 1)}{s^2(0.0824 s + 1) + 14.1 K_1(\tau_1 s + 1)}$$

The steady-state error for the input $\Theta_i(s) = \omega_i/s^2$ is

$$\lim_{t \to \infty} \epsilon(t) = \lim_{s \to 0} sE(s) = \lim_{s \to 0} \frac{s(0.0824 s + 1)}{s^2(0.0824 s + 1) + 14.1 K_1(\tau_1 s + 1)}$$

or

$$\lim_{t \to \infty} \epsilon(t) = 0$$

The specification for zero steady-state error for a constant velocity input is thus satisfied. Values of K_1 and τ_1 that will provide satisfactory performance must now be chosen.

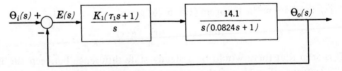

FIG. 13-27. Block diagram for the compensated positioning system.

At the crossover point it will be necessary for the lead term to contribute sufficient positive phase shift to provide a phase margin of at least $40°$. If crossover occurs near the original breakpoint of $\omega_m = 1/0.0824 = 12.12$ radians/sec, then the net negative phase shift is approximately $-225°$ (i.e., $-90°$ from each pole at the origin and approximately $-45°$ from the motor time constant). For a phase margin of 40°, the net phase must be on the order of −140°. Thus the lead term must contribute approximately 225° − 140° = +85°.

At the break point for the motor time constant, ω = 12.12, thus

$$\phi_1 = \tan^{-1} \frac{\tau_1(12.12)}{1} = 85°$$

and 12.12 τ_1 = 11.49 or τ_1 = 0.942.

In order to set the gain K_1, consider plotting the transfer function

$$\frac{G_c(j\omega)G(j\omega)}{K_1} = \frac{14.1\,(0.942j\omega + 1)}{(j\omega)^2\,(0.0824j\omega + 1)}$$

and adjust the decibel level of the curve upward or downward to provide a minimum 40° phase margin. The Bode diagram is given in Fig. 13-28. It is seen that the crossover frequency is ω = 10.2 radians/sec, and the phase margin is 44° at this point. Thus a gain K_1 of unity is sufficient for the specifications set forth in the problem.

The maximum magnitude ratio may be established by transferring the data from the Bode plot in Fig. 13-28 to the Nichols chart of Fig. 13-29. The plot is tangent to M = 2.4 db which corresponds to an M_m of approximately 1.3.

In compensating the system to eliminate error for a constant velocity input, the phase margin has been reduced, and the maximum magnitude ratio has been increased. Although the problem of a particular error has been solved, it should be noted that it is at the expense of having a slightly more oscillatory system. In general, however, the result here is a system which would normally have an acceptable frequency and transient response.   *Ans.*

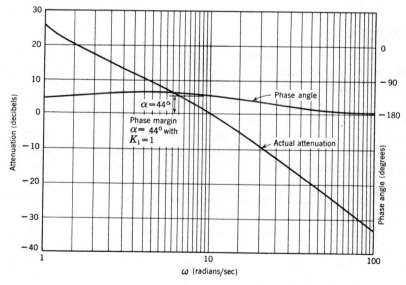

FIG. 13-28. Bode diagram for compensated system K_1 = 1.

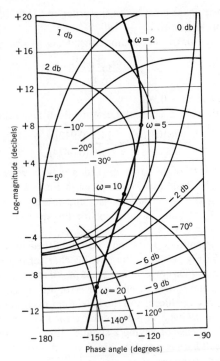

FIG. 13-29. Nichols chart for establishing M_m for compensated system.

13-9. Closure

The need for compensation of automatic control systems arises because a given set of system elements, chosen to achieve some desired end result, is not always capable of meeting the specifications for the control. The system may be inadequate from the standpoint of stability, speed of response, or accuracy. Naturally, a system that is unstable is useless. Similarly, a system that will not perform with some allowable minimum error is of little value. If a system fails to meet desired specifications, the alternatives at hand are to first, modify or replace the existing system elements so that the system will meet the requirements, or second, insert an additional element (a compensating element) into the system to alter the over-all behavior so that the system will behave as desired.

The use of compensating elements in a control system is a valuable technique because it is not always desirable or even possible to alter the basic nature of the system elements. Economically, it may be more desirable to use existing commercially available hardware for a system and design a suitable compensating device rather than set out to design and manufacture a completely new system. Compensation may be achieved by placing control components in series with the system (series compensation) or by placing elements in inner feedback paths around one or more of the system elements (feedback compensation). The choice is solely dependent on the nature of the system and the nature of the compensation required to alter the system to meet the specifications.

It must be pointed out that this treatment of compensation concepts and techniques is by no means conclusive. The synthesis of a compensation device may be carried out using frequency-response methods (Bode or Nyquist diagrams) or transient-response methods (root-locus plots). In this chapter it was decided to use attenuation and phase angle plots to facilitate solution of the example problems. It is quite possible that a particular problem is most effectively handled by one particular method.

Problems

13-1. A unity feedback control system has the forward transfer function

$$G(s) = \frac{K}{(0.1s + 1)(0.02s + 1)(0.006s + 1)}$$

It is desired to have not more than a steady-state error of 0.030 units for a unity step input and at the same time maintain a stable system.

a) By Routh's criterion, establish the maximum value of gain K allowable for stability of the closed-loop system. b) Evaluate the steady-state error $E_{ss}(t)$ for the value of gain obtained in part (a). c) Draw some conclusions regarding the design of this system to meet the stated specifications using gain adjustment alone. d) Recommend a series compensating device that will possess the necessary characteristics to allow the system to meet the stated specifications with the additional stipulation that a satisfactory phase margin be maintained. Explain.

13-2. Choose the parameters of a series compensating device that can be used to extend the open-loop frequency response of a system with a forward transfer function

$$G(s) = \frac{1}{(0.1s + 1)(0.02s + 1)(0.006s + 1)}$$

so as to give not more than −3db attenuation below 40 radians/sec and to provide −40 db or more attenuation at 500 radians/sec.

What is the effect of the compensating element on the closed-loop response?

13-3. A unity feedback control system has the forward transfer function

$$G(s) = \frac{K}{s(0.1s + 1)(2s + 1)}$$

It is necessary to design the system to meet the following specifications. The system must have zero steady-state error for a step input, and the closed-loop magnitude ratio must not exceed 1.3.

What compensation is necessary and what is the proper gain setting?

13-4. It is desired to design a series compensating element that will eliminate the steady-state error for a step input and provide a phase margin of 40° in a unity feedback control system with the transfer function

$$G(s) = \frac{K}{(0.1s + 1)(2s + 1)}$$

a) Specify the transfer function for a satisfactory compensating element and justify its use. b) Evaluate all parameters and constants of the compensated system. c) What is the effect of your compensating element on the frequency response of the closed-loop system?

13-5. The block diagram of a unity feedback electromechanical control system is shown in the accompanying illustration.

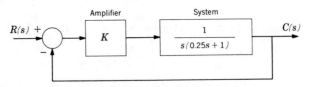

PROB. 13-5

a) Draw a new block diagram for the system incorporating a feedback compensating device around the system element that will permit improved speed of response. Show how the compensation will increase the speed of response. b) Calculate the parameters of the compensated system that will provide a 50% decrease in the original time constant and result in a maximum magnitude ratio of 1.3.

13-6. Sketch a mechanical system and specify the values of the parameters that will transmit a linear displacement with the transfer relationship

$$G_c(s) = \frac{0.5}{0.1s + 1}$$

The maximum input displacement encountered is 1 in. In the full range of operation, the spring force in the device should not exceed one-half pound. Specify any special restrictions on the application of the device that must be observed for proper operation.

13-7. Calculate the parameters of an electrical lag-lead network of the form shown in Fig. 13-24b. The current i_1 is not to exceed 10 ma for a potential difference of 100 volts. The transfer relationship desired is

$$G_c(s) = \frac{0.01s + 1}{0.5s + 1}$$

Specify any restrictions that must be placed on the coupled equipment to provide proper operation.

13-8. A system with a proportional controller is to be compared with the same system incorporating cascade control. a) For the system shown in the figure (Prob. 13-8*a*) *determine the gain* K_1 that will yield a system damping ratio $\zeta = 0.5$. For the gain obtained, determine the steady-state deviation in the controlled variable when a unit step disturbance is applied. Estimate the time for the system to respond completely. b) The same system elements are shown in Prob. 13-8*b* with cascade control. Determine K_1 and K_2 for $\zeta = 0.5$ and minimum steady-state deviation for a step disturbance. Determine the steady-state deviation of the controlled variable to a unit step disturbance and estimate the time for the system to completely respond. A constraint on the maximum values of the two controller gains is $K_{1\ max} = K_{2\ max} = 25$. c) Comment on the desirability of using cascade control for the system under study.

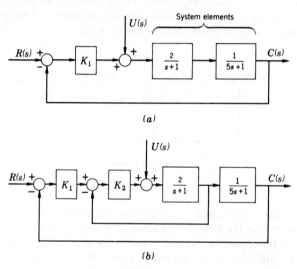

(a)

(b)

Prob. 13-8

13-9. A feedback control system is illustrated in the figure (Prob. 13-9) with feedforward compensation. Compare the response of the controlled variable to disturbances $u(t) = 50$ with and without the feedforward element. What is the optimum value of K?

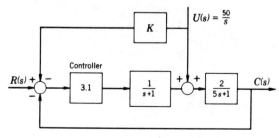

Prob. 13-9

13-10. A controller is to be designed to control a plant with the transfer function

$$P(s) = \frac{K}{\tau s + 1}$$

The overall closed-loop response for the system with unity feedback is to be

$$K_R(s) = \frac{1}{\dfrac{1}{\tau} s + 1}$$

a) Determine $G(s)$ and identify the control action. b) What is the closed-loop transfer function relating the controlled variable to noise? Determine the effect the controller will have on noise inputs $N(s)$. c) Determine the closed-loop transfer function relating the controlled variable to disturbance inputs. What effect will the controller have on step disturbances entering the system?

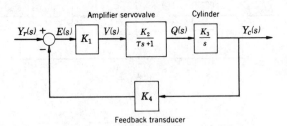

PROB. 13-11

13-11. A block diagram for an electrohydraulic positioning system is given. It is imperative that the steady-state error to a step input be zero. In reviewing the system, the servovalve manufacturer points out that his valve is subject to null shift, i.e., with zero input, the valve spool will not always center precisely. With null shift, the steady-state error will not be zero. a) Revise the block diagram to permit a disturbance input corresponding to spool shift. b) Modify the system to eliminate the effect of null shift. Amplifier choice is limited to proportional or integral.

Chapter 14

Control System Nonlinearities

14-1. Introduction

In the design of actual automatic control systems, the control analyst and the component designer must face the unfortunate truth that nature has not made all physical relationships strictly linear. One says, for example, that a spring is linear, meaning that the force-deflection curve for that spring is a straight line. In the actual test of a spring it may be found that the curve is slightly concave upward. The spring rate would not be a constant for all deflections in this case and the spring would be nonlinear. Another example in which a spring must be considered nonlinear is the case of a coil spring wherein the coils bottom after a certain deflection is reached. Over the initial range of deflections, the force-deflection curve might be sufficiently close to a straight line to call the spring linear. When the spring bottoms, however, the spring rate is greatly increased and there is an abrupt change in slope of the force-deflection curve. If this portion of the curve is to be used, the spring must be considered nonlinear. This simple illustration of linearity versus nonlinearity points out a basic concept—the behavior of all physical systems in nature is inherently nonlinear. There may be regions over which a linear approximation is satisfactory (i.e., a linear range), but in general, a nonlinear region may be reached.

From a system analysis point of view, nonlinearity may be generally identified by the fact that the principle of superposition does not hold. The converse of this statement was pointed out in Sec. 3-1 in that a linear differential equation was characterized by the fact that the principle of superposition does hold. A nonlinear dynamic system may respond to a driving function $f_1(t)$ with motion $y_1(t)$ and then respond to another driving function $f_2(t)$ with motion $y_2(t)$. If the driving function for the system is $f_1(t) + f_2(t)$, the responding motion is not, in general, $y_1(t) + y_2(t)$ as in the linear case.

The study of the solution of linear differential equations is facilitated by the fact that a specific set of mathematical techniques may be applied to a large class of equations. With nonlinear differential equations such is not the case. In fact, among the innumerable possibilities for nonlinear differential equations, only a relatively few analytical solutions have been found by presently known techniques. General solutions for classes of nonlinear differential equations are nonexistent.

How then are nonlinear control systems analyzed? One of two philosophies may be adopted. First, it is possible to *linearize* a nonlinear element and then solve the resulting linear problem. Second, one may solve the actual nonlinear problem by special techniques.

Linearization of a nonlinear element may be accomplished in several ways. The most obvious approach is to overlook the fact that an element is nonlinear. The "for small oscillations" approximation says, in effect, that for a limited range of change of a variable it is acceptable to replace the actual curve by the tangent to the curve at a particular point and to assume that variations occur along the tangent. For the spring discussed previously, a varying spring rate would be replaced by a constant spring rate for a limited range of the load-deflection curve.

A widely used technique for linearization is to replace a nonlinear function by a linear function which closely resembles the nonlinear function. A nonlinear relationship may be replaced by a series expansion. The solutions of linear differential equations with series terms are generally available, although they are frequently extremely involved. A common example of this approach is the use of the power series expansion for sin θ, · wherein

$$\sin \theta = \theta - \frac{\theta^3}{3!} + \frac{\theta^5}{5!} - \cdots$$

If θ is known to take on sufficiently small values, the θ^3, θ^5,... terms will be negligible and sin θ could be replaced by θ. There is a general method for replacing nonlinear functions by the fundamental component of an infinite series which is known as the describing function method. This approach will be discussed in Sec. 14-3.

If it is deemed necessary to solve the particular nonlinear problem without linearization, there are many specialized techniques which may be used to obtain a specific solution. They include:

1. Special analytical and tabulated solutions
2. Piecewise solutions
3. Phase-plane analysis
4. Analog computer solution
5. Solution using the digital computer

There is one further introductory remark that should be made regarding nonlinear analysis in automatic controls. The actual solution of the equations defining a system may not be necessary. It has been emphasized throughout this text that stability is a primary consideration in control system design. Thus, in the literature on nonlinear analysis, one finds considerable emphasis on the prediction of stability. In many cases it is possible to establish criteria for the stability of a solution where it is not possible to obtain an actual analytical solution.

This chapter has been included because of the importance of nonlinear analysis in the design of modern control systems. It is the purpose here to introduce some of the basic types of nonlinearities that may exist and to present an introduction to some of the methods that are available for analysis and solution. The application of these methods is illustrated by examples.

14-2. Common Control System Nonlinearities

The occurrence of nonlinear phenomena in real systems is so common and of such a varied nature that it is hardly worth trying to construct a system for classification. One approach to discussing the occurrence of nonlinearities is to consider basic physical elements that make up mechanical, electrical, and hydraulic systems and to investigate how nonlinearities may commonly enter into the problem. Once a knowledge of the presence of nonlinear phenomena is gained, attention may be turned to a study of the methods of solving problems where certain nonlinearities exist. The goal of this section is to look further into the source of nonlinear phenomena.

Mechanical Systems. The presence of nonlinear spring rates in mechanical systems was mentioned briefly in the introductory remarks.

Figure 14-1 shows the force-deflection curve for three different springs. The ideal linear spring is represented by a straight line; therefore, the derivative of force with respect to deflection is a constant and the spring rate is said to be a constant k. If the force-deflection curve is concave upward, the spring rate is constantly increasing with displacement. Thus, the spring rate is a function of x, $k = k(x)$, and the spring may be called a hardening spring. The remaining alternative shown in Fig. 14-1 is the case where the

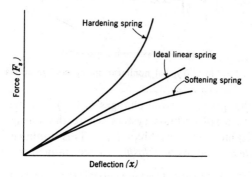

Fig. 14-1. Force-deflection curves.

stiffness decreases with displacement [here also, $k = k(x)$] and this is called a softening spring.

The spring rate of a mechanical system is not always represented by a continuous curve. Figure 14-2 illustrates four possibilities for spring

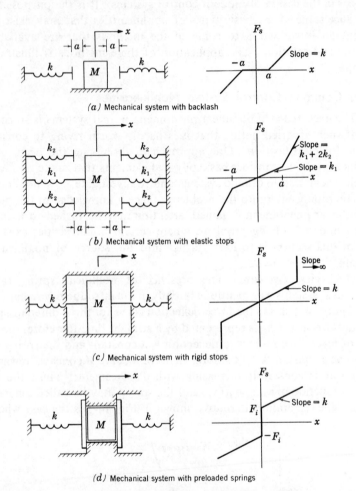

(a) Mechanical system with backlash

(b) Mechanical system with elastic stops

(c) Mechanical system with rigid stops

(d) Mechanical system with preloaded springs

FIG. 14-2. Typical nonlinear spring configurations.

combinations that possess abrupt changes in $k(x)$ at some position x. Figure 14-2a shows a spring-mass system where the springs do not contact the mass for displacements less than $\pm a$.[1] This nonlinearity is commonly referred to as *dead zone*, or *backlash*. In mechanical systems, backlash is

---

[1] The spring force is understood to act in a direction to restore the mass to its equilibrium position. The sign of the force may be accounted for on the free-body diagram of the mass.

frequently encountered in the analysis of the torque transmitted through a geared system.

A bilinear spring rate is obtained when there are two sets of springs, one set in contact with the mass for all displacements and a second set that makes contact for displacements greater than $\pm a$. Such a combination of springs and the corresponding force-deflection curve are given in Fig. 14-2b. As the second set of springs becomes stiffer, the case of a spring-mass system with stiff elastic stops is obtained. (The limit would be an infinite spring rate for a rigid wall, Fig. 14-2c.) A common example of a bilinear spring system is the suspension of an automobile. An overload device is generally used to protect the body from excessive axle deflections that may be occasionally encountered.

One further example of a nonlinear spring rate frequently encountered is that for a preloaded spring shown schematically in Fig. 14-2d. The force-deflection curve illustrates that an initial preload F_i must be overcome before the mass can be deflected.

In each case illustrated by Fig. 14-2 the differential equation obtained by applying Newton's law is nonlinear. Specifically, the differential equation of motion takes the form

$$M \frac{d^2x(t)}{dt^2} + k(x)x(t) = 0 \tag{14-1}$$

where the spring rate is a function of x as defined by the respective force-deflection curves.

Machine drives and power transmission equipment often require the use of a flexible coupling located between the driving and driven members. The flexible coupling is provided primarily to absorb shock and vibration and to permit limited misalignment without experiencing harmful loads on the shafting and bearings. Such devices provide flexibility in torsional systems which reduces stresses from shock and vibration and increases the resilience of a system by increasing its capacity to store energy. Further, some energy is lost in the coupling by internal hysteresis.

To the control designer this means that the use of a flexible coupling may necessitate consideration of one more spring rate and one more source of energy dissipation in the block diagram. In fact, the torque-deflection properties of some couplings are highly nonlinear. Figure 14-3 illustrates the torque-deflection curve for one cycle of loading of a commonly used commercial coupling. The closed contour which is formed by one complete torque-displacement cycle is referred to as a hysteresis loop. The area enclosed represents the energy lost in one cycle of loading. In this case, the motion of an inertia attached to the coupling is defined by a nonlinear differential equation of the form [26]

$$I\ddot{\theta} + cf_1(\dot{\theta}) + kf_2(\theta) = f_3(t) \tag{14-2}$$

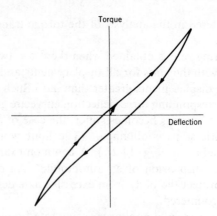

FIG. 14-3. Torque-deflection cycle
of a mechanical spring coupling.

Mechanical friction is another common source of system nonlinearity.
The nature of viscous friction was discussed in Sec. 2-2. Unfortunately,
viscous friction, which leads to a linear friction force proportional to
velocity, is an ideal which may only be approximated in most cases. It is
often used as a rough approximation in order to simplify an initial analysis.
At times a careful evaluation of the friction force acting on a mass reveals
that the friction force is a constant, independent of the magnitude of the
velocity and directed to oppose the direction of the velocity. Thus, the
coulomb friction force can be defined as[2]

$$F_f = (\text{sign of } \dot{x})\,\mu N \tag{14-3}$$

where (sign of $\dot{x}$) is the sign of the velocity, μ is the coefficient of sliding
friction (kinetic friction), and N is the normal force. Coulomb friction is
represented in Fig. 14-4a.

If a mass is known to come to rest in some portion of the cycle it may
be deemed necessary to consider the difference between the coefficient of
sliding friction and the coefficient of static friction, sometimes called breakaway friction. In general, the friction force at rest is greater than the friction force opposing some motion. This phenomenon is shown in Fig.
14-4b, where the friction force builds up to some maximum value F_{fs} before
the mass moves and then drops to some lower magnitude F_{fd} when the mass
has some velocity $\dot{x}$. Coulomb and static friction forces are nonlinear terms
which complicate the solution of equations of motion.

Electrical Systems. There are a great many nonlinearities associated
with electrical equipment. As an example, consider the two-phase a-c

---

[2]The friction force F_f is actually oriented in the negative direction for a positive velocity. The negative sign omitted in Eq. 14-3 may be accounted for by representing F_f properly on the free-body diagram of the mass.

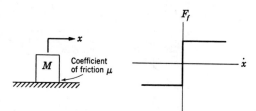

(a) Mechanical system with coulomb friction (static coefficient of friction equals coefficient of sliding friction)

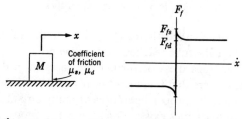

(b) Mechanical system with static and sliding friction
(breakaway friction coefficient is greater than sliding friction)

FIG. 14-4. Coulomb friction and static friction.

servomotor discussed in Sec. 7-9. The torque-speed curves for various control-field voltages and a constant reference field were represented as straight lines in Fig. 7-30*b*. Actually, these curves may be more accurately described as a family of nonlinear curves as shown in Fig. 14-5. Here, the rate of change of torque with speed is not a constant.

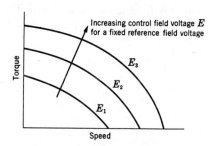

FIG. 14-5. A-c servomotor torque-speed characteristics.

A good example of an electrical element that exhibits a linear characteristic over some finite range is the electron tube. The typical characteristics of a triode are illustrated in Fig. 14-6. There is a finite range of grid voltage for which the plate current is a linear function of the grid voltage. To the left of a particular negative grid voltage, the plate current

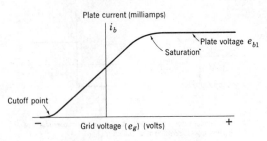

FIG. 14-6. Triode characteristics.

is zero. This point is called the *cutoff* point for the tube. Likewise, for positive grid voltage the tube reaches a *saturation* point (i.e., there is a limiting flow of current regardless of further increases in grid voltage). It is worthy of emphasis here that these nonlinearities in a triode are not always undesirable. In fact, the proper operation of many circuits depends on the fact that these nonlinearities exist. For example, a triode used in conjunction with a phototube in a relay circuit operates in the vicinity of the cutoff point. The nonlinearity in this case is essential. On the other hand, a triode in an audio amplifier must not be driven into the nonlinear range or the output will become distorted.

Hysteretic phenomena is not limited to mechanical devices. Electrical circuits with inductors (a common example is the magnetic amplifier) may exhibit sufficient hysteresis that it must be considered in the analysis. Figure 14-7 illustrates both hysteresis and saturation effects that are typical for the flux versus current curve of a coil with a soft iron core (commonly called a *choke*). The fact that the core permeability is not a constant results in some limiting of the flux for increasing current. The inductor is said to saturate. The hysteresis indicates that some energy is dissipated in each

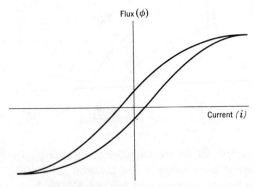

FIG. 14-7. Hysteresis and saturation for an inductance with a soft iron core.

cycle, and the area enclosed by the curve is a measure of the energy lost. This particular nonlinearity is important in controls because magnetic amplifiers are widely used power amplifiers. They are particularly well suited as power amplifiers for driving two-phase servomotors.

An electromechanical device known as a double-contact relay is an example of a device that is nonlinear and in practice is made even more nonlinear primarily because it works better and lasts longer. Figure 14-8a illustrates the operating characteristics of such a relay by showing a plot of the output (which may be the power supplied to a control element) of an ideal relay versus the input voltage to the relay coil. Any positive voltage e_i results in a constant positive output and conversely, any negative voltage results in a constant negative output. Because any small voltage fluctuations near zero would cause the relay to constantly oscillate between the two limits, and thus present a wear problem, actual relays are constructed with a dead zone $\pm e_d$. This is shown in Fig. 14-8b. As with mechanical systems, dead zone may have considerable effect on control system stability and performance. Nonlinearity in the double-contact relay may be carried one step further. Figure 14-8c illustrates a relay that must have an input voltage $e_i = e_d$ to make contact, but the voltage must return to a value e_h before the contact will break. This property is, again, a hysteretic phenomenon.

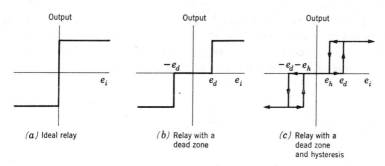

(a) Ideal relay (b) Relay with a dead zone (c) Relay with a dead zone and hysteresis

Fɪɢ. 14-8. Characteristics of a relay.

Hydraulic Systems. It is interesting to observe the similarity between the torque-speed curves for a two-phase servomotor and a hydraulic motor controlled by a servo valve. Figure 14-9 illustrates that the curves for this hydraulic element are not straight lines, and thus the derivative of torque with respect to speed is not a constant.

Figure 14-10 illustrates a saturation effect that exists in the flow versus input current relationship for a servo-valve. The rate of fluid flow through the valve, for a constant pressure drop across the valve, is proportional to the current for a finite range of current. If, however, the current becomes great enough, a linear relationship is no longer realistic.

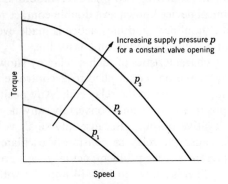

FIG. 14-9. Torque-speed curves for a servo-valve controlled hydraulic motor.

By now the reader should be reasonably convinced that nonlinearities do exist and that in many cases up to this point in the text nonlinear relationships have been replaced by handy straight lines in order to simplify the analysis. It must be pointed out that this is not necessarily bad, but really quite desirable in many instances. The important thing is that the analyst understand the implications and effects of the linearization process. The remaining sections of this chapter will illustrate further the concept of linearization and techniques for solving problems with nonlinearities present.

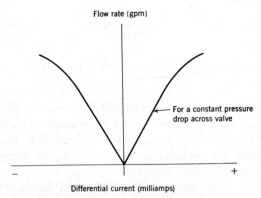

FIG. 14-10. Flow rate-differential current plot for a servo-valve.

14-3. The Linearization Approach in Analyzing Nonlinear Systems

The solution of a nonlinear problem may be simplified by replacing nonlinear functions by linear expressions that describe the behavior of a phenomenon sufficiently well for an engineering analysis. Two commonly used approaches to linearization will be presented here. First, replacement of a function by a tangent to that function, and second, replacement of an expression by the fundamental of a series expansion.

Linearization by the Tangent to the Curve. Consider as an illustration of the first technique for linearization, the functional relationship between $f(x)$ and x shown in Fig. 14-11. The function $f(x)$ in this case is not a linear expression of x. The value x_o is that of the independent variable from

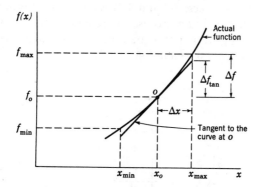

FIG. 14-11. Linearization of a nonlinear function in a specific operating range.

which a reference is taken. The independent variable then varies between x_{min} and x_{max} and $f(x)$ varies from f_{min} to f_{max} with $f(x_o) = f_o$. The first step is to construct a tangent to the curve at point o. (This corresponds to the reference point of the independent variable x_o.) Variations from the reference point f_o along the tangent to the curve are then defined by

$$\Delta f_{tan} = \frac{df}{dx}\bigg|_{x_o} \Delta x \cong \Delta f \tag{14-4}$$

where $df/dx \mid_{x_o}$ means the value of the derivative of $f(x)$ with respect to x evaluated at x_o. In other words, changes in f (i.e., Δf) are approximately equal to the slope at x_o times Δx. Picking values of $f(x)$ along the tangent rather than along the curve gives an exact result only when $x = x_o$. For the particular curve shown in Fig. 14-11, the value of Δf_{tan} at x_{min} evaluated by the linear approximation would be greater than the actual Δf, and at x_{max}, the linearized result would be less than that obtained from the actual curve. It is now up to the analyst to justify the acceptance of the error on the basis of the relative amount of error present and its physical significance. In general, if the percentage of error is small at the extremes, the approximation may be justified.

In the more general case, where the dependent variable is a function of more than a single independent variable, the derivative in Eq. 14-4 must be replaced by the partial derivative. For example, if $f = f(x,y,z)$, then[3]

$$\Delta f \cong \Delta f_{\tan} = \frac{\partial f}{\partial x}\bigg|_{(x_o,y_o,z_o)} \Delta x + \frac{\partial f}{\partial y}\bigg|_{(x_o,y_o,z_o)} \Delta y + \frac{\partial f}{\partial z}\bigg|_{(x_o,y_o,z_o)} \Delta z \qquad (14\text{-}5)$$

where $\partial f/\partial x \mid_{(x_o,y_o,z_o)}$ is the partial derivative of f with respect to x evaluated at the reference condition (x_o, y_o, z_o).

The analysis of a flyweight governor utilized in proportional and integral speed control (see Secs. 8-2 and 8-3) is an excellent example of the application of the linearization process as presented here. A schematic diagram of a flyweight governor that could be used in either a proportional or an integral controller is shown in Fig. 14-12.

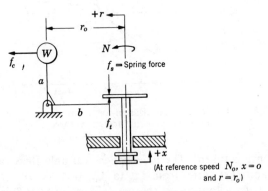

FIG. 14-12. Schematic drawing of flyweight governor.

The inertia force f_c is a function of two variables, the radius and the speed. Specifically,

$$f_c = \frac{W}{g} r\omega^2 = \frac{W}{g}\left(\frac{2\pi}{60}\right)^2 rN^2 = K_1 rN^2$$

where r is in inches and N is in revolutions per minute. The linear approximation which must be made at this point is that

$$\Delta f_c \cong \Delta f_{c\tan}$$

---

[3]This result may be verified by referring to any text on calculus and analytic geometry or advanced calculus.

with the tangent taken at the reference speed N_o. The corresponding reference radius is r_o. Thus,

$$\Delta f_{c\text{tan}} = \left.\frac{\partial f_c}{\partial r}\right|_{(r_o, N_o)} \Delta r + \left.\frac{\partial f_c}{\partial N}\right|_{(r_o, N_o)} \Delta N$$

Evaluating the partial derivatives and substituting gives

$$\Delta f_c \cong K_1 N_o^2(\Delta r) + 2K_1 r_o N_o(\Delta N)$$

From the geometry of Fig. 14-12, $\Delta r \cong (a/b)\Delta x$. Also, from $\Sigma F = 0$, $f_s = f_t = (a/b)f_c$. Therefore,

$$\Delta f_c = \frac{b}{a} \Delta f_s = \frac{b}{a} k_s \Delta x$$

where k_s is the spring rate (taken as a linear spring in this case). The final relationship between the motion x and governor speed N is then

$$\Delta x = \frac{2(a/b)K_1 r_o N_o}{k_s - K_1 N_o^2(a/b)^2} \Delta N$$

It is convenient to define the numerator as the ball-head rate K_b and the denominator as the effective spring rate K_o. Thus, for x representing changes in position taken from a zero reference position x_o,

$$x = \frac{K_b}{K_o}(\Delta N) \tag{14-6}$$

The linear-approximation approach has provided an expression for a position x (i.e., the valve-spool position) as a function of speed change ΔN, provided that the fluctuation in speed N is sufficiently small around N_o so that the plane tangent to the surface defining $f_c = K_1 r N^2$ gives a good approximation of the variation in inertia force.[4]

Describing Function Analysis. The material of Chapters 9, 10, 11, and 13 have illustrated the usefulness of the frequency-response approach in the study of linear systems. The describing function analysis is a frequency-response method for handling nonlinear systems. At the same time it must be placed in the category of a linearization technique.

The describing function method of analysis is based on the premises that the input to the nonlinear device is sinusoidal and that the oscillatory portion of the output is at the frequency of the input. Before progressing further to the mechanics of the method it is necessary to point out the restrictions on the type of nonlinear systems that may be handled by a describing function analysis.

The first restriction is that the system may have only one nonlinear element. If there is more than one nonlinear element, the system may be handled if the elements are arranged so that they may be lumped into a

[4]It should be pointed out that Eq. 14-6 is valid only in cases where the frequency of speed variations is small compared with the natural frequency of the governor.

single nonlinear element. A typical nonlinear control system block diagram is shown in Fig. 14-13. The nonlinear element is denoted by N and is

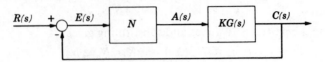

FIG. 14-13. Block diagram of a typical nonlinear control system.

shown in series with the system elements $KG(s)$. Except in very special cases, an attempt to consider more than one isolated nonlinearity leads to extreme complication.

A second restriction is that the nonlinearity not be time varying. That is, N may not be a function of time. In general, N is a function of the input $e(t)$ and may depend on both the frequency and magnitude of $e(t)$.

Figure 14-13 is a block diagram of a general nonlinear control system in which the nonlinear block need not be in the location shown. It could occur anywhere in the forward path or in the feedback path of the system. The procedure that is involved in the analysis includes the following steps:

1. Establish the actual output of the nonlinear element for a sinusoidal input.
2. Eliminate negligible quantities in the input-output relationship (this involves elimination of the higher harmonics in a harmonic series) to obtain the describing function N.
3. Make a polar plot (Nyquist diagram) of $KG(j\omega)$ and $-1/N$ and establish the nature of stability.

Each of these steps will be explained and illustrated in the following remarks.

Developing a Describing Function. The first step in describing function analysis is to establish the relationship between the output and the input of the nonlinear element when the input is a simple sinusoid. The expression for this transfer relationship that remains after neglecting small terms is the describing function N.

As an illustration of the procedure for determining the describing function, consider the previously discussed case of coulomb friction. The nonlinear block for this element is shown in Fig. 14-14a and the actual friction force as a function of the velocity $\dot{x}$ is given in Fig. 14-14b. Here, the velocity input to the nonlinear element is represented as a sine wave on a plot of $\dot{x}$ versus t. If projections are made vertically upward from the input function at various times, the output function may be obtained as illustrated. In this simple case any positive value of $\dot{x}$ results in a negative friction force while any negative value of $\dot{x}$ results in a positive friction force.

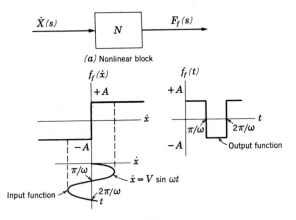

(a) Nonlinear block

(b) Coulomb friction with input and output

Fig. 14-14. A typical nonlinear element.

The negative sign may be accounted for in writing the differential equation. The resulting output is clearly a square-wave function. The output in this case is a periodic function and has the same frequency as the input. As frequency analysis is carried out with sinusoidal functions, it is now desirable to express the square wave in a different form. It is necessary then to call upon the theory of Fourier series.

Any periodic function that is sufficiently well behaved may be represented by a Fourier sine-cosine series of the form

$$f(t) = b_o + \sum_{n=1}^{\infty} (a_n \sin n\omega t + b_n \cos n\omega t) \qquad (14\text{-}7)$$

where b_o, a_n, and b_n are constant coefficients obtained by the following integrals

$$b_o = \frac{\omega}{2\pi} \int_0^{2\pi/\omega} f(t)\, dt \qquad (14\text{-}8)$$

$$a_n = \frac{\omega}{\pi} \int_0^{2\pi/\omega} f(t) \sin n\omega t\, dt \qquad (14\text{-}9)$$

$$b_n = \frac{\omega}{\pi} \int_0^{2\pi/\omega} f(t) \cos n\omega t\, dt \qquad (14\text{-}10)$$

The square wave of Fig. 14-14b may thus be expressed by a sum of sines and cosines where the constant coefficients are determined by Eqs. 14-8, 14-9, and 14-10. In this case,

$$f(t) = +A \qquad 0 \leq t \leq \frac{\pi}{\omega}$$

$$\qquad (14\text{-}11)$$

$$f(t) = -A \qquad \frac{\pi}{\omega} \leq t \leq \frac{2\pi}{\omega}$$

The term b_o represents the average value of the function over one cycle; thus, by symmetry, it is zero. To verify this,

$$b_o = \frac{\omega}{2\pi} \int_0^{\pi/\omega} A \, dt + \frac{\omega}{2\pi} \int_{\pi/\omega}^{2\pi/\omega} (-A) \, dt = 0 \qquad (14\text{-}12)$$

The coefficients of the sine terms are

$$a_n = \frac{\omega A}{\pi} \int_0^{\pi/\omega} \sin n\omega t \, dt - \frac{\omega A}{\pi} \int_{\pi/\omega}^{2\pi/\omega} \sin n\omega t \, dt$$

$$= \frac{-2A\omega}{\pi} \left(\frac{1}{n\omega}\right) (\cos n\pi - 1)$$

$$= \frac{2A}{n\pi} [1 - (-1)^n]$$

or $\qquad\qquad a_n = \dfrac{4A}{n\pi} \qquad$ for $\quad n = 1, 3, 5, \cdots \qquad\qquad (14\text{-}13)$

Evaluation of Eq. 14-10 for $f(t)$ between $0 \le t \le 2\pi/\omega$ gives

$$b_n = 0 \qquad\qquad\qquad (14\text{-}14)$$

The output function in Fig. 14-14 may then be represented by

$$f(t) = \frac{4A}{\pi} \left(\sin \omega t + \frac{1}{3} \sin 3\omega t + \frac{1}{5} \sin 5\omega t + \cdots\right) \qquad (14\text{-}15)$$

Equation 14-15 is a Fourier series representation of $f(t)$, and $\sin \omega t$ is the fundamental of the series. The approximation that is made in the describing function analysis is that this fundamental is a good representation of the output of the nonlinear element. Thus

$$f_f(t) = \frac{4A}{\pi} \sin \omega t \qquad\qquad (14\text{-}16)$$

This approximation is justified by two facts; first, each of the succeeding terms is reduced in magnitude by its respective n. The amplitude of the third harmonic, for example, is only one-third that of the fundamental. Second, each of the succeeding terms is at a frequency n times higher (the third harmonic is at frequency 3ω), and it is generally typical for the linear portions of the control system to attenuate higher frequencies. In the analysis process these factors must be weighed to justify the approximation made.

The describing function is defined as the ratio of the fundamental component of the output to the amplitude of the sinusoidal input. Thus, in this case

$$N = \frac{4A}{\pi V} \qquad\qquad (14\text{-}17)$$

It is interesting to point out what the describing-function approximation is actually doing. In the case of the coulomb friction nonlinearity the actual discontinuous force defined by Eq. 14-11 is being approximated by a linear "viscous" friction force with coefficient of viscous friction $c = 4A/\pi V$. The linearized friction force is then

$$F_f = \frac{4A}{\pi V} \dot{x}$$

Stability Analysis with the Describing Function. The stability of the system described by the block diagram of Fig. 14-13 may be investigated using the polar plot as discussed in Chapter 10. The criterion for stability must, however, be modified to consider the describing function N in the block diagram. The closed-loop transfer function of Fig. 14-13 is given by

$$\frac{C(j\omega)}{R(j\omega)} = \frac{NKG(j\omega)}{1 + NKG(j\omega)} \tag{14-18}$$

The stability criterion on the polar plot in this case is obtained in a fashion analogous to the arguments presented in Chapter 10 leading to Eq. 10-7. In Chapter 10 the result was that $KG(j\omega) = -1$ provided the necessary and sufficient condition for instability, and one was concerned with the encirclement of the point $-1 + j0$. In the case of the nonlinear system with describing function N, instability will result if

$$KG(j\omega) = -\frac{1}{N} \tag{14-19}$$

The value of N may be thought of as establishing the critical point which was $-1 + j0$ for the linear system.

The procedure for the analysis of stability of nonlinear systems where N is not a function of frequency involves the following:

1. Plot the frequency variant portion $KG(j\omega)$ on the polar diagram.
2. Plot the magnitude variant portion $-1/N$ on the polar diagram.
3. The relative position of the $-1/N$ locus and the $KG(j\omega)$ locus gives an indication of the stability condition of the nonlinear system. If $KG(j\omega)$ is by itself a stable transfer function, then the critical points from $-1/N$ must lie to the left of $KG(j\omega)$ (and thus not be enclosed by the Nyquist contour) for the nonlinear closed-loop system to be stable regardless of the oscillation amplitudes. Unstable operation of the closed-loop system can occur if the $KG(j\omega)$ locus touches or encloses any part of the $-1/N$ locus and oscillations are such that they correspond to enclosed points on the locus.

Figure 14-15 illustrates the concepts in point 3. These Nyquist diagrams are typical of the plots obtained for a unity feedback control system with the transfer function $K/s(\tau_1 s + 1)(\tau_2 s + 1)$ where the system is controlled by a relay contractor with dead zone, as shown in Fig. 14-8. Stable operation of the closed-loop system is insured for the gain setting K_a shown in Fig. 14-15a. The gain setting K_a is not sufficiently large to cause instability

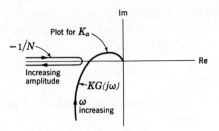

(a) Stable operation of nonlinear system

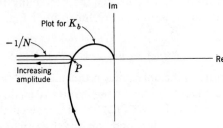

(b) Limitedly stable nonlinear system

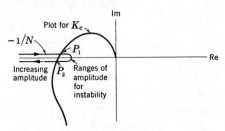

(c) Possibilities for an unstable nonlinear system

FIG. 14-15. Illustration of stability criterion for a nonlinear system.

for any magnitude of oscillation. In Fig. 14-15*b* the system gain is increased to K_b, and it is on the border of instability in that for a particular frequency and amplitude the system will have a sustained oscillation. Figure 14-15*c* illustrates the third possibility, that is, points which have amplitudes corresponding to $-1/N$ between P_1 and P_2 result in an unstable closed-loop system. Conversely, operation amplitudes to the left of points P_1 and P_2 represent stable operation. It may further be argued that P_1 is an unstable operating point while P_2 is a stable operating point.[5]

It should be pointed out that this discussion is intended as an introduction to the concepts of the describing function approach to nonlinear control problems and that no effort will be advanced to demonstrate the

[5]The reader is referred to a more advanced text on automatic controls for a detailed development of N and a discussion of these plots [19].

detail further. A further comment should be made to point out that a stability analysis with the describing function may be accomplished using either the Bode-plot techniques discussed in Chapter 11 or the root-locus concepts presented in Chapter 12.

14-4. Phase-Plane Analysis

One of the widely used engineering approaches to the analysis of the transient behavior of nonlinear systems is the semigraphical *phase-plane technique*. The method consists of constructing a plot of velocity versus displacement. Unlike the previous analysis of linear systems where it was found useful to transform the equations of motion into a new domain, the *s* domain, and further in the frequency-response methods where *s* was replaced by $j\omega$, the phase-plane analysis is performed in the time domain. In fact, time *t* may appear on the phase plane as a parameter of the curve. The curve obtained on the velocity versus displacements coordinates is frequently referred to as a *trajectory*. If one plots several trajectories on the phase plane for various initial states of the system the plot is termed the *phase portrait* of the system.

The most general treatment of this approach to the analysis of nonlinear systems involves the study of the behavior of a system in a *phase space*. If one is restricted to a *phase plane* then there are certain restrictions on the types and complexity of systems that may be studied. Because it is always valuable to be aware of the general restrictions on a method of analysis, they will be enumerated [32]:

1. The restriction to a phase plane limits the analysis to systems defined by first- or second-order equations.
2. The phase-plane approach is limited to the analysis of the transient behavior of systems with very simple inputs. Specifically, a system may have initial conditions and be otherwise unexcited. Step and ramp function inputs may also be handled by the phase-plane approach [32].

As an illustration of the phase-plane concept consider a spring-mass system defined by the differential equation

$$M\ddot{y} + ky = 0 \qquad (14\text{-}20)$$

with the initial conditions

$$y\,|_{t=0} = Y \qquad \text{and} \qquad \dot{y}\,|_{t=0} = 0$$

To construct a plot of $\dot{y}$ versus y, it is necessary to modify Eq. 14-20 so that $\dot{y}$ is expressed in terms of y. The first step is to recognize that

$$\ddot{y} = \frac{d(\dot{y})}{dt} = \frac{d\dot{y}}{dy}\frac{dy}{dt} = \dot{y}\frac{d\dot{y}}{dy}$$

Substituting for the second derivative in Eq. 14-20 and dividing by k gives

$$\frac{\ddot{y}}{\omega_n^2} \frac{d\dot{y}}{dy} + y = 0$$

or, multiplying by dy,

$$\frac{\dot{y}\,d\dot{y}}{\omega_n^2} + y\,dy = 0$$

The left-hand side of this equation may be integrated directly to give

$$\frac{\dot{y}^2}{2\omega_n^2} + \frac{y^2}{2} = C_1$$

where C_1 is a constant of integration. Multiplying through by 2 gives

$$\frac{\dot{y}^2}{\omega_n^2} + y^2 = C_o$$

The value C_o is now a constant that may be evaluated by substituting known values of y and $\dot{y}$ at some time t. The initial conditions give values of y and $\dot{y}$ at $t = 0$. In this case, $0 + Y^2 = C_o$. Thus,

$$\left(\frac{\dot{y}}{\omega_n}\right)^2 + y^2 = Y^2 \qquad (14\text{-}21)$$

Equation 14-21 is the equation of a circle with center at the origin and radius Y. It is often convenient to use $\dot{y}/\omega_n$ versus y on the phase-plane coordinates. The phase-plane plot, or trajectory, for this simple system is given in Fig. 14-16. Values of parameter t may be placed along the curve in this simple case as it is known that one cycle (traveling once around the circle) is completed in $t = 2\pi/\omega_n$ sec. Other values of initial deflection will give trajectories which are concentric circles on Fig. 14-16.

Introduction to Isoclinics. The next step is to investigate the more gen-

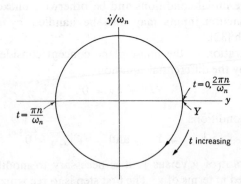

FIG. 14-16. Phase-plane plot for $M\ddot{y} + ky = 0$ with $\dot{y}\,|_{t=0} = 0$ and $y\,|_{t=0} = Y$.

eral case of the spring-mass-damper system. The differential equation of motion is

$$M\ddot{y} + c\dot{y} + ky = 0$$

with the initial conditions that $y\,|_{t=0} = Y$ and $\dot{y}\,|_{t=0} = 0$. Making the substitution $\ddot{y} = \dot{y}(d\dot{y}/dy)$ and rearranging gives

$$\frac{d\dot{y}}{dy} = -\frac{c}{M} - \frac{k}{M}\frac{y}{\dot{y}}$$

$$\frac{d(\dot{y}/\omega_n)}{dy} = -\frac{c}{M\omega_n} - \frac{y}{\dot{y}/\omega_n} = -2\zeta - \frac{y}{\dot{y}/\omega_n} \qquad (14\text{-}22)$$

In this case, the left-hand side of the final expression is the slope on the $\dot{y}/\omega_n$ versus y plane, and it may be evaluated for each point $P\,(\dot{y}_i/\omega_n,\, y_i)$. To facilitate plotting of the phase-plane trajectory, the left-hand side may be set equal to a constant, say, C_1, then

$$C_1 = -2\zeta - \frac{y}{\dot{y}/\omega_n}$$

or

$$\frac{\dot{y}}{\omega_n}(C_1 + 2\zeta) = -y$$

which is the equation of a straight line passing through the origin of the $\dot{y}/\omega_n$ versus y plane. This line represents the locus of all points where the trajectory has a slope C_1 and is called an *isoclinic*. In this case the isoclinics are straight lines as shown in Fig. 14-17. The proper slopes are drawn

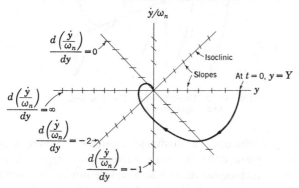

FIG. 14-17. Phase-plane plot for a spring-mass-damper system with less than critical damping ($\zeta = 0.5$).

along each isoclinic and then the trajectory is drawn in, starting at the initial state for $t = 0$ and following the proper slopes from isoclinic to isoclinic.[6] If, as in this case, the damping is less than critical ($\zeta < 1$) the

[6] For a detailed treatment of plotting techniques the reader is referred to reference [15].

trajectory will spiral toward the center as shown in Fig. 14-17. It may be noted that once the parameters of the system have been established, phase-plane trajectories for various initial conditions are all geometrically similar but with different starting points.

As the damping in the system is increased, there is less tendency for the trajectory to spiral toward the center (system oscillations are damped out more rapidly). For damping greater than critical ($\zeta > 1$) there will be two isoclinics that have the same slope as the derivative $d(\dot{y}/\omega_n)/dy$, which is the slope of the trajectory Thus, when this isoclinic is reached, the trajectory will follow that isoclinic back to the origin. These isoclinics may be found rapidly by setting $d(\dot{y}/\omega_n)/dy$ equal to the slope of the isoclinic $(\dot{y}/\omega_n)/y$ in Eq. 14-22. Thus,

$$\frac{\dot{y}/\omega_n}{y} = -2\zeta - \frac{y}{\dot{y}/\omega_n}$$

from which a value of $(\dot{y}/\omega_n)/y$ can be obtained for $\zeta > 1$. A typical family of trajectories for various initial conditions are shown in Fig. 14-18.

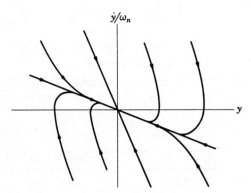

FIG. 14-18. Phase portrait for a second-order system with $\zeta = 1.4$.

Phase-Plane Techniques in Nonlinear Controls. Having seen the application of phase-plane concepts to the analysis of simple linear systems, we may extend the techniques to handle a simple nonlinear control system. Many nonlinear systems may be classified as piecewise linear systems. That is, the system behaves in a linear fashion for specific ranges. Thus, linear theory may be applied for each individual linear portion of the cycle. This is a powerful concept when used in connection with the phase-plane method.

Consider, for the purpose of illustration, the nonlinear feedback control system of Fig. 14-19. The nonlinearity in this case is an error detector with a dead zone. The transient response of the system will be investigated

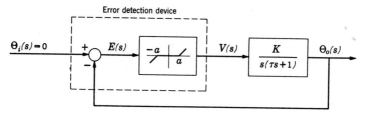

FIG. 14-19. Nonlinear control system with a dead zone in the error detector.

by letting the input $\theta_i(t)$ be zero and prescribing some initial output position $\theta_o(t)\,|_{t=0} = \Theta$. The equation of motion for the system under these conditions is

$$\frac{d^2\theta_o}{dt^2} + \frac{1}{\tau}\frac{d\theta_o}{dt} + \frac{K}{\tau}f(\theta_o) = 0$$

with $\qquad \theta_o(t)\,|_{\tau=0} = \Theta \qquad$ and $\qquad \dfrac{d\theta_o}{dt}\bigg|_{t=0} = 0$

This equation may be expressed in a form using previously defined general parameters

$$\frac{d^2\theta_o}{dt^2} + 2\zeta\omega_n\frac{d\theta_o}{dt} + \omega_n^2 f(\theta_o) = 0 \qquad (14\text{-}23)$$

The reader should recognize that this equation is the same as that which would be obtained by placing the spring of Fig. 14-2a in a classical spring-mass-damper system. In that case ω_n may be interpreted as the undamped natural frequency when the mass is in contact with the spring.

The control system is piecewise linear, and there are three distinct regions in which the motion may be defined by a linear equation. Substituting $\ddot{\theta}_o = \dot{\theta}_o(d\dot{\theta}_o/d\theta_o)$ in Eq. 14-23 and rearranging gives

$$\frac{d(\dot{\theta}_o/\omega_n)}{d\theta_o} = -2\zeta - \frac{f(\theta_o)}{\dot{\theta}_o/\omega_n} \qquad (14\text{-}24)$$

There are three regions in which Eq. 14-24 may be used to construct the phase-plane trajectory for this system. The first is defined by $\theta_o > a$. In this region,

$$\frac{d(\dot{\theta}_o/\omega_n)}{d\theta_o} = -2\zeta - \frac{\theta_o - a}{\dot{\theta}_o/\omega_n} \qquad \theta_o > a$$

Other regions are defined by

$$\frac{d(\dot{\theta}_o/\omega_n)}{d\theta_o} = -2\zeta \qquad -a \le \theta_o \le a$$

$$\frac{d(\dot{\theta}_o/\omega_n)}{d\theta_o} = -2\zeta - \frac{\theta_o + a}{\dot{\theta}_o/\omega_n} \qquad \theta_o < -a$$

Figure 14-20 presents the phase-plane trajectory for a case with $\zeta < 1$ and the corresponding nonlinear function $f(\theta_o)$. For the time interval $t_o \rightarrow t_1$, $\theta_o > a$ and the trajectory is constructed in the same manner as was the trajectory of Fig. 14-17. In the time interval $t_1 \rightarrow t_2$ the trajectory is a straight line with slope -2ζ. In terms of the spring-system analogy, during this interval the mass is coasting between the ends of the spring and is being decelerated by the viscous drag. Between t_2 and t_3, $\theta_o < -a$ and the motion is again defined by the linear spring-mass-damper relationship. The construction of the isoclinics for the two regions outside of the dead zone involves different origins. For the region of $+\theta_o$, the origin for the isoclinics is $+a$. For the region of $-\theta_o$, the origin for the isoclinics is $-a$.

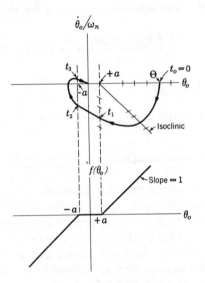

Fig. 14-20. Nonlinear function and transient response of a simple control system with a dead-zone error detector.

In the case where there is less damping or the dead zone is decreased, many cycles may occur before the system comes to rest. The segments of the trajectories obtained during the dead-zone interval result in parallel lines, all joined by decreasingly smaller spirals. It is important to recognize that the system may or may not come to rest with zero error. In fact, between the limits of $\pm a$, the system cannot detect that an error may exist because the output of the error detection device is zero.

Stability Analysis on the Phase Plane. For this introduction to phase-plane techniques, the concept of stability may be summarized in a few words. The phase-plane trajectory with parameter t is a time history of the behavior of the system when it is released from some initial disturbance. If the system is stable, the system will come to rest, that is, the velocity will go to zero and the displacement will reach some constant value. The phase trajectory must therefore proceed toward the equilibrium position. For example, in Fig. 14-17 the trajectory spiraled toward the origin.

There are two other possible alternatives for the nature of the trajectory. First, the trajectory could travel outward, or inward, to some limiting cycle on the phase plane (called a *limit cycle*) thus maintaining a sustained oscillation. Figure 14-21a shows a typical limit cycle which is approached from initial states either within (point A) or outside (point B) of the limit-cycle boundary.

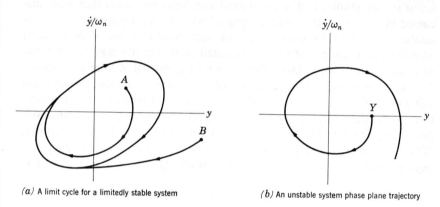

(a) A limit cycle for a limitedly stable system   *(b)* An unstable system phase plane trajectory

FIG. 14-21. Illustrations of phase-plane plots for limitedly stable and unstable systems.

A second possibility is that a system is unstable in which case the phase-plane trajectory spirals outward to infinity. This is illustrated in Fig. 14-21b. A phase plane that behaves in this fashion is easily visualized if one considers a second-order spring-mass system with negative damping. Nonlinear systems may also have a combination of a limit cycle that is reached from one area of initial conditions, but be unstable for other possible initial conditions.

The reader is again referred to one of many references on nonlinear analysis, in general, and on nonlinear control systems, in particular [15].

14-5. The Analog Computer in Nonlinear Analysis

The analog computer is an invaluable tool for the analysis of nonlinear systems. Most of the analytical and graphical techniques that are available for the analysis of nonlinear systems give the engineer a limited amount of information about a system's behavior. That is, they can be used to establish the nature of system stability, and they can, in general, give some information about system response for certain types of simple inputs. Frequently, the designer would like to be able to study the behavior of a nonlinear system for many settings of certain parameters, or investigate the response to unusual inputs, or as is often the case, investigate the behavior of a nonlinear system when it is operated in conjunction with another system that is actually available. The latter aspect falls in the category of computer simulation of portions of an actual system (e.g., a computer simulation of a gas turbine may be controlled by an actual governor). In these cases the analog computer is a flexible and effective design tool.

The main reason that analog computation is so well suited for nonlinear system studies is that the typical nonlinear functions that were discussed in Sec. 14-2 are easily generated with analog equipment. Simple nonlinearities may be easily obtained with combinations of operational amplifiers and diodes. More sophisticated nonlinearities may be obtained with nonlinear function fitters (which will be discussed in later comments).

As an example of the generation of a nonlinear function, the absolute value of a function offers an easily understood illustration. The function $y = |x|$ is shown in Fig. 14-22a. Figure 14-22b shows one possible computer setup for the generation of this function. The operation is as follows. When x is positive, diode 1 does not conduct; the output of amplifier 1 is negative and so diode 2 conducts. The resulting output of amplifier 2 is

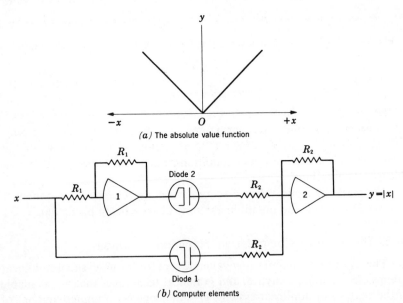

(a) The absolute value function

(b) Computer elements

Fig. 14-22. Generation of the absolute value function.

positive. (Recall from Chapter 4 that each amplifier changes the sign.) When x is negative, diode 1 conducts, but diode 2 does not. The resulting output of amplifier 2 is again positive. This is a particularly simple non-linearity to generate. A wide variety of computer circuits for various non-linearities including coulomb friction, dead zone, hysteresis, and so forth, are presented in most texts on analog computing including references [17, 18].

Example of Solution of a Nonlinear Differential Equation. A mechanical system with backlash is shown in Fig. 14-23a. The spring force acting on the mass is defined by the function $f(y - x)$ shown in Fig. 14-23b. The

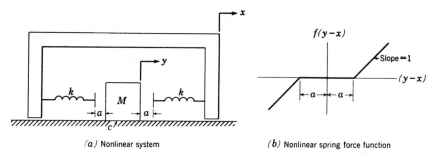

(a) Nonlinear system   *(b)* Nonlinear spring force function

Fɪɢ. 14-23. A mechanical system with backlash.

spring force is given by $f_s = kf(y - x)$. Applying Newton's law yields $M\ddot{y} = -c\dot{y} - kf(y - x)$. Thus the differential equation of motion for the system is $M\ddot{y} + c\dot{y} + kf(y - x) = 0$. The best form of this equation for computer solution is

$$\ddot{y} = -[+2\zeta\omega_n\dot{y} + \omega_n^2 f(y - x)] \tag{14-25}$$

The computer diagram for the solution of this equation is given in Fig. 14-24. The portion of the circuit that generates the dead zone is shown enclosed in the dashed rectangle of the figure. The output of amplifier 5, $f(y - x)$, is obtained by the superposition of a straight line and a limiter formed by diodes 1 and 2. The problem of scaling has not been discussed in the presentation of the solution. Generally, it is not a difficult task, but care must be taken in scaling the nonlinear element.

Nonlinear Function Generator. The nonlinear function generator is an electronic device that may be utilized to obtain a very good approximation of a single-valued nonlinear function. The concept of its operation is the following. It may be thought of as a "black box" composed of a multitude of line segments each of which has an adjustable break point and slope. By adjusting the breakpoints and slopes properly, a curve may be fitted to some desired nonlinear function. Figure 14-25 is a plot of a nonlinear spring force versus the deflection. A series of straight lines are adjusted to

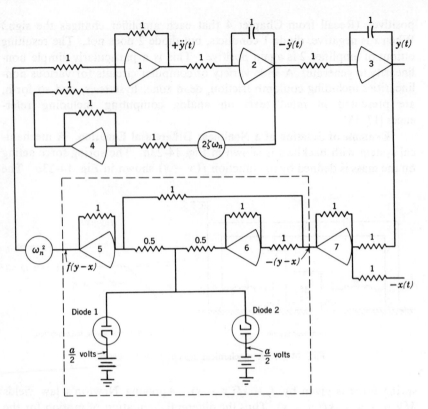

Fig. 14-24. Computer solution for the nonlinear mechanical system.

fit the curve, as illustrated in the figure. Although six straight lines have been shown, more than twice that many are generally available in the electronic devices used with most computers. To use this device, the variable x is fed in and the output $f(x)$ is multiplied by k to obtain the spring force.

These few comments on analog computing serve to illustrate the fact that analog computers are not at all limited to the study of linear systems. They are in many respects the engineer's best tool for the study of nonlinear phenomenon.

14-6. Closure

An introduction to some of the concepts underlying nonlinear system analysis has been presented. The analysis of a nonlinear control system begins with the identification of the nonlinearities present. The discussion in Sec. 14-2 considers how nonlinearities enter into some very common physical devices. The examples cited help to illustrate the point.

The foremost engineering approach to treating a nonlinear phenome-

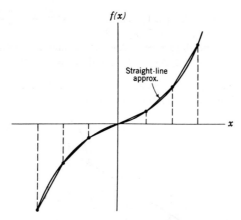

FIG. 14-25. Use of the nonlinear function generator.

non has been, in the past, to attempt to justify a linearization approach. The methods cited for linearization included a straight-line approximation of a curve for a limited change of a variable and the describing function analysis. The former is particularly valuable for initial estimates and rough analysis of a system's behavior. The latter is considerably limited in its use to frequency-response analysis.

The phase-plane method is a powerful tool, particularly when it is combined with the concept of piecewise linearity. It was pointed out that many systems are piecewise linear and may be solved with linear theory in a step-by-step fashion from one linear region to the next. For this latter approach one may turn to a digital computer.

The analog computer has also been discussed as a tool for handling nonlinear systems, particularly in the actual design process. Portions of systems may be simulated and combined with other actual hardware to circumvent initial prototype stages which could be prohibitively expensive.

In all cases, there is one comment that must not be overlooked. Each approach to handling nonlinear systems has certain advantages, limitations, and restrictions. Special procedures and methods are often necessary because nonlinear systems do not behave as nicely as linear systems. Specifically, the principle of superposition is no longer valid.

Problems

14-1. A spring rate is defined by the relationship $k(x) = 10^4 x^2$, where x is the deflection. The spring is to operate with a mean deflection of 0.1 in. Make a linear approximation for the incremental spring force Δf and calculate the maximum per cent error in the total spring force if the linear approximation is to be used between the limits of deflection $0.09 \leq x \leq 0.11$ in.

14-2. A function w is defined by $w(x,y,z) = \sqrt{x^2 + y^2 + z^2}$, where the reference points are $x_o = 1$, $y_o = 2$, $z_o = 3$.

Calculate the per cent difference in Δw, actual, and Δw, obtained by a linear approximation. if $\Delta x = -0.02$, $\Delta y = 0.01$, and $\Delta z = -0.09$.

14-3. A torsional pendulum system is shown in the accompanying illustration. a) Write the differential equation of motion for the system considering θ_i as

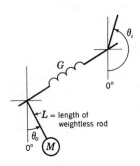

PROB. 14-3

the input. Identify the nonlinearity in the system. b) Linearize the differential equation for the condition that motions of the pendulum are on the order of a maximum of ± 0.1 radians from the vertical. What is the natural frequency of the pendulum in terms of the parameters of the system? c) If $L = 10$ in., $G = 1000$ lb-in./radian, $M = 1$ lb-sec^2/in., and $\theta_i = 0.05 \sin 5.6t$, would the linear approximation be valid for the solution of this problem? Explain. [The condition from part b is that the linear approximation is satisfactory for motions θ_o up to ± 0.1 radian.]

14-4. Discuss the importance of the footnote regarding the restriction of the validity of Eq. 14-6. In particular, what aspects of the dynamic problem have been omitted in this discussion of the flyweight governor.

14-5. Derive an expression for the describing function representing a deadzone nonlinear element in a control system as illustrated in the accompanying illustration.

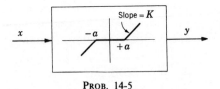

PROB. 14-5

14-6. Make an analog computer study of the nonlinear control system of Fig. 14-19. (It is valuable in this case to utilize the concepts presented in Appendix D.)

a) For this analysis let $\tau = 0.1$, $K = 5$, $0 \leq a \leq 5$, and $\theta_i = 10 \sin 7t$. Compare several solutions for values of a between the stated limits. b) Plot several trajectories on the phase plane for this system using values from part (a) and compare the results with phase plane trajectories which may be obtained directly from the analog computer solution.

14-7. Solve the following equations by the phase-plane technique (i.e., construct phase-plane plots).

a) $10\ddot{y} + 5000y = 0$ $\dot{y}\,|_{t=0} = 0$ $y\,|_{t=0} = 10$

b) $\ddot{y} + 2\dot{y} + 100y = 0$ $\dot{y}\,|_{t=0} = 10$ $y\,|_{t=0} = 50$

14-8. Make a phase-plane plot for the system of Fig. 14-2a with $M = 5$ lb-sec^2/in., $k = 5000$ lb/in., and $a = 1$ in. Place values of t along the plot and then construct a plot of x versus t. The initial conditions on the system are $\dot{x}\,|_{t=0} = 0$ and $x\,|_{t=0} = 2.0$ in.

14-9. For the system of Prob. 14-8 make a step-by-step "piecewise" solution for the motion of the mass. (Solve each linear equation over its linear range.) Plot a curve for x versus t and compare with the solution from Prob. 14-8.

Chapter **15**

Introduction to Advanced Topics

15-1. Introduction

Control theory, as we know it today, was almost nonexistent until about 1940 (although mathematical techniques had been applied to specific control problems long before this date). The body of knowledge has grown rapidly since then. During its development, control theory was influenced by the control problems encountered, and the human capabilities and physical facilities which were available to effect solutions. Thus, the development of control theory can be viewed as an evolutionary process.

During the late 1950's, a marked change in outlook occurred; one might say that evolution became revolution. The emerging theory became known as *modern control theory*. The most important factor which prompted the change was the general availability of the high-speed digital computer. This device not only offers a ready means for problem solution but can also be used as a part of the control system.

The use of the digital computer as an effective means for problem solution is unquestioned. However, in terms of control theory, an additional benefit is obtained. The digital computer requires that problems be formulated in a compatible way. This requirement, in turn, encourages the development of a unified approach to the treatment of control systems —a worthwhile result in its own right.

As a part of the control system, the digital computer is capable of receiving large quantities of data, performing sophisticated calculations, and determining suitable corrective actions. Its performance can be far superior to that obtained with conventional controllers. (Of course, the resulting improvement in control may or may not justify the additional expense of the computer.)

There are other factors which have influenced control theory. Competitive pressures, between companies and even nations, have resulted in more stringent performance requirements. Also, the systems which must be controlled have become increasingly more complex. Both factors have aggravated the control problem and have forced a re-evaluation of analytical and design techniques.

An increased level of research activity is yet another factor which has shaped control theory. Such activity has produced a vast body of knowledge, so much so, that the available theory is now well ahead of realistic practice.

The purpose of this chapter is to introduce a number of advanced topics, some of which are typically considered to be a part of modern control theory. A detailed treatment of the topics is, of course, beyond the scope of the book. The intention is to provide sufficient background so that the reader can understand some of the more recent concepts and can appreciate their potentialities.

15-2. State Variables [8, 23]

The *state* of a dynamic system is a quantitative description of the condition, or state, of the system at a particular instant in time. For example, the current state of some hypothetical process could be expressed by the statement that the pressure is 25 psi and the temperature is 150°F. In this case, pressure and temperature are the *state variables*, that is, the variables which can be used to define the state of the system. Of course, there must be a sufficient number of state variables to completely define the system.

The mechanical system of Fig. 15-1 will serve as a convenient exam-

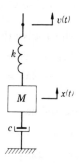

FIG. 15-1. Illustrative system.

ple for illustrating state-variable concepts. The defining equation is

$$M \frac{d^2x}{dt^2} + c \frac{dx}{dt} + kx = kv(t)$$

or

$$\frac{d^2x}{dt^2} + \frac{c}{M} \frac{dx}{dt} + \frac{k}{M} x = \frac{k}{M} v(t)$$

To permit specific solutions, let $c/M = 3$ and $k/M = 2$. Therefore, the system is defined by the equation

$$\frac{d^2x}{dt^2} + 3\frac{dx}{dt} + 2x = 2v(t) \qquad (15\text{-}1)$$

We now choose displacement and velocity as state variables and designate them as x_1 and x_2 respectively. The defining equations are now

$$\begin{aligned}\dot{x}_1 &= x_2 \\ \dot{x}_2 &= -2x_1 - 3x_2 + 2v(t)\end{aligned} \qquad (15\text{-}2)$$

The first equation results from the definition of velocity, namely, velocity is the time derivative of displacement. The second equation is Eq. 15-1 expressed in terms of acceleration. Equation 15-2 is in state-variable form and defines the system by two first-order differential equations instead of a single second-order equation, Eq. 15-1.

Note that the two state variables chosen are sufficient to define the system completely. Given values of displacement and velocity at some time $t = t_0$, it is possible to predict future states from a knowledge of the dynamics of the system and the input.

Equation 15-2 can be solved in a number of ways. One approach is to use matrix methods. Before proceeding with this technique, a brief review of matrix operations is presented.

Matrix Review. It is assumed that the reader is at least partially familiar with matrix manipulations and so only a short review (rather than an introduction to the subject) is in order. The review will be based on an example involving the solution of two simultaneous equations.

Let us suppose that we wish to solve the equations

$$\begin{aligned}2x_1 + x_2 &= 7 \\ 3x_1 - 2x_2 &= 0\end{aligned} \qquad (15\text{-}3)$$

by matrix methods.

Equation 15-3 can be expressed as

$$\begin{bmatrix} 2x_1 + x_2 \\ 3x_1 - 2x_2 \end{bmatrix} = \begin{bmatrix} 7 \\ 0 \end{bmatrix} \qquad (15\text{-}4)$$

The justification is that two matrices are equal if their corresponding elements are equal.

With the concept of matrix multiplication, it should be recognized that

$$\begin{bmatrix} 2 & 1 \\ 3 & -2 \end{bmatrix}\begin{bmatrix} x_1 \\ x_2 \end{bmatrix} = \begin{bmatrix} 2x_1 + x_2 \\ 3x_1 - 2x_2 \end{bmatrix}$$

Therefore, Eq. 15-4 can be expressed as

$$\begin{bmatrix} 2 & 1 \\ 3 & -2 \end{bmatrix} \begin{bmatrix} x_1 \\ x_2 \end{bmatrix} = \begin{bmatrix} 7 \\ 0 \end{bmatrix} \qquad (15\text{-}5)$$

Equation 15-5 can be written in matrix notation, the result being

$$AX = B \qquad (15\text{-}6)$$

where $\quad A = \begin{bmatrix} 2 & 1 \\ 3 & -2 \end{bmatrix} \quad X = \begin{bmatrix} x_1 \\ x_2 \end{bmatrix} \quad B = \begin{bmatrix} 7 \\ 0 \end{bmatrix}$

Here A is a 2×2 square matrix, X is a column matrix, or vector, and B is a column matrix.

The solution of Eq. 15-6 is facilitated by a form of matrix "division." We write

$$A^{-1}AX = A^{-1}B \qquad (15\text{-}7)$$

where A^{-1} is the inverse of A. Because

$$A^{-1}A = I \qquad (15\text{-}8)$$

and $\qquad\qquad\qquad IX = X \qquad (15\text{-}9)$

Eq. 15-7 becomes

$$X = A^{-1}B \qquad (15\text{-}10)$$

(We will verify Eqs. 15-8 and 15-9 at the conclusion of the example.)

The problem now is to determine the inverse matrix A^{-1}. First, the cofactor matrix is obtained through the use of the relationship

$$\text{cofactor } a_{ij} = (\text{minor})(-1)^{i+j}$$

Therefore, the cofactor matrix of A is

$$\text{cof } A = \begin{bmatrix} -2 & -3 \\ -1 & 2 \end{bmatrix}$$

The adjoint matrix is the transpose of the cofactor matrix, that is, the matrix obtained by interchanging the rows and columns of the cofactor matrix.

$$\text{adj } A = (\text{cof } A)^T$$

$$\text{adj } A = \begin{bmatrix} -2 & -1 \\ -3 & 2 \end{bmatrix}$$

The determinant of A is

$$|A| = a_{11}a_{22} - a_{12}a_{21}$$
$$|A| = 2(-2) - (1)(3) = -7$$

The inverse matrix is determined from

$$A^{-1} = \frac{\text{adj } A}{|A|}$$

Therefore,

$$A^{-1} = -\frac{1}{7} \begin{bmatrix} -2 & -1 \\ -3 & 2 \end{bmatrix} = \begin{bmatrix} 2/7 & 1/7 \\ 3/7 & -2/7 \end{bmatrix}$$

With Eq. 15-10,

$$X = A^{-1}B$$

$$X = \begin{bmatrix} 2/7 & 1/7 \\ 3/7 & -2/7 \end{bmatrix} \begin{bmatrix} 7 \\ 0 \end{bmatrix} = \begin{bmatrix} 2 \\ 3 \end{bmatrix} = \begin{bmatrix} x_1 \\ x_2 \end{bmatrix}$$

Thus, the required solution is $x_1 = 2$ and $x_2 = 3$.

Equations 15-8 and 15-9 are verified as follows.

$$A^{-1}A = \begin{bmatrix} 2/7 & 1/7 \\ 3/7 & -2/7 \end{bmatrix} \begin{bmatrix} 2 & 1 \\ 3 & -2 \end{bmatrix} = \begin{bmatrix} 1 & 0 \\ 0 & 1 \end{bmatrix} = I$$

$$IX = \begin{bmatrix} 1 & 0 \\ 0 & 1 \end{bmatrix} \begin{bmatrix} x_1 \\ x_2 \end{bmatrix} = \begin{bmatrix} x_1 \\ x_2 \end{bmatrix} = X$$

It is worth remarking at this stage that matrix operations are easily handled with the aid of a digital computer. This is one of the reasons for the current popularity of matrix methods.

Matrix Equations and Solutions. The state-variable equations (Eq. 15-2) for the spring-mass-damper system of Fig. 15-1 are

$$\dot{x}_1 = x_2$$

$$\dot{x}_2 = -2x_1 - 3x_2 + 2v(t)$$

Although these equations can be solved in a variety of ways, we will, for the moment, confine our attention to matrix methods. The equations can be written in matrix form as

$$\frac{d}{dt}\begin{bmatrix} x_1 \\ x_2 \end{bmatrix} = \begin{bmatrix} 0 & 1 \\ -2 & -3 \end{bmatrix} \begin{bmatrix} x_1 \\ x_2 \end{bmatrix} + \begin{bmatrix} 0 \\ 2 \end{bmatrix} v(t)$$

or

$$\frac{dX}{dt} = AX + Bv(t) \tag{15-11}$$

The solution of the matrix, or vector, equation (Eq. 15-11) will be deduced from that of the single first-order equation

$$\frac{dx}{dt} = ax + bv$$

With Laplace transformation,

$$sX(s) - x(0) = aX(s) + bV(s)$$

$$X(s) = \frac{x(0)}{s-a} + \frac{bV(s)}{s-a}$$

$$x(t) = x(0)e^{at} + \int_0^t e^{a(t-\tau)}bv(\tau)d\tau \tag{15-12}$$

The second term on the right side of Eq. 15-12 is, of course, the convolution integral.

The solution for a set of first-order equations expressed in matrix form as

$$\frac{dX}{dt} = AX + BV \tag{15-13}$$

is, by analogy from Eq. 15-12,

$$X = e^{At}X(0) + \int_0^t e^{A(t-\tau)}BV(\tau)d\tau \tag{15-14}$$

where V is an input vector. One notes that the matrix e^{At} appears in both terms of the solution. Because of its importance, this matrix is often called the *fundamental matrix*. It is also called the *transition matrix* because of its role in the determination of a system's transition from one state to the next. Two methods for obtaining the transition matrix are now presented.

Consider Eq. 15-13 with zero inputs, that is,

$$\frac{dX}{dt} = AX$$

With Laplace transformation,

$$sX(s) - X(0) = AX(s)$$
$$sIX(s) - AX(s) = X(0) = [sI - A]X(s)$$
$$X(s) = [sI - A]^{-1}X(0)$$
$$X(t) = \mathcal{L}^{-1}[X(s)] = \mathcal{L}^{-1}\left([sI - A]^{-1}\right)X(0) \tag{15-15}$$

A comparison of Eqs. 15-15 and 15-14 (with zero inputs) reveals that

$$e^{At} = \mathcal{L}^{-1}\left([sI - A]^{-1}\right) \tag{15-16}$$

Thus, Eq. 15-16 provides one means for the determination of the transition matrix.

A second means for obtaining the transition matrix is to use the relationship

$$e^{At} = I + At + \frac{A^2 t^2}{2!} + \frac{A^3 t^3}{3!} + \cdots \tag{15-17}$$

Due to convergence problems, Eq. 15-17 is of limited value (except for almost trivial cases) if one wishes the matrix to be a function of t. Equation 15-17 is, however, useful if one desires the numerical solution of a set of equations. More will be said about this when computer solutions are discussed.

The spring-mass-damper system of Fig. 15-1 provides a convenient example to illustrate the solution of vector equations. The defining equation, Eq. 15-11, is

$$\frac{dX}{dt} = AX + Bv(t) \tag{15-18}$$

where
$$A = \begin{bmatrix} 0 & 1 \\ -2 & -3 \end{bmatrix}$$

Let us consider the response of the system with zero input and initial conditions $x_1(0) = -1$ and $x_2(0) = 0$, that is,

$$X(0) = \begin{bmatrix} -1 \\ 0 \end{bmatrix} \tag{15-19}$$

The *shape* of the response curve will be the same as that which would result from subjecting the system, initially at rest, to a unit step input. Note that avoiding an input, through the strategic use of initial conditions, eliminates the complication of having to work with the convolution integral. From Eq. 15-14, the required solution is

$$X = e^{At} X(0) \tag{15-20}$$

We must now obtain the transition matrix in order to complete the solution.

With Eq. 15-16,

$$e^{At} = \mathcal{L}^{-1}\left([sI - A]^{-1}\right)$$

Now,

$$[sI - A] = s\begin{bmatrix} 1 & 0 \\ 0 & 1 \end{bmatrix} - \begin{bmatrix} 0 & 1 \\ -2 & -3 \end{bmatrix} = \begin{bmatrix} s & -1 \\ 2 & s+3 \end{bmatrix}$$

and
$$|sI - A| = s(s+3) - (-1)(2) = s^2 + 3s + 2$$
$$= (s+1)(s+2)$$

(Mathematically speaking, $s = -1$ and $s = -2$ are *eigenvalues* of matrix A. However, they are also the roots of the system characteristic equation.) Next,

$$[sI - A]^{-1} = \frac{\text{adj}[sI - A]}{|sI - A|} = \frac{\begin{bmatrix} s+3 & 1 \\ -2 & s \end{bmatrix}}{(s+1)(s+2)}$$

Therefore,

$$e^{At} = \mathcal{L}^{-1} \begin{bmatrix} \dfrac{s+3}{(s+1)(s+2)} & \dfrac{1}{(s+1)(s+2)} \\ \dfrac{-2}{(s+1)(s+2)} & \dfrac{s}{(s+1)(s+2)} \end{bmatrix}$$

or
$$e^{At} = F(t) = \begin{bmatrix} f_{11} & f_{12} \\ f_{21} & f_{22} \end{bmatrix} \tag{15-21}$$

where

$$f_{11} = \mathcal{L}^{-1}\left(\frac{s + 3}{(s + 1)(s + 2)}\right) = 2e^{-t} - e^{-2t}$$

$$f_{12} = e^{-t} - e^{-2t}$$

$$f_{21} = -2e^{-t} + 2e^{-2t}$$

$$f_{22} = -e^{-t} + 2e^{-2t}$$

The solution is written as

$$X = e^{At}X(0)$$

or with Eqs. 15-19 and 15-21,

$$\begin{bmatrix} x_1 \\ x_2 \end{bmatrix} = \begin{bmatrix} f_{11} & f_{12} \\ f_{21} & f_{22} \end{bmatrix} \begin{bmatrix} -1 \\ 0 \end{bmatrix}$$

from which

$$x_1 = -f_{11} = -2e^{-t} + e^{-2t} \tag{15-22}$$

$$x_2 = -f_{21} = 2e^{-t} - 2e^{-2t} \tag{15-23}$$

As a check, it is encouraging to note that x_2 is indeed equal to $\dot{x}_1$.

The approach just illustrated is certainly a tedious one and would become even more so for a higher-order system. However, the basic ideas involved are not without merit. A point to remember is that the eigenvalues of the A matrix are the roots of the system characteristic equation; this point is important because digital computer programs are available which can be used to determine the eigenvalues of a given matrix. Saying somewhat the same thing in a different way, it should also be remembered that

$$|sI - A| = 0 \tag{15-24}$$

is the system characteristic equation.

Computer Solutions. One advantage of the state-variable formulation of system equations is that this form is very suitable for use with both the analog and the digital computer. Consider the equations (Eq. 15-2)

$$\dot{x}_1 = x_2$$

$$\dot{x}_2 = -2x_1 - 3x_2 + 2v(t)$$

In terms of an analog computer, the state variables are integrator outputs, while the right-hand sides of the equations define the integrator inputs. (It is worth remarking at this point that, given an analog computer diagram, the integrator outputs provide a suitable set of state variables for the system.)

The digital computer approach to the solution of an n-th order differential equation is to solve n first-order equations. Thus, the state-variable formulation of system equations is, again, quite compatible.

There are a number of methods by which a set of first-order differential equations can be solved numerically. One way is to make use of the transition matrix as given in Eq. 15-17, namely,

$$e^{At} = I + At + \frac{A^2 t^2}{2!} + \cdots$$

Using the previous example for which

$$A = \begin{bmatrix} 0 & 1 \\ -2 & -3 \end{bmatrix}$$

and choosing a time increment of 0.5 seconds,

$$e^{0.5A} = \begin{bmatrix} 1 & 0 \\ 0 & 1 \end{bmatrix} + \frac{1}{2} \begin{bmatrix} 0 & 1 \\ -2 & -3 \end{bmatrix} + \cdots + \frac{1}{2^5} \frac{1}{5!} \begin{bmatrix} 30 & 31 \\ -62 & -63 \end{bmatrix}$$

where the series is terminated with the $A^5 t^5 / 5!$ term. The resulting transition matrix is

$$e^{0.5A} = \begin{bmatrix} b_{11} & b_{12} \\ b_{21} & b_{22} \end{bmatrix}$$

where
$$b_{11} = 0.84636$$
$$b_{12} = 0.23984$$
$$b_{21} = -0.47968$$
$$b_{22} = 0.12682$$

Now, with Eq. 15-20,

$$\begin{bmatrix} x_1(0.5) \\ x_2(0.5) \end{bmatrix} = \begin{bmatrix} b_{11} & b_{12} \\ b_{21} & b_{22} \end{bmatrix} \begin{bmatrix} x_1(0) \\ x_2(0) \end{bmatrix}$$

and
$$\begin{bmatrix} x_1(1) \\ x_2(1) \end{bmatrix} = \begin{bmatrix} b_{11} & b_{12} \\ b_{21} & b_{22} \end{bmatrix} \begin{bmatrix} x_1(0.5) \\ x_2(0.5) \end{bmatrix}$$

The process can be repeated for as many times as desired. For calculation purposes, the appropriate recursive equations are

$$x_1(n + 1) = b_{11} x_1(n) + b_{12} x_2(n)$$
$$x_2(n + 1) = b_{21} x_1(n) + b_{22} x_2(n)$$

These equations will yield values in close agreement with those obtained from the exact solutions, Eqs. 15-22 and 15-23 (for which $x_1(0) = -1$ and $x_2(0) = 0$). Note that all of the calculations are easily programmed for the digital computer.

15-3. Stability [8, 23]

The concept of stability for linear systems is, of course, quite straightforward. Unfortunately, the approach used for linear systems is not, in general, applicable for nonlinear systems. The purpose of this section is to introduce a method which provides a unified approach to both linear and nonlinear systems.

Stability is considered on the basis of system response after the system is released from non-zero (that is, non-equilibrium) initial conditions; all inputs are zero. Three types of stability are recognized. A system may be *unstable*, where the meaning of instability is the same as that for linear systems. A system may be *stable*; this classification is analogous to limited stability for a linear system or a limit cycle for a nonlinear system. Finally, a system may be *asymptotically stable*, meaning that it returns to the equilibrium position.

The stability condition can be modified by the adjectives *globally* or *locally*. For example, a system may be *globally unstable* while another may be *locally asymptotically stable*. The term globally refers to the entire phase space. (Recall that a second-order system can be defined by two state variables, and the response can be displayed on a phase plane. Third-order and higher-order systems require a *phase space*.) Thus, a globally unstable system is one which is unstable throughout the entire phase space, while a locally asymptotically stable system is one which, for a limited portion of the phase space, will return to the equilibrium position.

The Energy Viewpoint. Stability can be viewed in terms of energy relationships. Consider the system shown in Fig. 15-2. We know that the

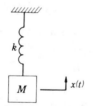

Fig. 15-2. Undamped system.

system will exhibit a persistent oscillation if released from a non-equilibrium position. Thus, it is stable in terms of our new definitions. Suppose $k/M = 1$. The defining equation, then, is

$$\ddot{x} + x = 0 \qquad (15\text{-}25)$$

We now choose displacement x_1 and velocity x_2 as state variables. Therefore,

$$\dot{x}_1 = x_2$$
$$\dot{x}_2 = -x_1 \qquad (15\text{-}26)$$

The system energy E is

$$E = \tfrac{1}{2}kx_1^2 + \tfrac{1}{2}M x_2^2$$

A function V, related to system energy, is formed as

$$V = x_1^2 + x_2^2 \tag{15-27}$$

Differentiation of Eq. 15-27 yields a new function W which is

$$W = \frac{dV}{dt} = 2x_1\dot{x}_1 + 2x_2\dot{x}_2$$

Substitution of Eq. 15-26 yields

$$W = 2x_1x_2 - 2x_1x_2 = 0 \tag{15-28}$$

What has been shown? Equation 15-27 reveals that V is positive for all non-zero values of x_1 and x_2, that is, there is energy in the system unless it is at rest. Equation 15-28 shows that $W = 0$, which means that the time rate of change of total system energy is zero. Since the energy remains in the system, one concludes that it will continue to oscillate and is, therefore, stable. There is, in fact, continuous periodic exchange of potential and kinetic energy within the system. For the case where $W < 0$, one would conclude that system energy is decreasing with time and so the system would come to rest; the system would be asymptotically stable.

The energy viewpoint will provide some insight into the criteria developed by Liapunov which follow next.

Liapunov's Stability Criterion. The stability criterion can be stated as follows: If there is a function $V(X)$ such that

$$V = 0 \text{ for } X = 0$$

$$V > 0 \text{ for } X \neq 0$$

$$V \rightarrow \infty \text{ for } |X| \rightarrow \infty$$

$$\frac{dV}{dt} = W$$

then, the system is stable if

$$W \leq 0 \text{ for } X \neq 0$$

and asymptotically stable if

$$W < 0 \text{ for } X \neq 0$$

(As a matter of terminology, the way in which the function V is defined makes it a *positive definite* function. If it were zero for $X = 0$ and negative for $X \neq 0$, it would be a *negative definite* function. Further,

$$|X| = (x_1^2 + x_2^2 + \cdots + x_n^2)^{1/2}$$

where $|X|$ can be visualized as the length of the vector in n-dimensional space.)

The system shown in Fig. 15-3 is known to be asymptotically stable and so it will serve as a convenient example for an exploration of the stability criterion. Using the same numerical values as with previous examples, the state-variable equations are (Eq. 15-2)

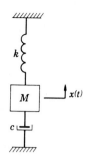

FIG. 15-3. Damped system.

$$\dot{x}_1 = x_2$$
$$\dot{x}_2 = -2x_1 - 3x_2$$

Let

$$V = x_1^2 + x_2^2$$

Then,

$$W = \frac{dV}{dt} = 2x_1\dot{x}_1 + 2x_2\dot{x}_2$$

With the state-variable equations,

$$W = 2x_1(x_2) + 2x_2(-2x_1 - 3x_2) = -6x_2^2 - 2x_1x_2$$

By inspection, W is *not* negative for all non-zero values of X.

Our predicament brings out the fact that Liapunov's criterion is a sufficient, but not a necessary, condition. Thus, we have (so far) determined nothing about the stability of the system. We must either try other V functions or learn some other approach.

The Instability Criterion. The instability criterion is stated as follows: With

$$W = \frac{dV}{dt} = 0 \text{ for } X = 0$$

$$W < 0 \text{ for } X \neq 0$$

$$W \to -\infty \text{ for } |X| \to \infty$$

then, the system is unstable unless

$$V > 0$$

This method can be used with the previous example in order to ascertain the stability condition. The approach is first to choose a quadratic form for V, that is,

$$V = \sum_{i=1}^{n} \sum_{j=1}^{n} a_{ij} x_i x_j \tag{15-29}$$

where $a_{ij} = a_{ji}$. Then, the W function is constrained to be negative definite. The resulting V function will establish the stability condition.

With Eq. 15-29, for $n = 2$, the V function becomes

$$V = a_{11}x_1^2 + 2a_{12}x_1x_2 + a_{22}x_2^2 \qquad (15\text{-}30)$$

from which

$$W = \frac{dV}{dt} = 2a_{11}x_1\dot{x}_1 + 2a_{12}x_1\dot{x}_2 + 2a_{12}\dot{x}_1x_2 + 2a_{22}x_2\dot{x}_2$$

The system equations are (Eq. 15-2)

$$\dot{x}_1 = x_2$$
$$\dot{x}_2 = -2x_1 - 3x_2$$

Upon substitution of these equations,

$$\begin{aligned}
W &= 2a_{11}x_1(x_2) + 2a_{12}x_1(-2x_1 - 3x_2) + 2a_{12}x_2(x_2) \\
&\quad + 2a_{22}x_2(-2x_1 - 3x_2) \\
&= -4a_{12}x_1^2 + (2a_{11} - 6a_{12} - 4a_{22})x_1x_2 \\
&\quad + (2a_{12} - 6a_{22})x_2^2 \qquad (15\text{-}31)
\end{aligned}$$

The function W is next constrained to be negative definite; let $W = -\alpha x_1^2 - \beta x_2^2$ where α and β are arbitrary real positive numbers. For this example, it will soon become apparent that a convenient choice is to make $\alpha = \beta = 4$. Thus,

$$W = -4x_1^2 - 4x_2^2 \qquad (15\text{-}32)$$

By equating coefficients of like terms in Eqs. 15-31 and 15-32,

$$-4a_{12} = -4 \qquad \text{and} \qquad a_{12} = 1$$
$$2a_{12} - 6a_{22} = -4 \qquad \text{and} \qquad a_{22} = 1$$
$$2a_{11} - 6a_{12} - 4a_{22} = 0 \qquad \text{and} \qquad a_{11} = 5$$

With Eq. 15-30,

$$V = 5x_1^2 + 2x_1x_2 + x_2^2$$

We must now determine whether or not function V is positive definite. By Sylvester's theorem, V is positive definite if the following determinants are all positive:

$$|a_{11}| \qquad \begin{vmatrix} a_{11} & a_{12} \\ a_{21} & a_{22} \end{vmatrix} \qquad \cdots$$

For our example,

$$|a_{11}| = |5| = 5$$
$$\begin{vmatrix} a_{11} & a_{12} \\ a_{21} & a_{22} \end{vmatrix} = \begin{vmatrix} 5 & 1 \\ 1 & 1 \end{vmatrix} = 5 - 1 = 4$$

Both determinants are positive and so V is positive definite.

From the instability criterion, we have determined that the system is not unstable. However, the stability criterion can be applied to the results to show that the system is asymptotically stable, that is,

$$V = 0 \quad \text{for} \quad X = 0$$
$$V > 0 \quad \text{for} \quad X \neq 0$$
$$W < 0 \quad \text{for} \quad X \neq 0$$

A Nonlinear Equation. Liapunov's stability criterion is applicable to nonlinear systems. As an example, consider the equation

$$\ddot{x}_1 + (2 + 3\dot{x}_1^2)\dot{x}_1 + x_1 = 0$$

from which

$$\dot{x}_1 = x_2$$
$$\dot{x}_2 = -x_1 - 2x_2 - 3x_2^3$$

With

$$V = x_1^2 + x_2^2$$
$$W = \frac{dV}{dt} = 2x_1\dot{x}_1 + 2x_2\dot{x}_2$$
$$= 2x_1(x_2) + 2x_2(-x_1 - 2x_2 - 3x_2^3)$$
$$= -4x_2^2 - 6x_2^4$$

The function W is clearly negative definite and so the system is asymptotically stable.

Closing Comment. The approach to stability presented in this section is commonly known as Liapunov's *second method*, the first method being to solve the equation. The approach is also known as the *direct method* because the nature of stability is sought directly without solving the equation.

15-4. Optimal Control [13, 24]

Everyone is familiar with the idea of seeking the most desirable, or best, alternative out of a number of possibilities. For example, some may wish to obtain the most return for a given investment, while others may desire to purchase the best used car for a given sum. With *optimal control*, the idea is to provide the best control for a particular system which must be controlled. The implication is that we must define what is meant by the best control and then devise a control scheme which will meet the objective. An example, of an exploratory nature, will be presented to help introduce the concepts of optimal control.

Suppose that the system represented in Fig. 15-4 is to be controlled. (As a matter of terminology, the system which must be controlled is often called the *plant*.) Assume that the objective is to minimize the error re-

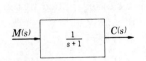

Fig. 15-4. System to be controlled.

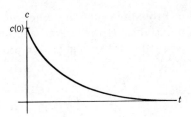

Fig. 15-5. Response of first-order
plant.

sulting from a non-zero initial value of the controlled variable c. We must
specify the manipulated variable $m(t)$ so that the objective is met. Saying
the same thing in another way, we must choose a *control law*, or *control
policy*.

The defining equation for the first-order plant shown in Fig. 15-4 is

$$\frac{dc}{dt} + c = m(t) \tag{15-33}$$

Let us investigate a negative-feedback control law. The manipulated
variable is specified as

$$m = -Kc \tag{15-34}$$

With Eq. 15-33,

$$\frac{dc}{dt} + (1 + K)c = 0 \tag{15-35}$$

The solution, assuming a non-zero initial value $c(0)$, is

$$c = c(0) e^{-(1+K)t} \tag{15-36}$$

A sketch of the response is shown in Fig. 15-5.

We are now in a position to define what is meant by the requirement
that the error be minimized. Because the value of the controlled variable
c is a measure of system error, the objective will be met by minimizing
the area under the response curve shown in Fig. 15-5. The requirement
can be expressed in terms of a *performance index PI* specified as

$$PI = \int_0^\infty c \, dt \tag{15-37}$$

where the performance index is to be minimized.

By inspection of Eq. 15-36, one recognizes that $K = \infty$ will provide
minimum error (but probably not a truly optimal system because infinite
correction is likely to be expensive). However, with $K = \infty$, the manipulated variable (Eq. 15-34) would be infinite for $c \neq 0$ and zero for $c = 0$.
This suggests that, for a finite maximum corrective signal M, an improved
control law would be

$$m = -M \quad \text{for} \quad c > 0$$
$$m = 0 \quad \text{for} \quad c = 0 \quad (15\text{-}38)$$
$$m = M \quad \text{for} \quad c < 0$$

Note that this control law will result in a nonlinear system.

The performance index can be used to establish the fact that the nonlinear control law is indeed superior. So that numerical values can be obtained, let $c(0) = 1$ and $M = 1$. For the linear case, Eq. 15-34 yields

$$m(0) = -Kc(0) = -K$$

The maximum corrective signal available is M; therefore,

$$m(0) = -M = -1 = -K \quad \text{and} \quad K = 1$$

The response, Eq. 15-36, is

$$c = c(0)\, e^{-(1+K)t} = e^{-2t}$$

With Eq. 15-37, the performance index is

$$PI = \int_0^\infty c\, dt = \int_0^\infty e^{-2t}\, dt = \left. -\frac{e^{-2t}}{2} \right|_0^\infty$$
$$PI = 0.5 \quad (15\text{-}39)$$

Thus, the performance index for the linear control law is 0.5.

Equations 15-33 and 15-38 are used to define the system with the nonlinear control law. For $c > 0$,

$$\frac{dc}{dt} + c = m(t) = -M = -1 \quad (15\text{-}40)$$

The solution, with $c(0) = 1$, is

$$c = 2e^{-t} - 1$$

We must now determine the time t_f at which $c = 0$. (At this time, $m = 0$ and the system remains at rest.) With the response equation,

$$0 = 2e^{-t_f} - 1$$
$$e^{-t_f} = 1/2 \quad \text{and} \quad t_f = 0.69$$

The performance index is calculated as

$$PI = \int_0^{t_f} c\, dt = \int_0^{0.69} (2e^{-t} - 1)\, dt$$
$$= (-2e^{-t} - t)\big|_0^{0.69} = -1 - 0.69 + 2$$
$$= 0.31 \quad (15\text{-}41)$$

A comparison of Eqs. 15-39 and 15-41 reveals that the nonlinear control law is superior because it yields a smaller performance index.

It is instructive to study the performance of the two systems on the phase plane. For the linear case with $K = 1$, Eq. 15-35 becomes

$$\frac{dc}{dt} + (1 + K)c = \frac{dc}{dt} + 2c = 0$$

from which

$$\dot{c} = -2c \tag{15-42}$$

Equation 15-42 defines the trajectory on the phase plane plot of Fig. 15-6a

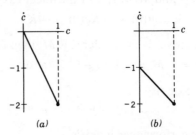

(a)   (b)

FIG. 15-6. Phase plane plots for the first-order plant with linear and non-linear control laws.

where the system starts from $c(0) = 1$. Note that, with a first-order system, it is possible to go instantaneously from zero velocity to a non-zero value—0 to -2 for the system shown.

The appropriate equation, Eq. 15-40, for the nonlinear case is

$$\frac{dc}{dt} + c = -1$$

from which

$$\dot{c} = -c - 1$$

The resulting trajectory, with $c(0) = 1$, is shown in Fig. 15-6b. Note that when $c = 0$, the manipulated variable is switched to zero and the system immediately comes to rest with $c = \dot{c} = 0$. It can be seen that the average velocity is greater with the nonlinear control law, which is another way of explaining its superiority.

In closing, two remarks are in order. First, the performance index as expressed in Eq. 15-37, although adequate for the purpose intended, is not generally applicable because it does not take into account both positive and negative values of c. For the general case, one would use the performance index

$$PI = \int c^2 \, dt$$

The second remark is that the nonlinear control law we devised has resulted in what is known as *bang-bang control*. This topic will be discussed next.

Bang-Bang Control. Bang-bang control is an approach which involves the strategic switching of maximum corrective actions. It provides optimal control in cases where the requirement is for either minimum error, as in the previous example, or minimum time. As an illustration of a minimum-time problem, consider an automobile at rest; we wish to bring it to rest a block away in the least amount of time. Obviously, full throttle followed by full braking is the solution. This is bang-bang control, that is, the strategic switching of maximum corrective actions.

The rotating inertia I shown in Fig. 15-7 is often called a *double-*

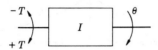

FIG. 15-7. Double-integration
plant.

integration plant because the transfer function involves $1/s^2$. We will investigate the control of displacement θ using bang-bang control. An example of an actual system of this type is a spacecraft which utilizes thrusters to apply torque T for attitude control.

The defining equation for the double-integration plant, with constant positive torque T, is

$$I \frac{d^2\theta}{dt^2} = T \qquad (15\text{-}43)$$

For convenience, let $T/I = 1$. Then,

$$\frac{d^2\theta}{dt^2} = 1 \qquad (15\text{-}44)$$

Equation 15-44 can be written as

$$\dot{\theta} \frac{d\dot{\theta}}{d\theta} = 1 \qquad \text{or} \qquad \dot{\theta}\, d\dot{\theta} = d\theta$$

Integration yields

$$\frac{\dot{\theta}^2}{2} = \theta + C_1$$

or

$$\dot{\theta}^2 = 2\theta + C \qquad (15\text{-}45)$$

To obtain a trajectory on the phase plane which passes through the origin, the constant C is made zero. Thus,

$$\dot{\theta}^2 = 2\theta \qquad (15\text{-}46)$$

This equation defines the phase-plane trajectory shown by the solid line in Fig. 15-8. Branch 1 of the curve represents driving the system from the origin with positive torque T. Branch 2 can be regarded as a "braking" trajectory, that is, for a set of initial conditions corresponding to a point on the curve, the application of positive torque would cause the system to reach $\theta = \dot{\theta} = 0$.

The trajectory in Fig. 15-8 shown by the dashed line is that obtained

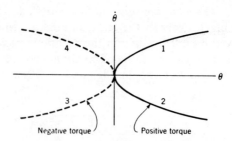

FIG. 15-8. Phase plane plot showing trajectories for full applied torque.

with negative torque applied. Branch 3 is the "driving" portion while Branch 4 is the "braking" portion.

Suppose that the inertia I of Fig. 15-7 exhibits an initial condition, $\theta_0 = -1$, and that we wish to return to $\theta = 0$ in minimum time. The appropriate trajectory is shown in Fig. 15-9a. The applied torque pattern

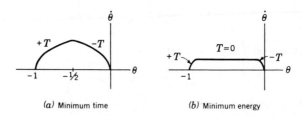

(a) Minimum time   (b) Minimum energy

FIG. 15-9. Phase plane trajectories for minimum time and minimum energy.

is first positive, then negative, and finally zero. It should be clear that the torque is switched from positive to negative at $\theta = -1/2$, and switched from negative to zero with $\theta = \dot{\theta} = 0$.

Figure 15-9b shows a trajectory where the requirement is for correction with minimum energy. A short pulse of positive torque, followed by a period of $T = 0$, and terminated by a short pulse of negative torque is the required pattern.

Switching is a fundamental part of bang-bang control. The *switching curve* for the double-integration plant is shown in Fig. 15-10. For initial conditions resulting in a point below and to the left of the switching curve, positive torque is applied until the switching curve is encountered. Then, negative torque is used to reach the origin. (See Fig. 15-9a for an example.) Negative torque is applied first if the starting point on the phase plane is above and to the right of the switching curve.

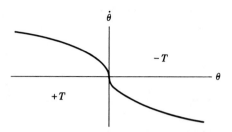

FIG. 15-10. Phase plane plot showing switching curve.

The switching curve for a second-order system becomes a switching surface for a third-order system. Third-order and higher-order systems aggravate the problem of the determination of switching points. However, the increased use of the digital computer as a part of the control system should tend to alleviate the problem because of the speed at which the computer can make calculations.

Control hardware poses somewhat of a problem to bang-bang control. For example, it is not possible to switch instantaneously from one corrective extreme to the other. However, hardware limitations would not seem to place serious restrictions on the use of bang-bang control concepts.

Calculus of Variations Approach. The calculus of variations is fundamental to the study of optimal control. First we will explore the variational approach and then use it to solve a problem.

Suppose that we are seeking a function $x(t)$ which will minimize the performance index

$$PI = \int_0^T F(x, \dot{x}, t) \, dt \qquad (15\text{-}47)$$

where F is a given function. Assume that $x(0)$, $x(T)$, and T are specified. The hypothetical optimal function $x(t)$ is shown in Fig. 15-11 as a solid

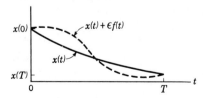

FIG. 15-11. The variational approach.

line. A neighboring function is shown as a dashed line and is defined, in terms of the optimal function, as

$$x(t) + \epsilon f(t)$$

where ϵ is small and $f(0) = f(T) = 0$. (The latter restriction is necessary so that both functions have the same values at $t = 0$ and $t = T$.) The performance index can be expressed in terms of the neighboring function. It is

$$PI = \int_0^T F(x + \epsilon f, \dot{x} + \epsilon \dot{f}, t)\, dt$$

In order that the neighboring function be optimal, the difference ΔPI between the two performance indices must be zero, that is

$$\Delta PI = \int_0^T [F(x + \epsilon f, \dot{x} + \epsilon \dot{f}, t) - F(x, \dot{x}, t)]\, dt = 0 \qquad (15\text{-}48)$$

With Taylor series expansion,

$$F(x + \epsilon f, \dot{x} + \epsilon \dot{f}, t) = F(x, \dot{x}, t) + \epsilon\left(f\frac{\partial F}{\partial x} + \dot{f}\frac{\partial F}{\partial \dot{x}}\right) + \begin{array}{l}\text{higher order}\\\text{terms in } \epsilon\end{array}$$

By neglecting the higher order terms in ϵ, Eq. 15-48 becomes

$$\Delta PI = \int_0^T \epsilon\left(f\frac{\partial F}{\partial x} + \dot{f}\frac{\partial F}{\partial \dot{x}}\right) dt = 0$$

The quantity ϵ is not zero; therefore,

$$\int_0^T \left(f\frac{\partial F}{\partial x} + \dot{f}\frac{\partial F}{\partial \dot{x}}\right) dt = 0 \qquad (15\text{-}49)$$

Through integration by parts with

$$u = \frac{\partial F}{\partial \dot{x}} \qquad du = \frac{d}{dt}\left(\frac{\partial F}{\partial \dot{x}}\right)$$

$$v = f \qquad dv = \frac{df}{dt}\, dt$$

the second integral in Eq. 15-49 is evaluated as

$$\int_0^T \dot{f}\frac{\partial F}{\partial \dot{x}}\, dt = f\frac{\partial F}{\partial \dot{x}}\Big|_0^T - \int_0^T f\frac{d}{dt}\left(\frac{\partial F}{\partial \dot{x}}\right)$$

Because $f(0) = f(T) = 0$,

$$\int_0^T \dot{f}\frac{\partial F}{\partial \dot{x}}\, dt = -\int_0^T f\frac{d}{dt}\left(\frac{\partial F}{\partial \dot{x}}\right)$$

and Eq. 15-49 becomes

$$\int_0^T f\left[\frac{\partial F}{\partial x} - \frac{d}{dt}\left(\frac{\partial F}{\partial \dot{x}}\right)\right] dt = 0$$

The function $f(t)$ is not zero. Therefore,

$$\frac{\partial F}{\partial x} - \frac{d}{dt}\left(\frac{\partial F}{\partial \dot{x}}\right) = 0 \qquad (15\text{-}50)$$

Equation 15-50 is called the Euler-Lagrange equation and states the requirement for an optimal solution.

As an example of the use of the calculus of variations, consider the first-order plant shown in Fig. 15-12. The defining differential equa-

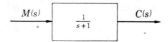

FIG. 15-12. First-order plant.

tion is

$$\frac{dc}{dt} + c = m$$

Let the controlled variable c be set equal to the state variable x_1. The appropriate state variable equation is

$$\dot{x}_1 = -x_1 + m \qquad (15\text{-}51)$$

The objective will be to minimize the error resulting from a non-zero initial condition. In addition, there is to be a penalty assigned to the use of correction, that is, we also wish to take into account the cost of control effort. The performance index can be expressed as

$$PI = \int_0^\infty (x_1^2 + Am^2)\, dt$$

where the constant A can be used to introduce the relative importance of the use of corrective action. For our case, let $A = 1$. The performance index becomes

$$PI = \int_0^\infty (x_1^2 + m^2)\, dt \qquad (15\text{-}52)$$

We must now determine the appropriate function F to be used in Eq. 15-50. A comparison of Eqs. 15-47 and 15-52 reveals that the function F is

$$F = x_1^2 + m^2 \qquad (15\text{-}53)$$

However, in the present case, x_1 is subject to the constraint expressed in the state variable equation, Eq. 15-51. Therefore, the function must be modified. Equation 15-51 can be written as

$$\dot{x}_1 + x_1 - m = g_1 = 0 \qquad (15\text{-}54)$$

A new function F' is constructed as follows:

$$F' = F + \lambda_1 g_1$$

where λ_1 is a Lagrangian multiplier. In the general case,

$$F' = F + \lambda_1 g_1 + \lambda_2 g_2 + \cdots \tag{15-55}$$

For the problem at hand,

$$F' = x_1^2 + m^2 + \lambda_1(\dot{x}_1 + x_1 - m) \tag{15-56}$$

There are two variables, namely x_1 and m, involved and so Eq. 15-50 must be used twice. Therefore, with Eq. 15-56,

$$\frac{\partial F'}{\partial x_1} - \frac{d}{dt}\left(\frac{\partial F'}{\partial \dot{x}_1}\right) = 2x_1 + \lambda_1 - \frac{d}{dt}(\lambda_1)$$

$$= 2x_1 + \lambda_1 - \dot{\lambda}_1 = 0 \tag{15-57}$$

$$\frac{\partial F'}{\partial m} - \frac{d}{dt}\left(\frac{\partial F'}{\partial \dot{m}}\right) = 2m - \lambda_1 = 0 \tag{15-58}$$

Substitution of Eq. 15-54 into Eq. 15-58 yields

$$2(\dot{x}_1 + x_1) - \lambda_1 = 0 \tag{15-59}$$

The simultaneous solution of Eqs. 15-57 and 15-59, to eliminate λ_1, yields the differential equation

$$\ddot{x}_1 - 2x_1 = 0$$

The solution is

$$x_1 = C_1 e^{\sqrt{2}t} + C_2 e^{-\sqrt{2}t} \tag{15-60}$$

Because $x_1(\infty) = 0$, $C_1 = 0$. With $x_1(0) = X_1$, $C_2 = X_1$. Thus,

$$x_1 = X_1 e^{-\sqrt{2}t} \tag{15-61}$$

is the optimal response.

We now determine the appropriate control law, that is, specify $m(t)$ such that the response of Eq. 15-61 is realized. With Eq. 15-54,

$$m = \dot{x}_1 + x_1$$

Using Eq. 15-61,

$$m = -\sqrt{2}X_1 e^{-\sqrt{2}t} + X_1 e^{-\sqrt{2}t}$$

$$= -\sqrt{2}x_1 + x_1 = -0.414x_1 \tag{15-62}$$

Equation 15-62 specifies a negative-feedback control law with a gain constant of 0.414. The system block diagram is given in Fig. 15-13. It is

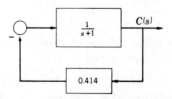

FIG. 15-13. First-order plant with optimal control.

interesting to note that the negative-feedback control law arose from the mathematical approach rather than from any presupposition on our part.

Although the variational approach to optimal control is appealing, there are very real problems involved in its general application to practical systems. Some of the problems are:

1. The performance index must be defined.
2. The dynamic equations for the plant must be obtained.
3. The control law must be computed.
4. The solution must be realized physically.

Any of these problems poses great difficulty for a complicated plant.

The scope of this chapter limits the topics which can be discussed. However, the reader should at least be aware of the fact that Pontryagin's *Maximum Principle* and Bellman's *Dynamic Programming* are related topics in optimal control [24, 31].

15-5. Adaptive Systems

An *adaptive system* is one which automatically varies one or more system parameters in response to changing conditions. The objective, of course, is to achieve effective control under all conditions. An early example of adaptive control occurred many years ago in the home-heating field. Control engineers postulated that, with a hot-water heating system, better control would result if the water temperature were related to the outside temperature. The idea was implemented by sensing outdoor temperature and using this information to maintain boiler water temperature inversely proportional to outdoor temperature. This application was a forerunner of adaptive systems because a system parameter was varied to meet a changing condition.

Three types of adaptive system can be identified. The first is *programmed adaptive* where the designer anticipates the difficulty and builds in a programmed change of parameters. The example just cited would fit into this classification. The second type is *optimal adaptive* where the performance index is calculated frequently and changes are made in an attempt to minimize a performance index. The third type is *learning adaptive* where the system is sufficiently sophisticated to determine for itself what to do under any circumstances. Very complex systems, even including the human being, fall into this category.

The first two types of adaptive system are now treated in more detail.

The Programmed-Adaptive Approach. We will explore an example for which the programmed-adaptive approach might be useful. Suppose that the nonlinearity shown in Fig. 15-14 is exhibited by one component in an otherwise linear closed-loop system. For a given reference input, the nominal value of the component input i might be i_o. An appropriate

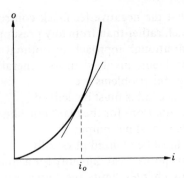

FIG. 15-14. Component nonlinearity.

gain for the component could be determined by linearization. It should be clear that the gain of the component changes with the value of i_o. If the system operates at levels corresponding to a wide range of i_o, we must take into account the gain change. One approach is to set the loop gain for the desired system response using the maximum value of gain to be expected from the nonlinear component. This approach has an obvious limitation; when the component is operated in the low-gain region, the loop gain will be small and the system response, sluggish.

The programmed-adaptive approach is to measure component input i and use this information to change another variable gain in the loop to compensate for the nonlinearity. In this fashion, we can maintain the loop gain essentially constant under all operating conditions. The beneficial result is not without cost; a small closed-loop system may well be required to accomplish the programmed-adaptive feature.

The Optimal-Adaptive Approach. With the optimal-adaptive approach, the idea is to constantly change certain system parameters in order to minimize the performance index. (Of course, the performance index should be maximized if it is expressed in terms of profit.) Given a suitable performance index and the capability to calculate it, the remaining step is to devise a strategy for changing the system parameters. Although a number of strategies are available, only a relatively simple one will be mentioned here to illustrate the idea.

Assume that only two manipulated variables influence the value of the performance index. The first variable is changed in an incremental manner while the second variable is held fixed. The performance index is calculated after each change. The process is continued until no further improvement is possible. Then, the first variable is held fixed while the second is changed incrementally. After each change, the performance index is calculated. When no further improvement is possible, the first variable is changed again, and so forth. The result is a system which modifies itself continuously in an attempt to produce an optimal output.

15-6. Closure

The degree to which the concepts presented in this chapter are useful will vary from topic to topic and from application to application. Nevertheless, the concepts are sound. The wise control engineer will keep them in mind and will seek to realize their potentialities. The creative control engineer will combine these ideas with others he has encountered to produce truly imaginative solutions to control problems.

Problems

15-1. For the liquid-level system shown in the figure: a) Determine the defining state variable equations where h_1 and h_2 are the state variables. b) Determine the defining state variable equations where $h_2 = x_1$ and $\dot{h}_2 = x_2$ are the state variables. (HINT: First obtain the second-order differential equation relating h_2 to q_i.) c) Write the equations from parts a) and b) in matrix notation. d) With the results of part c), determine the system characteristic equation from the relationship

$$| sI - A | = 0$$

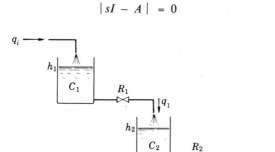

PROB. 15-1

15-2. For the electrical circuit shown in the figure, justify the following state variable equations:

$$\dot{i}_L = -\frac{R_2}{L} i_L + \frac{1}{L} e_C$$

$$\dot{e}_C = -\frac{1}{C} i_L - \frac{1}{R_1 C} e_C + \frac{1}{R_1 C} e_i$$

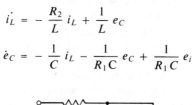

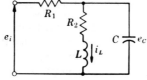

PROB. 15-2

15-3. A system block diagram is shown in the figure. a) Obtain the state variable equations where $c = x_1$, $\dot{c} = x_2$, and $\ddot{c} = x_3$ are the state variables. b) Express the results from part a) in matrix notation.

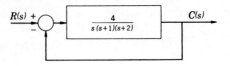

PROB. 15-3

15-4. With zero input, Eq. 15-1 is

$$\frac{d^2x}{dt^2} + 3\frac{dx}{dt} + 2x = 0$$

Solve the equation using Laplace transformation; the initial conditions are $x(0) = -1$ and $\dot{x}(0) = 0$. Compare the results with Eqs. 15-22 and 15-23 where $x_1 = x$ and $x_2 = \dot{x}$. Also, compare the amount of work involved in obtaining the solutions by the two methods.

15-5. Using Liapunov's second method, show that the system, with the following defining equations, is unstable.

$$\dot{x}_1 = x_2$$
$$\dot{x}_2 = 3x_2 - 2x_1$$

15-6. Investigate system stability; the defining equation is

$$\ddot{x} + A\,\dot{x}^2 + Bx = 0$$

15-7. The main propeller for a deep-sea, underwater search vehicle is driven by an electric motor. Motor controls permit only three speed levels—full forward propeller speed, full reverse propeller speed, and off. The corresponding forward thrust is 300 lb while the reverse thrust is 200 lb. You may assume that full thrust is instantaneously available. The vehicle mass M is 15,000 lb-sec^2/ft. Assume that the drag force behaves as viscous friction with a damping coefficient c of 150 lb-sec/ft. The search vehicle is to be moved forward by 10 ft. The vehicle starts from rest and is to be at rest in the new position.

a) Devise a bang-bang control scheme so that the distance is travelled in minimum time. Define switching points in terms of both distance travelled and time. b) Devise a bang-bang control scheme so that the distance is travelled with minimum energy expenditure. To increase controller and motor life, use no more than one pulse of forward thrust and no more than one pulse of reverse thrust. Define switching points in terms of both distance travelled and time.

15-8. A liquid-level system is shown in the figure. With the inlet valve 100% open, the level will ultimately rise to $h = H$. Assume that the inlet valve is linear with percentage opening and that the outlet valve is linear. We desire to fill the tank from $h = 0$ to $h = 0.5H$ in minimum time by strategically manipulating the inlet valve. You may assume that the valve opening can be changed to any desired value instantaneously. Devise a control scheme. *Sketch* plots of *h vs t* and valve opening *vs t*.

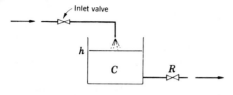

15-9. With the liquid-level system shown in the figure, the inlet valve is used to control level h_2. The inlet valve is linear and will provide a flow rate of 10 ft^3/min with 100% opening. The outlet valves are linear with $R_1 = R_2 = 1$ min/ft^2. Capacitance $C_1 = 5$ ft^2 and $C_2 = 10$ ft^2.

At the start of the problem, the system is in equilibrium with $h_2 = 2$ ft. We desire to make $h_2 = 6$ ft in minimum time by strategically manipulating the inlet valve. You may assume that the valve opening can be changed to any desired value instantaneously—a good assumption considering how slow the system is. Devise a bang-bang control scheme. Define switching points in terms of both level h_2 and time t.

NOTE: This problem is related to those involved in making a product grade change where the manufacturing process is continuous. The change must be made as rapidly as possible to reduce waste.

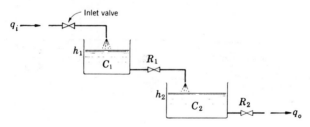

15-10. In Sec. 15-4, the calculus of variations was used to determine the optimal control law for the first-order plant shown in Fig. 15-12. The plant state variable equation is (Eq. 15-51)

$$\dot{x}_1 = -x_1 + m$$

and the performance index is (Eq. 15-52)

$$PI = \int_0^\infty (x_1^2 + m^2)\, dt$$

The resulting optimal system, Fig. 15-13, is shown in part *a* of the accompanying figure. (Recall that $c = x_1$.)

a) Compute the performance index for the optimal case using $c(0) = x_1(0) = 1$. b) Compute the performance index for the non-optimal case shown in part b of the figure. Again use $c(0) = x_1(0) = 1$. Compare with the result from part a.

NOTE: Comparisons of this type are easily carried out on an analog computer. The system is simulated and the performance index is calculated with the use of two multipliers and one integrator.

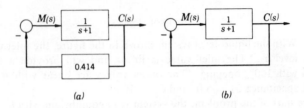

(a)   (b)

PROB. 15-10

15-11. A double-integration plant is given in part a of the figure. For a non-zero initial value of the controlled variable, the performance index to be mini-

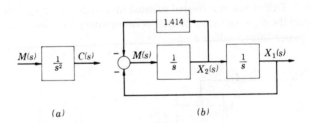

(a)   (b)

PROB. 15-11

mized is

$$PI = \int_0^\infty (x_1^2 + m^2)\, dt$$

where $x_1 = c$. Use the calculus of variations to show that the system given in part b of the figure is optimal.

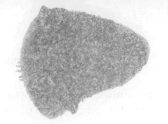

Chapter **16**

Digital
Computer Control

16-1. Introduction

During the 1960's the application of digital computers expanded from business and scientific computation systems to plant and process control installations. By 1966, the application of digital computers had become almost a standard approach to certain types of plant supervision and control [7]. Some leading early applications include computers for startup, shutdown, monitoring, and on-line control of power plant boiler-turbine-generator installations. In the petroleum and chemical industries, computer control has been applied to hydro-crackers, ethylene manufacture, and amonia plants. The monitoring and control of basic oxygen furnaces is an example from the steel industry. In other industries, examples of computer control applications of various degrees of sophistication include the control of cement kilns, paper-machines, glass furnaces, assembly lines, newspaper type setting, freight car classifications and railroad train makeup, environmental control, and textile finishing.

Not all of the early applications of digital computers to control were considered successful. Such problems as inadequate preliminary engineering, programming difficulties, manpower shortages, inadequate instrumentation, and hardware delivery delays caused considerable difficulties. In some instances, computers have been returned or projects have degenerated to a state of merely monitoring and data collection.

Implementation of computer control of a plant or process is a challenging problem in systems engineering. Detailed consideration must be given to cost analysis, system definition, control, design and strategy, and the installation and implementation. As in most well designed systems the most effective installations have sought the simplest strategy and techniques that will meet the performance needs. Where installations have sufficient financial backing, competent technical personnel, and sound management, computer control has and will continue to be exceptionally productive. Although management is dollar profit oriented, the benefits from the installation of digital computer control should be measured in broader terms including increased knowledge of the process, more uniform and consistent manufacturing guides, and improved technological capabilities of the engineering staff.

This chapter will present some of the fundamental concepts of digital computer control. Following an introduction to control configurations, an example of direct digital computer control will be presented. Formulation of control laws in the discrete sample time domain will then be developed. Analogous to the use of the Laplace transform for control systems treated in earlier chapters, the z-transform is introduced for sampled-data control systems. The chapter closes with some comments on the stability of sampled-data control systems.

The material in this chapter is an introduction to the area of digital computer control. The concept of working in the time domain with sampled functions is of particular importance. With an understanding of difference-differential equations, pulse sequences, and z-transforms, the reader will be aided in proceeding into the vast number of technical articles and texts in the field.

16-2. Concepts and Control Configurations

The digital computer provides a programmable, decision making device for control applications. Hardware may range from flexible, general purpose machines, operating in time-sharing with a multitude of tasks, to special purpose equipment designed to perform a particular task. The most important aspect of the digital computer is the ability to process and transform vast amounts of information in a prescribed way. In addition to calculation, the general purpose digital computer may store and retrieve information, make logical decisions, adapt to changing conditions, and ultimately learn.

In digital computer control applications the computer may be programmed to perform basic functions such as:
1. Collection of plant data
2. Identification of plant dynamics
3. Selection of control parameters
4. Implementation of control algorithms[1]
5. Implement optimization procedures to establish optimum control
6. Manipulation of plant input variables

The extent to which any of these functions are utilized in a given application depends on the system complexity, operating objectives and the engineering capabilities.

Figure 16-1 illustrates a computer with the potential input and out-

---

[1] An algorithm is a specification for calculation with a particular notation. The word flourished in 825 A.D. and was used by Persian mathematicians.

put functions. Command information may be fed into the computer through card or tape information, or operator manipulation of the con-

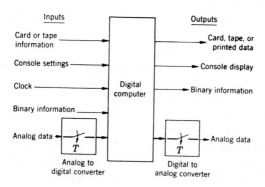

Fig. 16-1. Digital computer input-output functions.

sole. A clock input establishes a time reference and information from the process may be entered in binary form directly or in analog form through an analog-to-digital converter. Information output from the computer may be in the form of cards, tape, printed data, or console displays. Binary information may be read out or analog information may be obtained by digital-to-analog conversion. The control capability of the computer is governed essentially by computer speed or access time, and memory capacity or the number of words of a specific length that may be stored.

There are a variety of digital computer control concepts which have been employed. A number of the possibilities will be considered in increasing order of actual degree of computer control. The first approach may be termed off-line computer control and is illustrated in Fig. 16-2. In this case the computer is utilized as a computational device to assist

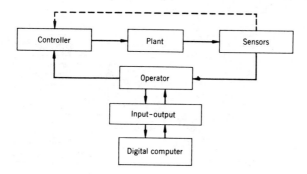

Fig. 16-2. Off-line computer control.

the operator in decision making. The operator may serve in one of two capacities; he may actually close the loop around the plant and provide the primary feedback by manipulating input functions, or he may adjust the set points of controllers with feedback signals coming directly from the sensors.

Two types of computer control represent a first attempt at incorporating the computer into a more direct control relationship. Programmed computer control places the computer in control of the process with the operator closing the loop. From observation, the operator submits operating data to a computer whereby the computer manipulates the process inputs through digital-to-analog conversion or by direct programmed pulse control. The inverse of this scheme is a system of data logging and operational guide formulation. In this arrangement, the computer monitors the process by sampling signals from sensors through an analog-to-digital converter. The computer serves two functions; it maintains an operating log of the process variables and it may periodically compute guidelines for the operator to initiate changes in process set points.

The most complete form of computer control is where the computer is used to close the loop, or many loops, in the process. Figure 16-3 represents schematically the concept of closed-loop computer control. Information is obtained from many sensors by a scanner which samples each sensor at set intervals. Information is fed through an analog-to-digital converter and then into the computer for storage. The computer is generally programmed for a multitude of functions, the primary one being implementation of digital algorithms which provide set point and direct manipulation signals. The operator observes the process and supplements the control by console, card, or tape input-output information. More sophisticated systems also involve automatic identification of process dynamics, controller tuning or algorithm modification and programmed change from one process state to another.

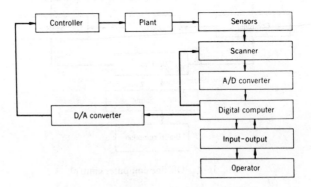

FIG. 16-3.   Closed-loop computer control.

16-3. An Example of Direct Digital Control

Direct digital control infers the direct manipulation of process functions by the digital computer. One form of direct digital control may be described as pulse duration modulation. With pulse duration modulation an element is manipulated by applying a fixed voltage for a predetermined length of time.

Figure 16-4 illustrates a valve controlling the flow rate in a line. The

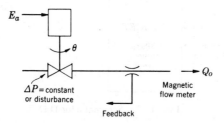

FIG. 16-4. A process valve for direct digital control.

valve is activated by a control voltage E_a which produces a rotation at velocity $\dfrac{d\theta}{dt}$. Therefore,

$$\theta = \alpha \int_0^{T_m} E_a \, dt \qquad (16\text{-}1)$$

where θ = valve position (rad)

α = valve rate constant $\left(\dfrac{\text{rad/sec}}{\text{volt}}\right)$

T_m = time duration of the fixed amplitude actuating pulse E_a

In the simplest case where the pressure drop across the valve is constant, the flow rate is given by

$$Q_o = K_v \theta = K_v \alpha \int_0^{T_m} E_a \, dt + Q_{\text{i.c.}} \qquad (16\text{-}2)$$

where K_v is the valve flow constant.

Closed-loop control based on the pulse duration modulation concept is shown schematically in Fig. 16-5. The reference or set point for flow is Q_r. Q_o is measured by a magnetic flow meter and the value of flow is sampled and stored in the computer. An algorithm is then implemented to determine the length of time the actuating voltage is to be turned on to correct for error E.

$$E = Q_r - Q_o \qquad (16\text{-}3)$$

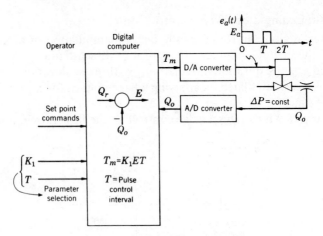

FIG. 16-5. Schematic for D.D.C.

The pulse duration T_m may be defined as a portion of the product of the interval T and the amount of error. Thus

$$T_m(nT) = K_1 \mid E(nT) \mid T \qquad (16\text{-}4)$$

where K_1 is the pulse duration coefficient. K_1 physically represents the ratio of the pulse length to the control interval T per unit error in flow rate. The performance of the control will be shown to depend on the selection of control parameters by studying a series of responses of flow rate to unit step set point changes. The analysis is most easily handled in a stepwise fashion where in each interval

$$\left. \begin{array}{l} T_m(nT) = K_1 \mid E(nT) \mid T \\ Q_o[nT + T_m(nT)] = K_1(K_v \alpha E_a T) E(nT)^* + Q_{nT} \end{array} \right\} \qquad (16\text{-}5)$$

In Eq. 16-5, n is an integer and the value of Q at time $(nT + T_m)$ is held until the time $(n + 1)T$ as no valve change occurs following the calculated duration T_m. The following summarizes the calculation of the response:

        Interval $0 \rightarrow T$; (for the unit step $E = 1$)

$$T_m(0) = K_1 T$$

$$Q_o[T_m(0)] = K_1(K_v \alpha E_a T) + 0$$

        Interval $T \rightarrow 2T$;

$$E(T) = 1 - Q_o(T)$$

$$T_m(T) = K_1 \mid 1 - Q_o(T) \mid T$$

$$Q_o[T + T_m(T)] = K_1(K_v \alpha E_a T)\,[1 - Q(T)] + Q_o(T)$$

    The response is seen to depend on two quantities, namely K_1 and $(K_v \alpha E_a T)$. For example, consider $K_1 = \frac{1}{2}$ and $K_v \alpha E_a T = 1$, then

---

*In this equation, E_a is taken as positive and the absolute value bars have been eliminated to preserve the sign of the incremental flow rate of Q_o.

Interval $0 \to T$

$$T_m(o) = 1/2\ T$$

$$Q_o(\tfrac{1}{2}T) = 1/2\ (1) + 0 = 1/2$$

Interval $T \to 2T$

$$T_m(T) = 1/2\ (1/2)\ T = T/4$$

$$Q_o(1.25\,T) = 1/2\ (1)\ (1/2) + 1/2 = 3/4$$

Interval $2T \to 3T$

$$T_m(2T) = 1/2\ (1/4) = T/8$$

$$Q_o(2.125\,T) = 1/2\ (1)(1/4) + 3/4 = 7/8$$

$$\vdots$$

A plot for the case of $K_1 = 1/2$ and $(K_v \alpha E_a T) = 1$ is shown in Fig. 16-6a.

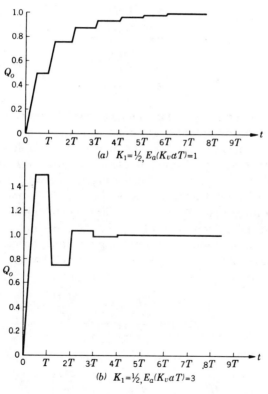

FIG. 16-6. Response of pulse duration modulated system with pure integration.

A plot for the case of $K_1 = 1/2$ and $(K_v \alpha E_a T) = 3$ is shown in Fig. 16-6b.

These response curves are typical of the behavior of a pure integration system with pulse duration modulation. The closed-loop system may behave in concept analogous to an overdamped, underdamped, or unstable system depending on the parameter selection.

16-4. Difference Differential Equations

As the digital computer is only capable of performing numerical calculations on discrete pieces of data, it is necessary to visualize a functional relationship as a series of discrete values of a dependent variable at specified intervals of the independent variable. Figure 16-7 illustrates a

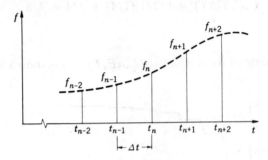

FIG. 16-7. Discrete values of a continuous function at intervals Δt.

continuous function represented by points at discrete intervals Δt.

In dealing with differential equations, it is necessary to develop expressions for the rates of change of the dependent variable. There are a number of alternatives in selecting an expression for the slopes of curves represented by discrete points. Considering the derivative of the curve in the vicinity of a point t_n on the curve of Fig. 16-7, one approach to obtaining an approximate expression is to fit the neighboring curve with a parabolic curve and differentiate at the point. An alternative is to use the Taylor series expansion and neglect the higher order terms. In general, the Taylor series expansion may be used to express the value of a function to the right and to the left of a point t_n. To the right,

$$f(t_n + \Delta t) = f_n + \frac{df}{dt}\bigg|_n (\Delta t) + \frac{\frac{d^2f}{dt^2}\big|_n (\Delta t)^2}{2!} + \frac{\frac{d^3f}{dt^3}\big|_n (\Delta t)^3}{3!} + \cdots \quad (16\text{-}6)$$

Similarly, to the left,

$$f(t_n - \Delta t) = f_n - \frac{df}{dt}\bigg|_n (\Delta t) + \frac{\frac{d^2f}{dt^2}\big|_n (\Delta t)^2}{2!} - \frac{\frac{d^3f}{dt^3}\big|_n (\Delta t)^3}{3!} + \cdots \quad (16\text{-}7)$$

Subtracting Eq. 16-7 from Eq. 16-6 and neglecting terms above the first derivative yields,

$$\frac{df}{dt}\bigg|_n = \frac{f(t_n + \Delta t) - f(t_n - \Delta t)}{2 \Delta t}$$

which from Fig. 16-7 is equivalent to

$$\frac{df}{dt}\bigg|_n = \frac{f_{n+1} - f_{n-1}}{2 \Delta t} \tag{16-8}$$

This expression, commonly termed the central difference approximation of the derivative at a point, is geometrically interpreted as the slope of a line adjoining the points that straddle f_n at t_n. The central difference approximation to the second derivative may readily be obtained by adding Eqs. 16-6 and 16-7 to obtain

$$\frac{d^2f}{dt^2}\bigg|_n = \frac{f_{n+1} - 2f_n + f_{n-1}}{(\Delta t)^2} \tag{16-9}$$

In general, the central difference technique is the most accurate technique for numerical differentiation. However, there are other alternatives that may be developed from the Taylor series. When data to the right of t_n is not available the backward-difference approach is used. The backward difference approximation to the derivative at t_n in Fig. 16-7 is given by

$$\frac{df}{dt}\bigg|_n = \frac{f_n - f_{n-1}}{\Delta t} \tag{16-10}$$

Similarly, when data to the left of t_n is not known the forward-difference technique is useful. Expressing the approximate derivative at t_n in terms of data to the right of t_n yields

$$\frac{df}{dt}\bigg|_n = \frac{f_{n+1} - f_n}{\Delta t} \tag{16-11}$$

For a more complete review of the formulation of finite difference equations, the reader is referred to a text on advanced calculus or numerical computer methods [28].

A finite difference equation is formulated by substituting the approximate derivatives expressed in Eqs. 16-8 through 6-11 into the original differential equation.

EXAMPLE 16-1. Formulate a finite difference equation for the differential equation

$$\tau \frac{dx}{dt} + x = f(t)$$

Solution: Using the first backward difference technique

$$\frac{\tau(x_n - x_{n-1})}{\Delta t} + x_n = f_n$$

or solving for x_n

$$\tau x_n + \Delta t x_n = \Delta t f_n + \tau x_{n-1}$$

$$x_n = \frac{\Delta t f_n + \tau x_{n-1}}{(\tau + \Delta t)}$$

Utilizing the forward difference technique yields

$$\frac{\tau(x_{n+1} - x_n)}{\Delta t} + x_n = f_n$$

$$x_{n+1} = \frac{\Delta t}{\tau}\left[f_n + x_n\left(\frac{\tau}{\Delta t} - 1\right)\right] \qquad\qquad Ans.$$

Solutions of equations in the form illustrated by Example 16-1 are readily implemented on the digital computer. In principle equations of these forms are similar to the digital algorithms utilized to implement control actions as will be discussed in the following section.

16-5. Formulation of a Digital Controller

A straightforward approach to developing a digital algorithm for computer control is to work directly from the continuous system and utilize the finite difference approximation for the desired control action. Suppose, for example, that a process controller $G_c(s)$ is composed of proportional, integral and derivative control actions.

$$\frac{M(s)}{E(s)} = G_c(s) = K_p + \frac{K_i}{s} + K_d s$$

The differential equation representing the controller is thus

$$\frac{dm(t)}{dt} = K_d \frac{d^2 e(t)}{dt^2} + K_p \frac{de(t)}{dt} + K_i e(t)$$

Utilizing first backward differences where

$$\left.\frac{df}{dt}\right|_n = \frac{f_n - f_{n-1}}{\Delta t}$$

and

$$\left.\frac{d^2 f}{dt^2}\right|_n = \frac{f_n - 2f_{n-1} + f_{n-2}}{(\Delta t)^2}$$

The finite difference equation becomes

$$\frac{m_n - m_{n-1}}{\Delta t} = \frac{K_d(e_n - 2e_{n-1} + e_{n-2})}{(\Delta t)^2} + K_p \frac{e_n - e_{n-1}}{\Delta t} + K_i e_n$$

m_n represents the value of the manipulated variable which is to be applied to the process at time t_n. Solving for m_n

$$m_n = \left(\frac{K_d}{\Delta t} + K_p + K_i \Delta t\right)e_n$$

$$- \left(\frac{2K_d}{\Delta t} + K_p\right)e_{n-1} + \frac{K_d}{\Delta t}e_{n-2} + m_{n-1} \qquad (16\text{-}12)$$

which is a control algorithm expressing the nth value of the manipulated variable in terms of the previous value, m_{n-1}, and past and present values of the error e.

Digital control algorithms may be established by applying a combination of controller design for a desirable closed loop response as discussed in Sec. 13-6, and finite difference approximations.

It should be pointed out that the anticipated degree of stability for a continuous system may be reduced by the sampling process. Consider the design of a controller for a plant represented by a pure dead time.

$$\frac{C(s)}{M(s)} = P(s) = e^{-sD}$$

The problem is to formulate a $G_c(s)$ such that

$$\frac{C(s)}{R(s)} = \frac{G_c(s)\, P(s)}{1 + G_c(s)\, P(s)} \qquad (16\text{-}13)$$

has some desirable response. First consider designing a controller $G_c(s)$ such that

$$\frac{C(s)}{R(s)} = K_R(s) = \frac{\lambda}{s + \lambda} \qquad (16\text{-}14)$$

Equating Eqs. 16-13 and 16-14 yields

$$G_c(s) = \frac{1}{P(s)} \frac{\dfrac{\lambda}{s + \lambda}}{1 - \dfrac{\lambda}{s + \lambda}} = e^{sD}\frac{\lambda}{s}$$

The control law is then

$$M(s) = \frac{\lambda}{s}\, E(s)\, e^{+sD}$$

which cannot be inverted to the time domain directly. Physically, the control is not realizable because it is necessary to integrate values of a function which do not yet exist.

For a realistic system consider

$$K_R(s) = K'(s)\, e^{-sD}$$

that is, the closed-loop system will have the same dead time as the open-loop plant. Then,

$$G_c(s) = \frac{K'(s)\, e^{-sD}}{P(s)\, (1 - K'(s)\, e^{-sD})}$$

For steady-state analysis an approximation at low frequencies ($e^{-j\omega D} \to 1 \underline{/0^\circ}$ as $\omega \to 0$)

$$\frac{C(s)}{N(s)} = \frac{1}{1 + G_c(s)\, P(s)} = 1 - K'(s)\, e^{-sD} \cong 1 - K'(s)$$

In order to suppress low frequency noise let

$$\frac{C(s)}{N(s)} = \frac{s}{s + \lambda}$$

Then,

$$K'(s) = \frac{\lambda}{s + \lambda}$$

and the controller transfer function becomes

$$G_c(s) = \frac{\dfrac{\lambda}{s + \lambda} e^{-sD}}{e^{-sD}\left(1 - \dfrac{\lambda}{s + \lambda} e^{-sD}\right)} = \frac{\lambda}{s + \lambda - \lambda e^{-sD}}$$

Next, $G_c(s)$ may be expressed in the time domain and then approximated by finite difference expressions. Taking the inverse transform, an equivalent differential equation for the controller is

$$\frac{dm(t)}{dt} + \lambda m(t) - \lambda m(t - D) = \lambda e(t)$$

where $m(t - D) = 0$ for $t < D$. In finite difference form this becomes

$$\frac{m_n - m_{n-1}}{T} + \lambda m_n - \lambda m_{n-\Gamma} = \lambda e_n$$

where Γ is the number of sample intervals of length T in the dead time.

$$\Gamma = \frac{D}{T}$$

Γ may be rounded to the next highest integer of the ratio D/Γ. The values of $m_{n-\Gamma}$ from $m_{-\Gamma}$ to m_0 are zero, providing that the initial conditions on the controller are zero.

The control algorithm is then given by

$$m_n = \frac{1}{1 + \lambda T}\left[m_{n-1} + \lambda T m_{n-\Gamma} + \lambda T e_n\right]$$

The control action is thus a function of the previous values of the manipulated variable and the present error e_n. Note should be made of the fact that nothing has been said concerning the sample time T and its relationship to the process time constant τ or the closed loop break frequency λ. It will suffice here to say that T must be smaller than τ or λ to approach the control characteristics of a continuous system controller of the same form. If it is not, oscillation and instability may result.

16-6. Impulse Modulation Concepts

The process of sampling as treated in sampled-data control system theory involves the transformation of data from a continuous analog form to data in a discrete value form. The process is easily visualized by referring to Fig. 16-8. Continuous information, Fig. 16-8a, is passed through a sampling switch, Fig. 16-8b, which is closed momentarily every T seconds. The sampled form of the data is shown in Fig. 16-8c as points on the curve.

$$f(0), f(T), f(2T), \ldots, f(nT), \ldots$$

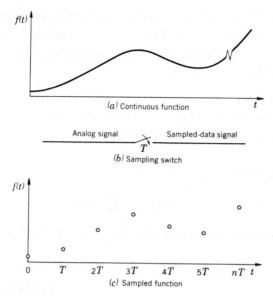

FIG. 16-8. Graphic interpretation of the sampling process.

The function may be represented mathematically as a modulated pulse train. The true, or ideal, samples may be thought of as a series of impulse functions. The area under each impulse represents the value of the function $f(t)$ at each of the sampling instants. The sampled-data form of $f(t)$ is typically described by a star notation $f^*(t)$. Thus,

$$f^*(t) = f(t)\, \delta^*(t)$$

where $\delta^*(t)$ is a unit impulse carrier, or train of unit impulses.

$$\delta^*(t) = \sum_{n=0}^{\infty} \delta(t - nT) \tag{16-15}$$

Thus,

$$f^*(t) = f(t) \sum_{n=0}^{\infty} \delta(t - nT) \tag{16-16}$$

or

$$f^*(t) = \sum_{n=0}^{\infty} f(nT)\, \delta(t - nT) \tag{16-17}$$

Equation 16-17 for the starred function is the sampled-data form of the continuous function $f(t)$. It is generally Laplace transformable, that is,

$$\mathcal{L}\{f^*(t)\} = \mathcal{L}\left\{\sum_{n=0}^{\infty} f(nT)\, \delta(t - nT)\right\}$$

$$F^*(s) = \sum_{n=0}^{\infty} f(nT)\, e^{-nTs} \tag{16-18}$$

The starred function in s denotes the Laplace transform of the sampled time function. Equation 16-18 may be expanded into a sequence

$$F^*(s) = f(0) + f(T)e^{-Ts} + f(2T)e^{-2Ts} + \cdots + f(nT)e^{-nTs} + \cdots \quad (16\text{-}19)$$

Equation 16-19 is the basic relationship from which the z-transform is defined as will be illustrated in the following section.

16-7. z-Transforms

In the sequence of $F^*(s)$, Eq. 16-19, e^{-nTs} appears as a multiplier. A new variable z may be introduced where

$$z = e^{Ts}$$

Therefore, substituting into Eq. 16-18,

$$F^*(s)\Big|_{s=\frac{1}{T}\ln z} = F(z) = \sum_{n=0}^{\infty} f(nt)z^{-n} \quad (16\text{-}20)$$

The inverse transform of $F(z)$ is given by Eq. 16-16, or in expanded form

$$f^*(t) = f(o)\delta(t) + f(T)\delta(t - T) + f(2T)\delta(t - 2T) + \cdots$$
$$+ f(nT)\delta(t - nT) + \cdots$$

From a geometric point of view, the z variable introduced carries the index n which identifies the relative position of a pulse with respect to the pulse train.

EXAMPLE 16-2. Derive the z-transform for the impulse train

$$\delta^*(t) = \delta(t) + \delta(t - T) + \delta(t - 2T) + \cdots + \delta(t - nT) + \cdots$$

Solution:

$$\delta^*(t) = \sum_{n=0}^{\infty} \delta(t - nT)$$

$$\Delta^*(s) = \sum_{n=0}^{\infty} e^{-nTs} = 1 + e^{-Ts} + e^{-2Ts} + \cdots + e^{-nTs} + \cdots$$

which is the geometric progression

$$\Delta^*(s) = \frac{1}{1 - e^{-Ts}}$$

Thus,

$$\Delta(z) = \frac{1}{1 - z^{-1}}$$

The z-transform of the unit step function $u(t)$ is readily obtained by realizing that $u^*(t)$ is equivalent to $\delta^*(t)$. That is, $u^*(t) = u(t)\delta^*(t) = \delta^*(t)$, thus,

$$U(z) = \frac{1}{1 - z^{-1}} \quad\quad Ans.$$

The general form of the transformation given by Eq. 16-20 may be applied directly to sampled-data functions.

EXAMPLE 16-3. Find the z-transform of the sampled-data form of the continuous function $f(t) = t$.

Solution:

$$F(z) = \sum_{n=0}^{\infty} (nT)z^{-n}$$

$$= Tz^{-1} + 2Tz^{-2} + 3Tz^{-3} + \cdots + nTz^{-n} + \cdots$$

$$= Tz^{-1}(1 + 2z^{-1} + 3z^{-2} + \cdots + (n + 1)z^{-n} + \cdots)$$

The bracketed quantity on the right is recognized as the geometric progression

$$\frac{1}{(1 - z^{-1})^2} = 1 + 2z^{-1} + 3z^{-2} + \cdots + (n + 1)z^{-n} + \cdots$$

Thus,

$$F(z) = \frac{Tz^{-1}}{(1 - z^{-1})^2} \qquad \qquad Ans.$$

EXAMPLE 16-4. Develop the z-transform of the sampled-data form of the continuous function $f(t) = e^{-bt}$ where $F(s) = 1/(s + b)$.

Solution:

Thus

$$F(z) = \sum_{n=0}^{\infty} e^{-bnT} z^{-n}$$

$$F(z) = 1 + e^{-bT}z^{-1} + e^{-b2T}z^{-2} + \cdots + e^{-bnT}z^{-n} + \cdots$$

$$F(z) = \frac{1}{1 - e^{-bT}z^{-1}} \qquad \qquad Ans.$$

Ordinarily, the sampling process results in a linear operation, therefore,

$$[\alpha f_1(t) + \beta f_2(t)]^* = \alpha f_1^*(t) + \beta f_2^*(t)$$

As with Laplace transforms the uniqueness of the z-transform permits the development of tables of transform pairs. Relatively complete tables are available in most texts on sampled-data control systems. Additional manipulative skills will be developed in the following section.

16-8. Block Diagrams and Transfer Functions with z-Transforms

The *pulse transfer function* $G(z)$ for a sampled-data control system relates the z-transform of the sampled output to the z-transform of the sampled input. Figure 16-9 illustrates a typical block representation of a continuous element with a sampler at the input and a fictitious sampler at the output. The fictitious sampler is utilized to illustrate that in sampled-data systems, information is described at the sample instants. The z-transform relates the sampled output to the sampled input. Although information may be constructed between the sample instants by special

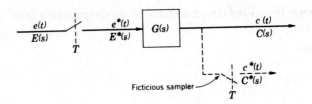

FIG. 16-9.   Sampled-data control system block.

techniques, these techniques will not be presented. The reader is referred
to one of the many texts on sampled-data control systems.

$E^*(s)$ is the Laplace transform of the sampled signal $e^*(t)$. $E(z)$
is the z-transform of $E^*(s)$. With the introduction of the fictitious sampler, the sampled output $c^*(t)$ may be obtained. $C^*(s)$ is the Laplace
transform of the sampled output and $C(z)$ is the z-transform of $C^*(s)$.

For a unit impulse function input $e(t) = \delta(t)$, $e^*(t)$ is also a unit
impulse. The impulse response of the continuous element $G(s)$ is defined
by

$$g(t) = \mathcal{L}^{-1}\{G(s)\}$$

The sampled output for a single impulse at $t = 0$ is

$$c^*(t) = g^*(t) = \sum_{n=0}^{\infty} g(nT)\delta(t - nT)$$

The *pulse transfer function* is defined as the Laplace transform of the
sampled impulse response

$$G^*(s) = \sum_{n=0}^{\infty} g(nT)e^{-nTs} \qquad (16\text{-}21)$$

$g(nT)$ in Eq. 16-21 is defined as the weighting sequence of the element
$G(s)$.

In the case where $e(t)$ is a general input, the input to $G(s)$ is a train
of pulses $e^*(t)$. Therefore the response at any time nT is, by superposition,
the sum of the contributions of the individual impulse responses initiated
at all previous sample instants. Figure 16-10 illustrates the formation of
the output at $t = 2T$ for an input $e(t)$.

$$C(2T) = e(o)\,g(2T) + e(T)\,g(T) + e(2T)\,g(o)$$

In general terms at time nT, this relationship may be expressed as

$$C(nT) = e(o)\,g(nT) + e(T)\,g[(n - 1)\,T] + e(2T)\,g[(n - 2)\,T]$$
$$+ \cdots + e[(n - 1)\,T]\,g(T) + e(nT)\,g(o) \qquad (16\text{-}22)$$

multiplying both sides of Eq. 16-22 by e^{-nTs} and summing from $n = 0$
to $n = \infty$,

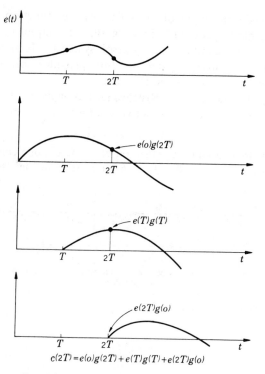

$$c(2T) = e(o)g(2T) + e(T)g(T) + e(2T)g(o)$$

FIG. 16-10. Super position of the impulse responses.

$$\sum_{n=0}^{\infty} C(nT) e^{-nTs} = \sum_{n=0}^{\infty} e(o) g(nT) e^{-nTs} + \sum_{n=0}^{\infty} e(T) g[(n-1) T] e^{-nTs}$$

$$+ \cdots + \sum_{n=0}^{\infty} e[(n-1) T] g(T) e^{-nTs} + \sum_{n=0}^{\infty} e(nT) g(o) e^{-nTs}$$

Then, expanding and regrouping,

$$\sum_{n=0}^{\infty} C(nT) e^{-nTs} = [e(o) + e(T) e^{-Ts} + e(2T) e^{-2Ts} + \cdots] \sum_{n=0}^{\infty} g(nT) e^{-nTs}$$

$$= \sum_{n=0}^{\infty} e(nT) e^{-nTs} \sum_{n=0}^{\infty} g(nT) e^{-nTs}$$

or, by definition of the Laplace transform of the sampled function,

$$C^*(s) = E^*(s) G^*(s)$$

from which the z-transform is

$$C(z) = E(z) G(z) \tag{16-23}$$

With the development of the pulse transfer function $G(z) = C(z)/E(z)$ it is possible to proceed in investigating techniques for manipulating block diagrams of sampled-data control systems. It should be re-emphasized that the pulse transfer function provides information at the sample instants and is undefined elsewhere.

Elements Cascaded with Synchronous Samplers. First consider a continuous element cascaded with two samplers. From Eq. 16-23

$$C(z) = G(z) E(z)$$

EXAMPLE 16-5. Find the z-transform of the output of a continuous element $G(s) = 1/s$ when it is bracketed by two samplers as in Fig. 16-11. The input

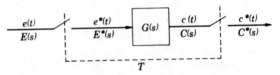

FIG. 16-11. A continuous element cascaded with samplers.

is $e(t) = e^{-at}$ and $E(s) = 1/(s + a)$.

Solution:

$$G(s) = \frac{1}{s}, \quad \text{therefore} \quad G(z) = \frac{1}{1 - z^{-1}}$$

For the input function $E(s) = 1/(s + a)$ which has a corresponding z-transform $E(z) = 1/(1 - e^{-Ta}z^{-1})$. Thus,

$$C(z) = \frac{1}{1 - z^{-1}} \frac{1}{1 - e^{-Ta}z^{-1}}$$

$$C(z) = \frac{1}{1 - (1 + e^{-Ta})z^{-1} + e^{-Ta}z^{-2}} \qquad Ans.$$

$C*(t)$ may be obtained by inverting $C(z)$ to a pulse sequence form by long division or by the method of residues.

Continuous Elements Separated by Synchronous Samplers. Figure 16-12 illustrates two continuous elements separated by synchronous samplers. The overall z-transfer function may be derived in the following manner.

$$C_2(z) = G_2(z) C_1(z)$$
$$C_1(z) = G_1(z) E(z)$$

Thus,

$$C_2(z) = G_1(z) G_2(z) E(z)$$

or where

$$G(z) = G_1(z) G_2(z)$$
$$C_2(z) = G(z) E(z) \qquad (16\text{-}24)$$

EXAMPLE 16-6. Find the z-transform of the output $C_2(z)$ for two cascaded time constant elements separated by synchronous samplers as in Fig. 16-12. The

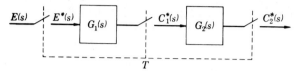

FIG. 16-12. Continuous elements separated by synchronous samples.

input $e(t)$ is a unit step.

$$G_1(s) = \frac{1}{s + a}, \qquad G_2(s) = \frac{1}{s + b}$$

Solution: From a z-transform table

$$G_1(z) = \frac{1}{1 - e^{-Ta}z^{-1}}, \qquad G_2(z) = \frac{1}{1 - e^{-Tb}z^{-1}}$$

and

$$E(z) = \frac{1}{1 - z^{-1}}$$

Therefore,

$$C_2(z) = \frac{1}{(1 - z^{-1})(1 - e^{-Ta}z^{-1})(1 - e^{-Tb}z^{-1})} \qquad Ans.$$

Continuous Elements Cascaded by a Continuous Connection. The case of two continuous elements cascaded directly with a continuous connection is illustrated in Fig. 16-13. In this instance, $G_2(s)$ is driven

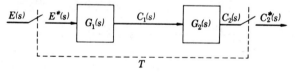

FIG. 16-13. Continuous elements separated by a continuous connection.

not only by signals at the sample instants T, but by the continuous output of $G_1(s)$. Therefore $G_1(s)$ and $G_2(s)$ are first lumped together in the conventional manner and there the z-transform is

$$C(z) = G_1 G_2(z) E(z)$$

where $G_1 G_2(z)$ is the notation commonly used to indicate the z-transform of the product $G_1(s) G_2(s)$. This is *not* the same as the product of the z-transforms of $G_1(s)$ and $G_2(s)$.

EXAMPLE 16-7. Two continuous elements cascaded as in Fig. 16-13 have the transfer functions $G_1(s) = 1/(s + a)$ and $G_2 = 1/(s + b)$. For a unit step input, find $C_2(z)$.

Solution:

$$G_1(s)\,G_2(s) = \frac{1}{(s+a)(s+b)} = \frac{\frac{1}{b-a}}{s+a} - \frac{\frac{1}{b-a}}{s+b}$$

$$= \frac{1}{b-a}\left(\frac{1}{s+a} - \frac{1}{s+b}\right)$$

$$G(z) = G_1 G_2(z) = \frac{1}{(b-a)}\left(\frac{1}{1-e^{-Ta}z^{-1}} - \frac{1}{1-e^{-Tb}z^{-1}}\right)$$

$$= \frac{1-e^{-Tb}z^{-1} - 1 + e^{-Ta}z^{-1}}{(b-a)(1-e^{-Ta}z^{-1})(1-e^{-Tb}z^{-1})}$$

$$= \frac{-z^{-1}}{b-a}\frac{(e^{-Tb}-e^{-Ta})}{(1-e^{-Ta}z^{-1})(1-e^{-Tb}z^{-1})}$$

Therefore,

$$C_2(z) = \frac{(e^{-Tb}-e^{-Ta})z^{-1}}{(1-z^{-1})(a-b)(1-e^{-Ta}z^{-1})(1-e^{-Tb}z^{-1})} \qquad \textit{Ans.}$$

Sampled-Data Feedback Systems. In sampled-data feedback systems, the form of the closed loop transfer function or the form of the z-transform of the output depends on the location of the samplers. Con-

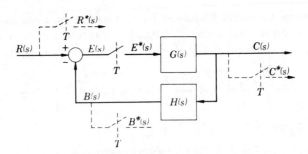

FIG. 16-14.   The error-sampled feedback system.

sider as an example the error-sampled feedback system of Fig. 16-14.

$$E(s) = R(s) - B(s)$$

as the error is sampled, and the sample of a difference is the difference of the samples,

$$E^*(s) = R^*(s) - B^*(s)$$

Thus,
$$E(z) = R(z) - B(z)$$

Also
$$B(z) = GH(z)\,E(z)$$

and
$$C(z) = G(z)\,E(z)$$

Thus,

$$\frac{C(z)}{R(z)} = \frac{G(z)}{1 + GH(z)}$$

It may be shown also that

$$E(z) = \frac{1}{1 + GH(z)}$$

In computer control a possible configuration could involve a continuous plant and other elements bracketed by synchronous samplers as

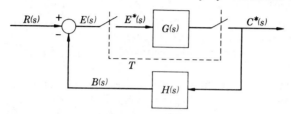

FIG. 16-15. A feedback system with plant bracketed by samplers.

shown in Fig. 16-15. In this case,

$$E(z) = R(z) - B(z)$$
$$B(z) = C(z) H(z)$$
$$C(z) = E(z) G(z)$$

Therefore,

$$\frac{C(z)}{G(z)} = R(z) - C(z) H(z)$$

Thus

$$\frac{C(z)}{R(z)} = \frac{G(z)}{1 + G(z) H(z)}$$

It should be apparent that a multitude of forms may exist depending on the location of the samplers throughout the system. It is not always possible to solve for an overall closed-loop transfer function explicitly as inputs and elements not separated by samplers must be transformed as a product.

In sampled-data systems, particularly with digital computer control, data may or may not be fed directly to control elements from a sampler. Rather, it is typical to incorporate some form of data reconstruction or "data hold" device to maintain signals between samples. This aspect of the problem will be discussed in the next section.

16-9. Data Reconstruction

In an actual control system where continuous analog elements are manipulated by sampled-data information it is generally desirable to first "smooth" the data before it is utilized. Naturally, if the sampling time is short relative to the systems ability to respond, the system acts as a filter and the smoothing process is a natural result. When digital computers are utilized for control, the computer is generally controlling a multitude of loops of a given process, or even a number of processes. If this is the case, economics dictate that the sample times be as long as possible. Therefore the output of samplers must be modified by elements that transform the pulse train to analog information. The process of obtaining continuous or at least piecewise continuous data from sampler outputs is termed *data reconstruction*. Devices to accomplish this task are called "data holds," "desampling filters," or "data extrapolators." Physically, the process of data reconstruction may be accomplished in a number of ways.

Filtering. It is possible to recover a sampled signal by passing the sampled signal through a filter which has a "flat" frequency response up to one half of the sampling frequency. An ideal filter with unity transmission from zero frequency up to one half of the sampling frequency, and zero transmission elsewhere is called a *cardinal data hold*. Clearly, this filter may only be approximated by a physical system. Furthermore, in real filters the phase shift generally becomes apparent at frequencies much lower than the cut-off frequency. The resulting phase shift causes signal phase distortion, and in closed-loop feedback system adds to the problem of system stability.

Polynomial Extrapolation. The problem of data reconstruction of a sampled signal may be viewed as a problem of extrapolating between

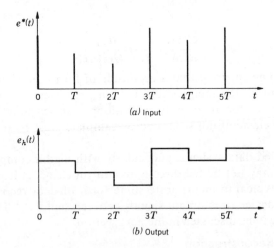

FIG. 16-16. Input and output of the zero order data hold.

sample instants. Extrapolation can be accomplished by polynomial expressions using a variety of mathematical techniques. Various orders of polynomial extrapolation may be achieved by combinations of analog and digital circuitry.

Zero-Order Data Hold. The simplest form of extrapolation involves maintaining or holding the value of the last pulse until the next pulse occurs. This is often called a "staircase" or "boxcar" data system. For example, the input to a zero-order hold $e^*(t)$ is shown in Fig. 16-16a. The output $e_h(t)$ or staircase function is shown in Fig. 16-16b.

To obtain the transfer function of the zero-order hold it is necessary to refer to the impulse response of the hold element. The unit impulse response $e_h\delta(t)$ of the zero-order hold is shown in Fig. 16-17a. This function is equivalent to a plus and minus step function as illustrated in Fig. 16-17b. Thus,

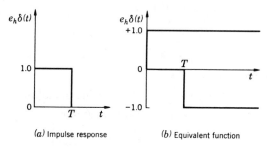

(a) Impulse response (b) Equivalent function

FIG. 16-17. Unit impulse response of a zero-order hold.

$$e_h\delta(t) = u(t) - u(t - T)$$

In transform form this yields

$$G_h(s) = \frac{1}{s} - \frac{e^{-Ts}}{s} = \frac{1 - e^{-Ts}}{s} \tag{16-25}$$

In analyzing a control system utilizing a zero-order hold for data reconstruction, the transfer function of Eq. 16-25 is placed in a separate cascade block ahead of the plant or controllers as seen in Fig. 16-18.

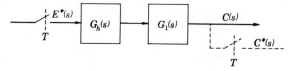

FIG. 16-18. Cascade connection of the zero-order hold.

From the work in Sec. 16-8 it is apparent that it must be lumped with the plant transfer function to obtain the z-transform.

$$G(z) = G_h G_1(z) = \frac{C(z)}{E(z)}$$

With the transfer function of Eq. 16-25 the hold element is simply treated as another analog element with the addition of dead time equal to the sample interval.

16-10. Stability Concepts

In the s domain, stability was characterized by the nature of the roots of the characteristic equation. A system is unstable if the roots of the characteristic equation have positive real parts (i.e., $\sigma > 0$ for $s = \sigma + j\omega$).

On the s-plane one is concerned with identifying roots on the right hand half of the plane. The z-transform is defined through the relationship, $z = e^{sT}$, where

$$s = \sigma + j\omega$$

Thus,

$$z = e^{(\sigma + j\omega)T} = e^{\sigma T} e^{j\omega T}$$

Then

$$|z| = |e^{\sigma T}||e^{j\omega T}| = e^{\sigma T}$$

Therefore the following may be summarized

$$\sigma = 0, |z| = 1 \qquad \{\text{Borderline of stability}$$
$$\sigma < 0, |z| < 1 \qquad \{\text{System stable}$$
$$\sigma > 0, |z| > 1 \qquad \{\text{System unstable}$$

As a result, for stability of the sampled data control, the roots of the z-transform of the characteristic equation or zeros of the characteristic function must lie within a unit circle on the z-plane. In other words, the unit circle of the z-plane in Fig. 16-19 is analogous to the imaginary axis of the s-plane.

It is logical to expect at this point that many techniques have been developed for analyzing sampled-data control systems along lines parallel to the techniques available for analog systems. Many techniques are presented in texts on sampled-data control [20, 25].

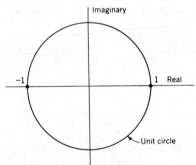

Fig. 16-19. Stability on the z-plane.

16-11. Closure

The intent in this chapter has been to present an introduction to the concepts underlying computer control. In particular, emphasis has been placed on the mathematics of sampled-data control systems. The reader should appreciate that the theory underlying this broad and rapidly expanding topic is the subject of many individual textbooks and papers. As computer control finds more widespread successful application, control engineers and the designers of processes will need a greater degree of familiarization with the principles and application techniques.

Problems

16-1. a) For the flow control problem of Sec. 16-3, plot responses to a unit step input for the following:

$$K_1 = 1/2 \quad \begin{cases} K_v \alpha E_a T = 2 \\ \quad '' \quad = 4 \\ \quad '' \quad = 5 \end{cases}$$

$$K_1 = 1/4 \quad \begin{cases} K_v \alpha E_a T = 2 \\ \quad '' \quad = 4 \\ \quad '' \quad = 8 \end{cases}$$

$$K_1 = 3/4 \quad \begin{cases} K_v \alpha E_a T = 1 \\ \quad '' \quad = 4/3 \\ \quad '' \quad = 2 \end{cases}$$

b) What general conclusions could be reached regarding the selection of K_1 and T for optimum computer control response? Consider that $K_v E_a \alpha$ is fixed by the valve selection.

16-2. Using backward differences, formulate a finite difference equation to numerically solve

$$\frac{d^2 y}{dt^2} + 4.8 \frac{dy}{dt} + 64y = 0$$

$$y \big|_{t=0} = 10$$

$$\frac{dy}{dt}\bigg|_{t=0} = 0$$

Illustrate how the problem would be initialized. Analyze the problem of selecting a Δt for the problem solution and recommend a value. If a digital computer is available, solve the equation for different values of Δt and compare the results with the exact continuous solution.

16-3. Derive a control algorithm for a proportional plus integral plus derivative controller where the controller is expressed by

$$m(t) = K_p e(t) + K_i \int e(t)\, dt + K_d \frac{de(t)}{dt}$$

Discuss the relative merits between the resulting form and that expressed by Eq. 16-12.

16-4. Design a digital controller (digital control algorithm) that will control a plant $P(s) = Ke^{-sD}/(\tau s + 1)$ such that

$$\frac{C(s)}{R(s)} = \frac{\lambda}{s + \lambda}$$

$$\frac{C(s)}{N(s)} = \frac{s}{s + \lambda}$$

Express $G_c(s)$ in finite difference form.

16-5. Derive the z-transform for the sampled-data form of the following continuous functions:

a) $f(t) = t^2$ c) $f(t) = \sin at$

b) $f(t) = \dfrac{1}{(b - a)} (e^{-at} - e^{-bt})$ d) $f(t) = te^{-at}$

16-6. Find $G(z)$ for the following:

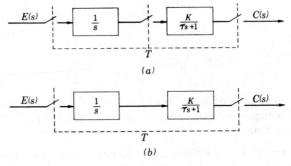

(a)

(b)

PROB. 16-6

16-7. Derive expressions for $C(z)$ and $E(z)$ in the following:

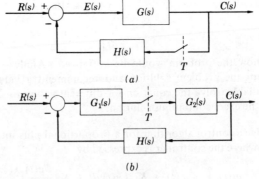

(a)

(b)

PROB. 16-7

16-8. Find $C(z)$ for the following:

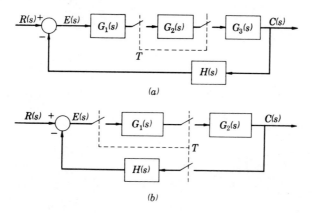

(a)

(b)

PROB. 16-8

16-9. Find the overall pulse transfer function for a control system with a zero-order data hold and the sampler configuration shown.

PROB. 16-9

16-10. The block diagram of a computer controlled system is shown below.

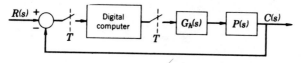

PROB. 16-10

If the control algorithm is based on a continuous function

$$G_c(s) = \frac{\lambda}{k} \frac{\tau s + 1}{s}$$

and the plant is defined by $K/(\tau s + 1)$, develop an expression for $C^*(t)$ if the input to the system is a unit step.

Appendix **A**

Digital Computing for
Control System Analysis

Many engineering groups in industry and an ever-increasing number of engineering students have access to modern computing facilities. These facilities generally include analog and digital computers of various types. Chapter 4 and Appendix D discuss the use of an analog computer in control-system analysis. The purpose of this appendix is to discuss the types of problems in control analysis that are suited to digital computation. Some programs are presented to illustrate the points made. In the latter regard, a basic understanding of computer programming in one of the many compiler languages is assumed.[1]

The digital computer is extremely useful for the solution of problems involving many repeated computations. In particular, iteration procedures may be handled most economically using a digital machine. In evaluating roots of a system characteristic equation, it may be necessary to factor a higher-degree polynomial. Appendix B describes two procedures, either of which may require considerable computation. Lin's method, described in Appendix B, is the type of iteration procedure that lends itself well to digital programming. A detailed analysis and a computer program will be presented later.

In frequency response analysis using either the polar plot (Nyquist diagram) or rectangular plot (Bode diagram), considerable computation is required to evaluate the transfer function magnitudes and phase angles for a desired range of frequency. This is another good example of a computation that is suited for digital computation. Sample programs will be presented to illustrate the principles here.

In Chapter 14, it was pointed out that many nonlinear control systems are piecewise linear; that is, their behavior may be described by linear equations for particular regions of their operation. The complete solution may be obtained by evaluating a linear equation (i.e., the solution to a linear differential equation) over a specific range and then using the final values of the variables as the initial conditions of the following range.

[1]The programs presented in this appendix are written in the FORTRAN language.

The digital computer may be used to obtain the numerical solution of a differential equation using the Runge-Kutta integration method or another numerical approach [29]. Here, however, the digital machine is in competition with the analog computer. For the purpose of analysis and synthesis the analog computer has certain advantages. The utility of a digital computer for the solution of linear and nonlinear differential equations in an engineering problem is a matter of availability, convenience, and cost. It should be mentioned here that certain digital programming schemes have been developed, such as DYSAC (Digitally Simulated Analog Computer), which facilitate the solution of a class of engineering problems which are most easily visualized or described by block diagrams. To utilize programming of this nature requires a large, high-speed, digital computer, and the cost of computer time may become prohibitive.

Factoring Higher-Degree Polynomials

The quadratic factors of a higher-degree polynomial may be extracted by the use of Lin's method described in Appendix B. The method is an iterative procedure wherein one guesses a trial quadratic factor, divides it into the polynomial, and then looks at the remainder. Once the quadratic factors of the polynomial have been determined to a satisfactory accuracy, the real and complex conjugate roots are readily established. The following is a development of a computer program for obtaining the quadratic factors by Lin's method.

The nth degree equation may be arranged in the form

$$a_1 p^n + a_2 p^{n-1} + \cdots + a_{n-1} p^2 + a_n p + a_{n+1} = 0 \qquad (A-1)$$

where the a's are constant coefficients. The order of subscripting is chosen here for convenience in indexing the computer program. Figure A-1 is a general flow diagram that illustrates the computation and decisions that are incorporated into the program for iteration. The input data read in consists of N, the degree of the polynomial on the left in Eq. A-1, and the coefficients $a_1, a_2, \ldots, a_{n+1}$. In every case the equation must be divided by the coefficient of the highest power of p so that $a_1 = 1$.

Special note may be made of the test made on the size of the remainder factor, which is of the form $b_1 p + b_0$. Both the linear portion b_1 and the constant portion b_0 must be tested. Further, it is more meaningful to test the ratio of the numbers subtracted to obtain the relative size of the remainder terms rather than the absolute values of the remainder terms themselves. The quantity ϵ chosen for the test then indicates a percentage difference between the last two terms of the last two rows of the division process.

The last test illustrated in Fig. A-1 establishes how many quadratic factors are yet to be determined. An even-degree polynomial will factor into all quadratic factors. An odd-degree polynomial will factor into all

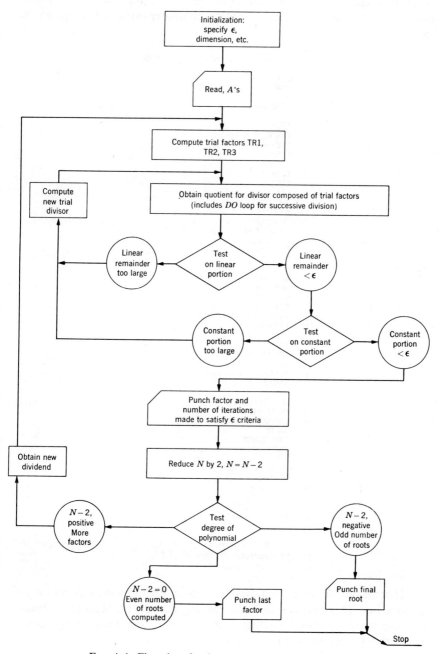

Fig. A-1. Flow chart for the program shown in Fig. A-2.

quadratic factors plus one real factor of the first degree. Both these possibilities are considered. The complete computer program is given in Fig. A-2.

```
C     QUADRATIC FACTORS BY LIN'S METHOD
      READ,N
      EPSIL=0.0001
      DIMENSION A(100),QUO(100)
      N1=N+1
      READ,(A(I),I=1,N1)
1     L=0
      TR1=1.
      TR2=A(N1-1)/A(N1-2)
      TR3=A(N1)/A(N1-2)
2     L=L+1
      T1=A(1)
      T2=A(2)
      T3=A(3)
      K=1
      QUO(K)=T1
      S2=T1*TR2
      S3=T1*TR3
      DO 10 I=4,N1
      T1=T2-S2
      T2=T3-S3
      T3=A(I)
      K=K+1
      QUO(K)=T1
      S2=T1*TR2
10    S3=T1*TR3
      TT1=ABSF(1.-S2/T2)
      IF(EPSIL-TT1) 12,12,11
11    TT2=ABSF(1.-S3/T3)
      IF(EPSIL-TT2) 12,12,15
12    TR2=(TR2+T2/T1)/2.
      TR3=(TR3+T3/T1)/2.
      GO TO 2
15    PUNCH,TR1,TR2,TR3,L
      N=N-2
      N1=N+1
      IF(N-2) 19,18,16
16    DO 17 I=1,N1
17    A(I)=QUO(I)
      GO TO 1
18    TR1=QUO(1)
      TR2=QUO(2)
      TR3=QUO(3)
      PUNCH,TR1,TR2,TR3
      GO TO 20
19    TR1=QUO(1)
      TR2=QUO(2)
      PUNCH,TR1,TR2
20    STOP
      END
```

FIG. A-2. Program for obtaining quadratic factors by Lin's method.

There is one difficulty that may be encountered in such a program. If Lin's method does not provide rapid convergence, the problem may require excessive computer time.[2] Further, if the solution does not converge at all, no solution will be obtained.

[2]In the program shown in Fig. A-2 the new trial divisor is obtained by a modification of Lin's method, which, in general, was found to give faster convergence.

Evaluation of Transfer Functions for Frequency Response Analysis

Frequency response analysis requires the evaluation of the magnitude and angle of a transfer function. If one has evaluated any number of complex transfer functions, it is needless to say that considerable labor is involved. This may be the case regardless of the many shortcuts available for plotting. Nyquist- and Bode-plot data are readily obtained using the digital computer. Figures A-3 and A-4 illustrate the types of programs that may be written for these computations.

```
C       TRANSFER FUNCTION EVALUATION INT AND LAG
        READ,FK,T
        READ,WI,DW,FN
        W=WI
        PUNCH,FK,T
1       FM=FK/(W*SQRTF((T*W)**2+1.))
        FMDB=20.*.4342944819*LOGF(FM)
        ANG = -90.-57.295778*ATANF(T*W)
        PUNCH,W,FM,FMDB,ANG
        IF(W-FN*WI)2,3,3
2       W=W+DW
        GO TO 1
3       STOP
        END
```

Fig. A-3. Program for evaluation of
the transfer function $K/[s(Ts + 1)]$
for frequency response analysis.

The program shown in Fig. A-3 is for evaluation of a single lag with one integration, or

$$G(s) = \frac{K}{s(Ts + 1)}$$

```
C       TRANSFER FUNCTION EVALUATION LEAD LAG DOUBLE INT
        READ,FK,T1,T2
        READ,WI,DW,FN
        W=WI
        PUNCH,FK,T1,T2
1       FM=(FK/W**2)*SQRTF(((T1*W)**2+1.)/((T2*W)**2+1.))
        FMDB=20.*.4342944819*LOGF(FM)
        ANG = -180.+57.295778*(ATANF(T1*W)-ATANF(T2*W))
        PUNCH,W,FM,FMDB,ANG
        IF(W-FN*WI)2,3,3
2       W=W+DW
        GO TO 1
3       STOP
        END
```

Fig. A-4. Program for evaluation of the transfer function $[K(T1s + 1)]/[s^2(T2s + 1)]$ for frequency response analysis.

The input data consists of one card containing the gain FK and the time constant T, and a second card with the initial frequency $W1$, the increment in frequency for each computation DW, and the number of increments FN. The program computes the magnitude ratio $M(\omega)$, its value in decibels, and the phase angle at each desired frequency.

Figure A-4 shows a similar program for computing the same output as the program in Fig. A-3 for the combination of a double integration and a lead-lag compensator of the form:

$$G(s) = \frac{K(T1\,s + 1)}{s^2(T2s + 1)}$$

Computer programs for frequency response analysis, for factoring polynomials for evaluation of the roots of a characteristic equation, and for the actual numerical solution of linear and nonlinear differential equations are extremely useful in control system analysis. The digital computer is particularly well suited for numerical evaluation of mathematical expressions and problems involving iterative procedures. The brief presentation here is intended to provide a starting point—it is by no means conclusive. While digital computer programming is not a prerequisite for control-system studies, it is easy to see that the ability to program will be a time-saving asset.

Solution of Equations

In control work it is sometimes necessary to determine the roots of an equation or to factor a polynomial. Since the approach is the same in either case, only root solving will be discussed here.

Consider the general equation

$$a_n p^n + a_{n-1} p^{n-1} + \cdots + a_1 p + a_0 = 0 \qquad \text{(B-1)}$$

in which the coefficients are real numbers. The coefficient of p^n can be made unity by division:

$$p^n + \frac{a_{n-1}}{a_n} p^{n-1} + \cdots + \frac{a_1}{a_n} p + \frac{a_0}{a_n} = 0$$

or

$$p^n + b_{n-1} p^{n-1} + \cdots + b_1 p + b_0 = 0 \qquad \text{(B-2)}$$

Several observations can be made regarding the roots of Eq. B-2:

1. The number of roots is equal to the degree of the equation n.
2. The roots may be real or complex. However, complex roots occur in pairs. This means that if n is odd, there must be at least one real root.
3. The roots r are related to certain coefficients by the expressions

$$r_1 + r_2 + \cdots + r_n = -b_{n-1} \qquad \text{(B-3)}$$

and

$$r_1 r_2 \cdots r_n = (-1)^n b_0 \qquad \text{(B-4)}$$

There are literally dozens of methods for root solving. One popular method for seeking real roots is the use of synthetic division. This approach will be illustrated by an example. Suppose that the roots of the equation

$$f(p) = p^3 + 5p^2 + 11p + 15 = 0$$

are required. The first step is to assume a real root, say, -1 (an integer for simplicity). Then the coefficients of the equation are written down. The steps are as follows: Multiply the first coefficient by -1, add the result to the second coefficient, multiply the sum by -1 and add to the third coefficient, and so forth. Thus,

1	$+5$	$+11$	$+15$	$\lfloor -1$
	-1	-4	-7	
1	$+4$	$+7$	$+8$	

413

The value of $f(-1)$ is 8; therefore -1 is not a root of the equation. Try -4.

$$
\begin{array}{ccccc}
1 & +5 & +11 & +15 & \underline{\,|-4} \\
  & -4 & -4 & -28 & \\
\hline
1 & +1 & +7 & -13 &
\end{array}
$$

Thus, $f(-4) = -13$. Not only is -4 not a root, but the sign change [remember that $f(-1) = +8$] indicates the presence of a root between -1 and -4. Next, try -3.

$$
\begin{array}{ccccc}
1 & +5 & +11 & +15 & \underline{\,|-3} \\
  & -3 & -6 & -15 & \\
\hline
1 & +2 & +5 & 0 &
\end{array}
$$

The value of $f(-3)$ is zero, and so -3 is a root. Once a real root is found by synthetic division, the complete solution is easily obtained. Division of $f(p)$ by the factor $(p + 3)$ yields $p^2 + 2p + 5$ from which $p = -1 \pm j2$. Therefore, the three roots are $r_1 = -3$, $r_2 = -1 + j2$, $r_3 = -1 - j2$. Equations B-3 and B-4 can be used as a check. From Eq. B-3,

$$
-3 - 1 + j2 - 1 - j2 = -5
$$
$$
-5 = -5
$$

From Eq. B-4,

$$
-3(-1 + j2)(-1 - j2) = (-1)^3\, 15
$$
$$
-15 = -15
$$

A popular approach to solving equations in which n is even and greater than 2 is known as Lin's method[1] [22]. The method involves repeated division by assumed quadratic factors until the remainder is sufficiently small. In this manner, a polynomial can be factored into quadratic factors from which the roots of the equation are easily obtained. Lin's method is best illustrated by an example. Suppose that the roots of the equation

$$
f(p) = p^4 + 6p^3 + 18p^2 + 30p + 25 = 0
$$

are to be determined. In using Lin's method, the polynominal $f(p)$ is divided by an assumed quadratic factor in an attempt to express $f(p)$ as a product of two quadratic factors. The last three terms of $f(p)$ are used to obtain the first trial divisor.

---

[1]It should be pointed out that Lin's method is not restricted to equations in which n is even. However, in this discussion it is assumed that at least one real root has been determined by synthetic division if n is odd.

$$p^2 + \frac{30}{18}p + \frac{25}{18} = p^2 + 1.7p + 1.4$$

Division yields

$$
\begin{array}{r}
p^2 + 4.3p + 9.3 \\
p^2 + 1.7p + 1.4 \overline{\smash{\big)}\ p^4 + 6.0p^3 + 18.0p^2 + 30.0p + 25.0} \\
\underline{p^4 + 1.7p^3 + 1.4p^2} \\
4.3p^3 + 16.6p^2 + 30.0p \\
\underline{4.3p^3 + 7.3p^2 + 6.0p} \\
(9.3p^2 + 24.0p + 25.0) \\
\underline{9.3p^2 + 15.8p + 13.0} \\
8.2p + 12.0
\end{array}
$$

The remainder is somewhat large and so a second trial divisor is obtained from the expression enclosed in parenthesis in the foregoing step. It is

$$p^2 + \frac{24.0}{9.3}p + \frac{25.0}{9.3} = p^2 + 2.6p + 2.7$$

Division with this assumed factor produces a remainder $3.9p + 7.4$, which is smaller than the previous remainder. The process can be continued by using a third trial divisor determined in the same manner. The remainder is then $2.8p + 5.2$. After eight trials, the remainder is $-0.12p - 0.10$, which is reasonably small. (In later trials, after the remainder became small, a second decimal place was carried. The procedure of carrying progressively more decimal places as the remainder gets smaller results in less computational labor without sacrifice in accuracy.) The resulting factors are $(p^2 + 4.01p + 4.98)(p^2 + 1.99p + 5.04)$ which are close to the actual factors $(p^2 + 4p + 5)(p^2 + 2p + 5)$. The roots of the equation are easily obtained since each factor can be equated to zero.

The methods described can be used for solving equations. However, the remark must be made that these methods are not necessarily the best approaches to any given equation. The interested reader may wish to consult other sources of information [29].

Routh's Criterion

In 1877, E. J. Routh developed a method for determining whether or not an equation has roots with positive real parts without actually solving for the roots. When used with the characteristic equation of a control system, Routh's criterion offers a simple means for detecting system instability, because roots with positive real parts indicate transients which increase, rather than decay, with time.

The usual characteristic equation can be written in general form as

$$a_0 p^n + a_1 p^{n-1} + \cdots + a_{n-1} p + a_n = 0 \qquad (C-1)$$

in which the coefficients typically are positive numbers. An unstable system, or one with limited stability (roots with zero real parts), is immediately indicated if any coefficient is either zero or negative.

The first step in applying Routh's criterion is to form the Routh array. For simplicity of illustration, a characteristic equation with $n = 5$ will be used.

$$a_0 p^5 + a_1 p^4 + a_2 p^3 + a_3 p^2 + a_4 p + a_5 = 0 \qquad (C-2)$$

The array is

$$
\begin{array}{ccc}
a_0 & a_2 & a_4 \\
a_1 & a_3 & a_5 \\
b_1 & b_3 & \\
c_1 & c_3 & \\
d_1 & & \\
e_1 & &
\end{array}
\qquad (C-3)
$$

The first two rows are composed of the coefficients of the characteristic equation. The remaining entries are obtained by computation with the following pattern:

$$b_1 = \frac{\begin{array}{c} a_0 \quad a_2 \\ a_1 \quad a_3 \end{array}}{a_1} = \frac{a_1 a_2 - a_0 a_3}{a_1} \qquad (C-4)$$

$$b_3 = \frac{a_1 \quad a_5}{a_1} = \frac{a_1 a_4 - a_0 a_5}{a_1} \tag{C-5}$$

Similarly,

$$c_1 = \frac{b_1 a_3 - a_1 b_3}{b_1} \qquad c_3 = \frac{b_1 a_5 - a_1(0)}{b_1} \tag{C-6}$$

Entries d_1 and e_1 are determined in a similar manner.

Routh's criterion states that the number of sign changes in the first column of the array is equal to the number of roots with positive real parts.

Consider, as an example, the characteristic equation

$$p^5 + 3p^4 + 7p^3 + 20p^2 + 6p + 15 = 0$$

The Routh array is

1	7	6
3	20	15
$\frac{1}{3}$	1	
11	15	
$\frac{6}{11}$		
15		

The system in question is stable. There are no sign changes in the first column of the array, and so there are no roots with positive real parts. However, the Routh criterion gives no indication of the degree of stability.

Next, consider the characteristic equation

$$p^4 + 2p^3 + 3p^2 + 8p + 2 = 0$$

The Routh array is

1	3	2
2	8	
−1	2	
12		
2		

There are two sign changes (plus to minus and minus to plus) in the first column showing that there are two roots with positive real parts. Therefore, the system is unstable.

It can be shown that the entries in any row can all be multiplied or divided by a constant without affecting the sign changes in the first column. This procedure can be used to reduce the mathematical labor involved in forming the array.

Two special cases are worthy of note. Should the first entry in a row be zero while the other entries are not zero, the procedure is to replace the zero with a small positive number ϵ. The array can then be continued. Sign changes in the first column can be ascertained by letting ϵ approach zero. If a row of zeros occurs, the system is either unstable or possesses limited stability—it is *not* stable.

Transfer-Function Simulation on the Analog Computer

The analysis of automatic control systems is more often than not carried out in the s domain, where the differential equations defining the system have been transformed, and the transfer functions of the various elements have been inserted in blocks making up a complete block diagram. It is possible to represent the block diagram on the analog computer and to simulate the performance of the system without returning to the original differential equations. This procedure may be termed *transfer-function simulation.*

To understand the procedure for this method of analog computing, it is necessary to return to the basic computing element—the operational amplifier with its associated input and feedback impedances as shown in Fig. D-1. Certain properties of the operational amplifier should be recalled. They are as follows:

    1. The amplifier input voltage e_g is essentially 0.

    2. The amplifier input current is essentially 0 and therefore $i_1 \cong i_2$.

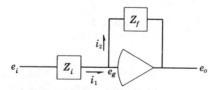

FIG. D-1. The basic analog computer element.

Suppose, now, that the input and feedback impedances are not restricted to individual resistors or capacitors. Figure D-2 illustrates one possibility where the feedback impedance is a parallel combination of a resistor and a capacitor. The derivation for the *transfer function* of this computer element is as follows:

$$i_1 \cong i_2 = i_2' + i_2''$$

but
$$i_1 = \frac{e_i}{R_i} \qquad i_2' = \frac{-e_o}{R_f}$$

and
$$-C_f \frac{de_o}{dt} = i_2''$$

If these equations are transformed with all initial conditions zero and substituted, the result is

$$\frac{E_i(s)}{R_i} = - \frac{E_o(s)}{R_f} - C_f s E_o(s)$$

or
$$\frac{E_o(s)}{E_i(s)} = - \frac{R_f}{R_i} \frac{1}{R_f C_f s + 1}$$

This is recognized as the transfer function for a simple lag, which may be expressed in the form

$$\frac{E_o(s)}{E_i(s)} = \frac{-K}{\tau s + 1}$$

where $K = R_f/R_i$ and $\tau = R_f C_f$.

Thus a combination of computer elements as shown in Fig. D-2 provides a single lag with gain K and a break point of $\omega = 1/\tau$.

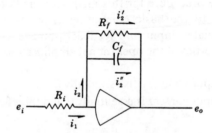

FIG. D-2. A complex feedback
impedance—the simple lag.

In a similar fashion, it is possible to derive relationships for other common transfer relationships, including lag-lead, lead-lag, proportional-plus-integral control, and many others. Table D-1 is a compilation of some computer elements for simulation of transfer functions. Complete tables of transfer relationships are available in most analog computing texts.[1]

There is one important restriction when analog computer elements are used to simulate transfer functions, that is, the transfer functions are ob-

[1]The general approach in computing texts is to consider the input and feedback impedances as short-circuited impedances. One may then use the basic relationship $e_o = (-Z_f/Z_i) e_i$ and a table of short-circuited impedances to construct complex elements. The reader is referred to reference [18].

TABLE D-1
Computer Elements for Transfer Function Simulation

Description	Component Symbol	Transfer Relationship
Constant	$e_i \longrightarrow \boxed{K} \longrightarrow e_o$	$\dfrac{E_o}{E_i} = K \text{ for } K < 1$
Constant		$\dfrac{E_o}{E_i} = -\dfrac{R_f}{R_i} \quad \begin{matrix} \text{generally for} \\ 0.1 \le \dfrac{R_f}{R_i} \le 10 \end{matrix}$
Integration		$\dfrac{E_o(s)}{E_i(s)} = -\dfrac{K}{s} \qquad K = \dfrac{1}{R_i C_f}$
Simple lag		$\dfrac{E_o(s)}{E_i(s)} = -\dfrac{K}{\tau s + 1} \quad K = \dfrac{R_f}{R_i}$ $\tau = R_f C_f$
Lag-lead or lead-lag		$\dfrac{E_o(s)}{E_i(s)} = \dfrac{-K(\tau_1 s + 1)}{(\tau_2 s + 1)}$ $K = \dfrac{R_f}{R_i}$ $\tau_1 = R_i C_i$ $\tau_2 = R_f C_f$
Lag-lead		$\dfrac{E_o(s)}{E_i(s)} = -\dfrac{K(\tau_2 s + 1)}{(\tau_1 s + 1)}$ $K = \dfrac{R_{f_1}}{R_i}$ $\tau_1 = (R_{f_1} + R_{f_2})C_f$ $\tau_2 = R_{f_2} C_f$
Proportional-plus-integral control		$\dfrac{E_o(s)}{E_i(s)} = -\dfrac{K(\tau s + 1)}{s}$ $K = \dfrac{1}{R_i C_f}$ $\tau = R_f C_f$

tained by assuming that all initial conditions are zero. If a problem calls for initial conditions, then this technique is not generally applicable. Quadratic or higher-order transfer function terms may be handled by cascading single factors if the term is factorable into real factors. Other special complex circuits are available [18]. It is often desirable, however, to expand these portions of a block diagram back into their differential equation form and solve the equation. It is logically possible to use a combination of computer techniques in one complete computer problem on the problem board.

As a simple example of the use of transfer-function simulation, consider constructing a computer diagram for the simulation of the simple speed-control system of Fig. D-3. The governor-controlled gas turbine is considered not to have any torque disturbance.

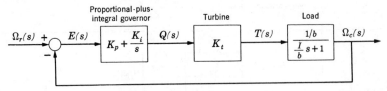

Fɪɢ. D-3. Speed-control system.

The control element in this system has the transfer function

$$K_p + \frac{K_i}{s} = \frac{K_i[(K_p/K_i)s + 1)]}{s} = K_i\frac{(\tau_1 s + 1)}{s}$$

The computer elements "proportional-plus-integral control" of Table D-1 will provide this relationship if (see Fig. D-4)

$$K_i = \frac{1}{R_1 C_1} \qquad \text{and} \qquad \tau_1 = \frac{K_p}{K_i} = R_2 C_1$$

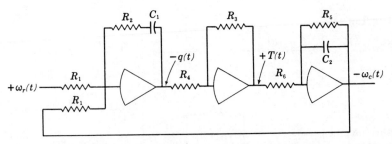

Fɪɢ. D-4. Computer diagram for the speed-control system.

The turbine is defined by a constant $K_t = R_3/R_4$. The load is simulated in this instance by a simple lag, where

$$\frac{1}{b} = \frac{R_5}{R_6} \quad \text{and} \quad \frac{I}{b} = R_5 C_2$$

The computer diagram is given in Fig. D-4.

It may be recalled from Chapter 4 that scaling is an important part of successful analog computing. No attempt will be made to scale the example given here. When the simulation approach to analyzing control systems is used, scaling is most easily accomplished directly on the analog computer diagram. If one has mastered the scaling concepts presented in Chapter 4, scaling a problem of this nature will come as a matter of course.

TABLE OF $1 + j\omega\tau = A \, \underline{/\theta°}$

$\omega\tau$	A	$\theta°$	$\omega\tau$	A	$\theta°$	$\omega\tau$	A	$\theta°$
0.05	1.00	2.9	2.30	2.51	66.5	6.1	6.18	80.7
0.10	1.00	5.7	2.35	2.55	66.9	6.2	6.28	80.8
0.15	1.01	8.5	2.40	2.60	67.4	6.3	6.38	81.0
0.20	1.02	11.3	2.45	2.65	67.8	6.4	6.48	81.1
0.25	1.03	14.0	2.50	2.69	68.2	6.5	6.58	81.3
0.30	1.04	16.7	2.55	2.74	68.6	6.6	6.68	81.4
0.35	1.06	19.3	2.60	2.79	69.0	6.7	6.77	81.5
0.40	1.08	21.8	2.65	2.83	69.3	6.8	6.87	81.6
0.45	1.10	24.2	2.70	2.88	69.7	6.9	6.97	81.8
0.50	1.12	26.6	2.75	2.93	70.0	7.0	7.07	81.9
0.55	1.14	28.8	2.80	2.97	70.3	7.1	7.17	82.0
0.60	1.17	31.0	2.85	3.02	70.7	7.2	7.27	82.1
0.65	1.19	33.0	2.90	3.07	71.0	7.3	7.37	82.2
0.70	1.22	35.0	2.95	3.11	71.3	7.4	7.47	82.3
0.75	1.25	36.9	3.00	3.16	71.6	7.5	7.57	82.4
0.80	1.28	38.7	3.1	3.26	72.1	7.6	7.67	82.5
0.85	1.31	40.4	3.2	3.35	72.6	7.7	7.76	82.6
0.90	1.35	42.0	3.3	3.45	73.1	7.8	7.86	82.7
0.95	1.38	43.5	3.4	3.54	73.6	7.9	7.96	82.8
1.00	1.41	45.0	3.5	3.64	74.1	8.0	8.06	82.9
1.05	1.45	46.4	3.6	3.74	74.5	8.1	8.16	83.0
1.10	1.49	47.7	3.7	3.83	74.9	8.2	8.26	83.0
1.15	1.52	49.0	3.8	3.93	75.3	8.3	8.36	83.1
1.20	1.56	50.2	3.9	4.03	75.6	8.4	8.46	83.2
1.25	1.60	51.3	4.0	4.12	76.0	8.5	8.56	83.3
1.30	1.64	52.4	4.1	4.22	76.3	8.6	8.66	83.4
1.35	1.68	53.5	4.2	4.32	76.6	8.7	8.76	83.4
1.40	1.72	54.5	4.3	4.41	76.9	8.8	8.86	83.5
1.45	1.76	55.4	4.4	4.51	77.2	8.9	8.96	83.6
1.50	1.80	56.3	4.5	4.61	77.5	9.0	9.06	83.7
1.55	1.84	57.2	4.6	4.71	77.7	9.1	9.15	83.7
1.60	1.89	58.0	4.7	4.81	78.0	9.2	9.25	83.8
1.65	1.93	58.8	4.8	4.90	78.2	9.3	9.35	83.9
1.70	1.97	59.5	4.9	5.00	78.5	9.4	9.45	83.9
1.75	2.02	60.3	5.0	5.10	78.7	9.5	9.55	84.0
1.80	2.06	60.9	5.1	5.20	78.9	9.6	9.65	84.1
1.85	2.10	61.6	5.2	5.30	79.1	9.7	9.75	84.1
1.90	2.15	62.2	5.3	5.39	79.3	9.8	9.85	84.2
1.95	2.19	62.9	5.4	5.49	79.5	9.9	9.95	84.2
2.00	2.24	63.4	5.5	5.59	79.7	10.0	10.05	84.3
2.05	2.28	64.0	5.6	5.69	79.9	12.0	12.04	85.2
2.10	2.33	64.5	5.7	5.79	80.0	14.0	14.04	85.9
2.15	2.37	65.1	5.8	5.89	80.2	16.0	16.03	86.4
2.20	2.42	65.6	5.9	5.98	80.4	18.0	18.03	86.8
2.25	2.46	66.0	6.0	6.08	80.5	20.0	20.02	87.1

Appendix **F**

DECIBEL CONVERSION TABLE ($m = 20 \log_{10} M$)

M	0	1	2	3	4	5	6	7	8	9
0.0	$m =$	−40.00	−33.98	−30.46	−27.96	−26.02	−24.44	−23.10	−21.94	−20.92
0.1	−20.00	−19.17	−18.42	−17.72	−17.08	−16.48	−15.92	−15.39	−14.89	−14.42
0.2	−13.98	−13.56	−13.15	−12.77	−12.40	−12.04	−11.70	−11.37	−11.06	−10.75
0.3	−10.46	−10.17	−9.90	−9.63	−9.37	−9.12	−8.87	−8.64	−8.40	−8.18
0.4	−7.96	−7.74	−7.54	−7.33	−7.13	−6.94	−6.74	−6.56	−6.38	−6.20
0.5	−6.02	−5.85	−5.68	−5.51	−5.35	−5.19	−5.04	−4.88	−4.73	−4.58
0.6	−4.44	−4.29	−4.15	−4.01	−3.88	−3.74	−3.61	−3.48	−3.35	−3.22
0.7	−3.10	−2.97	−2.85	−2.73	−2.62	−2.50	−2.38	−2.27	−2.16	−2.05
0.8	−1.94	−1.83	−1.72	−1.62	−1.51	−1.41	−1.31	−1.21	−1.11	−1.01
0.9	−0.92	−0.82	−0.72	−0.63	−0.54	−0.45	−0.35	−0.26	−0.18	−0.09
1.0	0.00	0.09	0.17	0.26	0.34	0.42	0.51	0.59	0.67	0.75
1.1	0.83	0.91	0.98	1.06	1.14	1.21	1.29	1.36	1.44	1.51
1.2	1.58	1.66	1.73	1.80	1.87	1.94	2.01	2.08	2.14	2.21
1.3	2.28	2.35	2.41	2.48	2.54	2.61	2.67	2.73	2.80	2.86
1.4	2.92	2.98	3.05	3.11	3.17	3.23	3.29	3.35	3.41	3.46
1.5	3.52	3.58	3.64	3.69	3.75	3.81	3.86	3.92	3.97	4.03
1.6	4.08	4.14	4.19	4.24	4.30	4.35	4.40	4.45	4.51	4.56
1.7	4.61	4.66	4.71	4.76	4.81	4.86	4.91	4.96	5.01	5.06
1.8	5.11	5.15	5.20	5.25	5.30	5.34	5.39	5.44	5.48	5.53
1.9	5.58	5.62	5.67	5.71	5.76	5.80	5.85	5.89	5.93	5.98
2.	6.02	6.44	6.85	7.23	7.60	7.96	8.30	8.63	8.94	9.25
3.	9.54	9.83	10.10	10.37	10.63	10.88	11.13	11.36	11.60	11.82
4.	12.04	12.26	12.46	12.67	12.87	13.06	13.26	13.44	13.62	13.80
5.	13.98	14.15	14.32	14.49	14.65	14.81	14.96	15.12	15.27	15.42
6.	15.56	15.71	15.85	15.99	16.12	16.26	16.39	16.52	16.65	16.78
7.	16.90	17.03	17.15	17.27	17.38	17.50	17.62	17.73	17.84	17.95
8.	18.06	18.17	18.28	18.38	18.49	18.59	18.69	18.79	18.89	18.99
9.	19.08	19.18	19.28	19.37	19.46	19.55	19.65	19.74	19.82	19.91
	0	1	2	3	4	5	6	7	8	9

NOTE: Table is easily extended. If $M' = M(10)^n$, $m' = m + 20n$. As an example: If $M' = 80 = 8(10)$, $m' = 18.06 + 20 = 38.06$.

Sequencing Control

The term *sequencing control* as used herein refers to that type of control by which a system is caused to go through a set of sequential operations. A very common example is an automatic washing machine. Operations such as filling with water, agitating, draining, rinsing, and drying take place in a sequential fashion. An industrial example is the machining of automobile engine blocks. Here, raw castings are fed into an automated line of machine tools. Once a particular machining operation is completed, the castings are automatically transferred to the next machining station, until the castings finally emerge from the line completely machined. The two examples cited point out the extreme importance of sequencing control in an automated society.

The purpose of this appendix is to introduce the basic concepts of sequencing control. Attention is first turned to control systems which utilize electromechanical components. Commonly used components are introduced, and then are combined into circuits which will produce the desired sequence of operations. Following a presentation of *logic functions*, the application of *Boolean Algebra* to switching circuits is discussed. Finally, the subject of *static switching* (switching which requires neither contacts nor moving parts) is introduced.

A comment is in order at this point having to do with the symbols used to represent various components. While symbols tend to change with time, the symbols used in this appendix are, in general, compatible with those developed by the National Electrical Manufacturers Association and the National Fluid Power Association.

Electromechanical Components and Circuits

The use of electromechanical components in sequencing control is so widespread that this subject deserves first attention. The approach will be to introduce a number of basic components, and then to illustrate how they can be integrated into useful sequencing control circuits. It should be understood that only a limited number of the many available components can be discussed here and that, for those discussed, only the most basic configurations will be treated. Further information is readily available from manufacturers' catalogs.

Pushbutton Switches. Switches of various sorts are required to make and break electrical circuits. First to be considered is the *momentary contact pushbutton switch.* Switch symbols are shown in Fig. G-1.

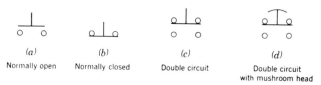

(a)	*(b)*	*(c)*	*(d)*
Normally open	Normally closed	Double circuit	Double circuit with mushroom head

Fig. G-1. Symbols for momentary contact pushbutton switches.

The *normally-open* switch of Fig. G-1*a* will make the circuit between the two terminals when the button is pushed manually, and will break (or open) the circuit when the button is released. The term *normally*, then, refers to the condition of the switch when it is in the normal, or unactuated, position.

The *normally-closed* switch of Fig. G-1*b* will break the circuit between the two terminals when pushed, and will make (or close) the circuit when released.

Multiple circuit pushbuttons can be purchased with various combinations of normally-open and normally-closed contacts. These switches permit the actuation of one pushbutton to initiate or interrupt a number of signals. The symbol for a double-circuit switch is given in Fig. G-1*c*.

The symbol of Fig. G-1*d* is for a double-circuit switch with a *mushroom head,* that is, an enlarged head with the general shape of a mushroom. Most pushbuttons can be purchased with mushroom heads. These heads are typically used to permit easy actuation in emergency situations, although, for example, they would also be specified for applications which require an operator to wear heavy gloves.

Because most circuit diagrams involve more than one pushbutton switch, some means of identification is required. One method is to designate a particular switch by a number followed by the letters PB, for example, 1PB, 2PB, and so forth. An alternate method is to label the switch by its function, for example, Start, Stop, and so forth.

Limit Switches. A second important class of switches is the *limit switch.* These switches are actuated mechanically through the use of some form of cam which contacts an actuating arm.

Limit-switch symbols are shown in Fig. G-2. The *normally-open* switch of Fig. G-2*a* will complete the circuit between the two terminals when the switch is actuated, and will open the circuit when the switch is deactuated. Figure G-2*b* symbolizes a limit switch which has the opposite function, that is, it will break the circuit when actuated, and re-make it when released. Thus, it is normally-closed.

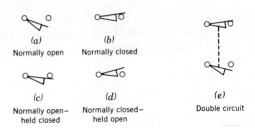

Fig. G-2. Symbols for limit switches.

There are circumstances in which a circuit diagram is clearer if a switch is shown in the *held* position. The symbols of Fig. G-2c and d then apply. An example will be introduced at a later point.

The double-circuit switch shown in Fig. G-2e is very useful in sequencing control applications. The actuation of a single switch will interrupt one operation and immediately initiate a second operation. The dashed line between the individual switch symbols denotes that the switches are mechanically connected, that is, they are physically located within the same switch housing.

Limit switches are typically designated by a number followed by the letters LS, for example, 1LS, 2LS, and so forth.

Control Relays. A *control relay* is an electromagnetic device which provides for a multiplicity of switching functions based on the energization of the relay coil. One way of describing a relay is to say that it is a switch (or switches) which is actuated by an electrical input (as opposed to, say, the manual input required by a pushbutton switch). However, and more fundamentally, control relays are *information handling devices* and, as such, provide the intelligence for a system. Sometimes control relays are classified as *logic devices*, which is saying the same thing in a different way. This latter designation will take on more meaning after the discussion of logic functions which is presented in the next major section.

A schematic drawing of a relay is given in Fig. G-3a. The relay is

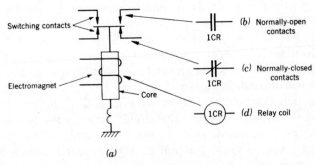

Fig. G-3. Control relay schematic drawing and related symbols.

composed of two basic parts. The first is an electromagnet. With the relay coil deenergized, the spring moves the magnetic core downward; the core moves upward with the coil energized. The second part of the relay contains switching contacts which make and break with core displacement.

The symbol for *normally-open contacts* is shown in Fig. G-3*b*. The term normally open refers to the switching condition existing with the relay coil *deenergized. Normally-closed contacts* are represented as shown in Fig. G-3*c*. The relay coil is represented by a circle.

A control relay is designated by a number followed by the letters CR. The designation is placed in the circle representing the relay coil. All contacts controlled by that coil are labelled the same as the coil.

In general, control relays can be purchased with as many contacts as the application demands.

Timer Relays. A *timer relay* is quite similar to a control relay except that provision is made for an adjustable delayed switching action. In Fig. G-4*a* the symbols for a timer relay, designated as 1TR, are

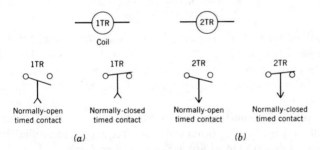

FIG. G-4. Timer relay symbols. Contact action is delayed in the direction of the arrowheads.

given. If the coil is energized, the normally-open timed contact will delay and then close. At that same time, the normally-closed timed contact will open. Upon coil deenergization, both contacts immediately revert to their original states. *The key to understanding the action is that the delay takes place in the direction of the arrowheads.*

The timer relay of Fig. G-4*b* is such that switching takes place immediately upon energization of the coil, while a delay in switching occurs following deenergization.

It is possible to purchase a timer relay with untimed contacts in addition to timed contacts. The untimed contacts behave in the same way as those on control relays.

So far, we have discussed examples of two broad classes of components. Pushbutton switches and limit switches are *input devices* because they provide inputs to a control system. Control and timer relays are *logic devices* because they can provide a form of intelligence for the system. Finally, we need *output devices* so that useful operations can be performed. Attention is now turned to two common output devices.

Motor Starters. *Motor starters,* as the name suggests, are components used to start and stop electric motors. The devices are basically relays with contacts capable of handling high currents. In addition, motor starters provide motor overload protection, a point which will not be pursued here.

Appropriate symbols for motor starters are shown in Fig. G-5a. In

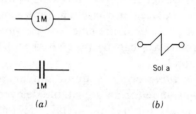

FIG. G-5. Symbols for motor starters and solenoids.

our discussion, we will be concerned only with the energization and de-energization of the motor starter coil; we will not consider the power wiring to the motor itself. Energization of the motor starter coil causes the motor to start while deenergization of the coil causes the motor to stop.

Solenoids. *Solenoids* are electromagnetic actuators. The symbol used is that of Fig. G-5b. Solenoids are frequently identified by lower-case letters and designated as Sol a, Sol b, and so forth.

Our interest in solenoids will be confined to their use in actuating the hydraulic and pneumatic valves to be discussed next.

Hydraulic and Pneumatic Valves. The term *hydraulics* refers to power transmission through the use of a pressurized liquid—typically oil. On the other hand, *pneumatics* refers to systems which utilize a pressurized gas—typically air.

Hydraulic and pneumatic cylinders are the power devices to be considered. Cylinder actuation requires that the direction of fluid flow to the cylinder be controlled. Valves which perform this function are called *direction-control* valves. Although valves can be actuated in a variety of ways, only solenoid actuation will be considered here.

As the first step to understanding valve symbols, consider a valve which has only two fixed positions, corresponding to open and closed. The appropriate symbol involves two squares as shown in Fig. G-6a, one square for each possible valve position. Next, the porting configuration for each position is indicated as shown in Fig. G-6b; in one position, flow

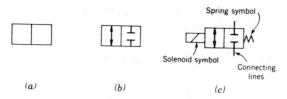

FIG. G-6. Hydraulic and pneumatic valve symbols.

is permitted, while in the other, flow is blocked. Figure G-6*c* gives a complete symbol. Solenoid actuation is indicated by the symbol shown. Connecting lines are also shown. The spring symbol indicates that the valve is spring loaded in one position.

The symbol in Fig. G-6*c* is interpreted as follows. With the solenoid deenergized, the square adjacent to the spring symbol is in effect, that is, the valve is closed (connecting lines are blocked). When the solenoid is energized, the square adjacent to the solenoid symbol is in effect and the valve is open. One can visualize the action by mentally sliding the symbol one square to the right such that the arrowed line lies between the connecting lines. Connecting lines to the valve symbol always are drawn to the square which is in effect with the solenoid (or solenoids) deenergized.

The valve symbolized in Fig. G-6*c* can be described as a *solenoid-actuated, spring-offset, 2-position* valve. Further, it is a *normally-closed* valve because it is closed with the solenoid deenergized. It is also a *2-way* (or *2-connection*) valve because two lines can be connected to the valve.

A 2-position, 3-way valve is shown in Fig. G-7*a*. An air supply to the

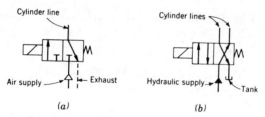

FIG. G-7. Symbols for 2-position, 3-way and 4-way valves.

valve is denoted by a line, with an open arrowhead, entering the valve. There is one cylinder line (which can be connected to a cylinder) and a provision for exhausting air. With the solenoid deenergized, the air supply is blocked and the cylinder line is ported to exhaust. With the solenoid energized, air is ported to the cylinder line and exhaust is blocked. An example of the use of this valve will be presented at a later point.

A 2-position, 4-way valve is shown in Fig. G-7*b*. A hydraulic supply is indicated by a line, with a filled arrowhead, entering the valve. The *tank* symbol denotes that exhaust oil is returned to the pump tank. Two cylinder lines are provided. Solenoid energization reverses the flow through the cylinder lines.

Valve-controlled cylinder actuation is illustrated in Fig. G-8. A 3-

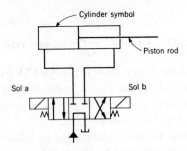

Fig. G-8. Control of a hydraulic cylinder using a double-solenoid, 3-position, 4-way valve.

position valve is used so that it is possible to stop the piston rod in any desired position. With both solenoids deenergized, the centering springs provide the center porting configuration. The cylinder lines are blocked and hydraulic fluid is returned to the tank. Energization of Sol a causes the porting in the adjacent square to be in effect, and so the piston rod will move to the right. Energization of Sol b (alone) causes rod motion to the left. (As a practical matter, both solenoids should never be energized at the same time; if they are, one or both will be damaged electrically.)

Electrical Circuit Diagrams. All electrical circuit diagrams start by drawing two vertical lines (see Fig. G-9). These lines represent electrical

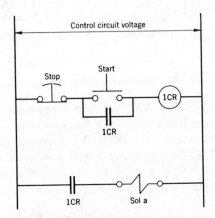

Fig. G-9. Illustrative electrical circuit diagram.

leads with the control circuit voltage applied between them. Horizontal lines appear on the diagram as various components are added. The result is sometimes called a *ladder diagram* because of its appearance.

An illustrative electrical circuit diagram is shown in Fig. G-9. When the control circuit voltage is applied, nothing happens because there is no current path to either the control relay coil or to the solenoid. If the Start button is pushed, coil 1CR will become energized and the two normally-open contacts will close. One of these energizes the solenoid, while the other establishes a parallel path around the Start button. This contact is called a *holding contact* because it holds the circuit in after the Start button is released. Solenoid a will remain energized until the Stop button is depressed.

The start-stop arrangement shown in Fig. G-9 is widely used. A key feature is that if a temporary power failure occurs, the circuit will remain inactive on power resumption, that is, Sol a will remain deenergized until the Start button is again pushed. This arrangement removes the hazard of a sudden unexpected return to operation.

In the discussion of the double-solenoid valve shown in Fig. G-8, it was stated that both solenoids should never be energized at the same time. This point can be amplified while, at the same time, further illustrating the nature of electrical circuit diagrams.

A partial electrical circuit diagram is shown in Fig. G-10. Relays

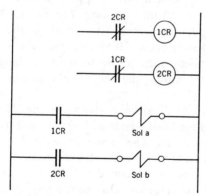

Fig. G-10. Partial electrical circuit diagram illustrating the interlocking of two solenoids.

1CR and 2CR control solenoids a and b through normally-open contacts. In addition, normally-closed contacts are wired in series with the relay coils so that the energization of one coil opens the circuit to the other coil. Thus, energization of the second solenoid is impossible while the first is energized. With this arrangement, the solenoids are said to be *interlocked*. In general, the term interlocking refers to an arrangement whereby certain conditions must be met before actuation is permitted. For example, would it be a good idea to permit the landing gear to be retracted while the airplane is standing on the ground?

Illustrative Examples. Two illustrative examples are now presented to help clarify the use of the various components and their symbols.

Figure G-11 shows an electrical circuit diagram for a motor starter.

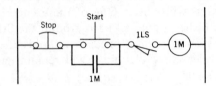

FIG. G-11. Control circuit for motor starter.

The start-stop arrangement is the same as the one discussed earlier; one of the normally-open contacts on the motor starter is being used as a holding contact. Note also that, not only need the Start button be depressed but, limit switch 1LS must be actuated before coil 1M can be energized. Coil 1M will remain energized until either the Stop button is pushed or switch 1LS is released.

A more complex system is shown in Fig. G-12. Shown in the pneumatic circuit, Fig. G-12a, are a spring-returned cylinder and the 2-position, 3-way valve first introduced in Fig. G-7a. With the solenoid deenergized, the cylinder line is exhausted to atmosphere and the internal spring retracts the piston rod. However, with the solenoid energized, pressurized air is applied to the piston to advance the piston rod. Note that, in the retracted position, the cam on the piston rod actuates limit switch 1LS.

The electrical circuit diagram is shown in Fig. G-12b. First of all, note that the symbol for limit switch 1LS shows a normally-open switch in the held-closed position. The choice of symbol is based on the fact that, when the system is at rest, switch 1LS is actuated.

System operation can be described as follows. With the Start button depressed, relay 1CR is energized. One normally-open contact is used as a holding contact. The second normally-open contact energizes timer relay coil 1TR through switches 1LS and 2LS. An untimed 1TR contact sets up a hold around switch 1LS (which will open as soon as the piston rod begins to advance). Following a timed delay, the timed 1TR contact will energize Sol a, with the result that the valve will shift and the piston rod will advance toward switch 2LS. Actuation of switch 2LS will interrupt the circuit to coil 1TR with the result that Sol a will be immediately deenergized and the piston rod will retract. Upon full retraction, switch 1LS will be actuated again. After a time delay, the piston rod will advance

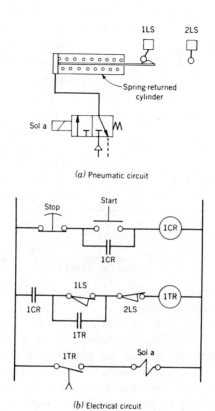

(a) Pneumatic circuit

(b) Electrical circuit

FIG. G-12. Circuits for pneumatic
cylinder application.

again. Piston reciprocation, with a time delay at switch 1LS, will continue until the Stop button is pushed, after which the piston rod will retract and then remain at rest.

Logic Functions

In the previous section, relays were said to be logic devices because they provide a form of intelligence for a system. As such, relays offer a convenient starting point for a discussion of *logic functions*. Logic functions will be introduced and illustrated by using various combinations of relay contacts (although these functions can be obtained through the use of hardware other than relays, as will be developed later). For simplicity, relay coils and their contacts will be designated by letters such as A, B, . . . instead of 1CR, 2CR, ...

Two normally-open contacts in series are shown in Fig. G-13a. This combination produces an AND function, so called because there must be an input at coil *A* AND coil *B* in order to close the circuit and, thus,

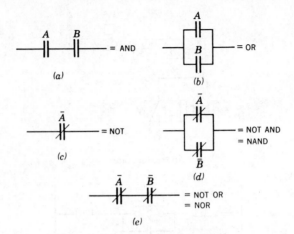

FIG. G-13. Contact configurations to yield
various logic functions.

provide an output. In general, an AND function provides an output if, and only if, all inputs are present.

In Fig. G-13*b*, a parallel arrangement of two normally-open contacts is given. This is an OR function because an input at coil *A* OR coil *B* (OR both) completes the circuit and provides an output. Thus, an OR function provides an output with one or more inputs present.

A single normally-closed contact is shown in Fig. G-13*c*. This is a NOT function because an input to coil *A* results in an open circuit (no output) while deenergization of (no input to) the coil provides a closed circuit (an output). Thus, there is an opposition between input and output. Normally-closed contacts are designated by a bar over the contact letter, such as $\overline{A}$, which is pronounced NOT *A*.

In Fig. G-13*d*, a parallel arrangement of two normally-closed contacts is given. This is a NOT AND or NAND function because its switching action is the opposite of that obtained from an AND function. For example, inputs to both coils of the NAND function opens the circuit while with the AND function, the circuit would be closed.

The series arrangment of two normally-closed contacts shown in Fig. G-13*e* produces a NOT OR or NOR function.

The section on static switching will introduce other physical means for obtaining logic functions. At that point, the subject of logic functions will be re-examined.

Boolean Algebra

George Boole (1815–1864) developed a branch of mathematics which is applicable to switching circuits; it is called *Boolean Algebra*. This section will illustrate the use of Boolean Algebra as applied to relay contact circuits. The section on static switching will extend the ideas to circuits which do not utilize relays.

Boolean Algebra permits the expression of contact configurations in a mathematical form. One use is that, through mathematical manipulation, one seeks to simplify the configuration, that is, to reduce the number of contacts required. The objective is to reduce cost and increase reliability. Reliability is extremely important because the failure of a single pair of relay contacts can shut down an entire automated process.

Contacts (and circuits) are either closed, represented by 1, or open, represented by 0.

Contacts in *series* (AND and NOR functions) are written as the *product* of the contact designations. For example, the contacts of Fig. G-13a can be expressed as $A \cdot B$ or simply AB.

Contacts in *parallel* (OR and NAND functions) are represented as the *sum* of the contact designations. For example, the contacts of Fig. G-13d can be expressed as $\overline{A} + \overline{B}$.

A number of Boolean Algebra theorems are given in Table G-1.

TABLE G-1
BOOLEAN ALGEBRA THEOREMS

1. $A + B = B + A$	10. $A\overline{A} = 0$
2. $AB = BA$	11. $AB + AC = A(B + C)$
3. $A + 0 = A$	12. $(A + B)(A + C) = A + BC$
4. $A \cdot 0 = 0$	13. $A + AB = A$
5. $A + 1 = 1$	14. $A + \overline{A}B = A + B$
6. $A \cdot 1 = A$	15. $A(A + B) = A$
7. $A + A = A$	16. $\overline{AB} = \overline{A} + \overline{B}$
8. $AA = A$	17. $\overline{A + B} = \overline{A}\overline{B}$
9. $A + \overline{A} = 1$	18. $\overline{\overline{A}} = A$

These theorems can be proved in several ways. The first is to sketch the contact configurations represented by both sides of an equation and to see that they are equivalent. For example, Theorem 3 is represented in Fig. G-14a. Obviously, a contact in parallel with an open circuit can be re-

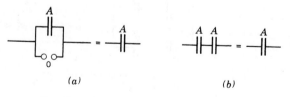

(a)   *(b)*

FIG. G-14. Equivalent contact configurations.

placed by the contact alone. Theorem 8 is represented in Fig. G-14b. The equivalence should be obvious.

A second type of proof involves algebraic manipulation and simplification. With Theorem 13, for example,

$$A + AB = A(1 + B)$$
$$= A(1)$$
$$= A$$

Note that Theorems 11, 5, and 6 (in that order) were used in the proof.

A third type of proof utilizes tabulations known as *truth tables*. This approach will be illustrated by proving Theorem 16. The first two columns of the table in Fig. G-15a show all possible open-and-closed combinations of contacts A and B. The entries in Column 3 are computed from those in the first two columns. Column 4 entries are the opposite of those in Column 3, that is $\overline{1} = 0$ and $\overline{0} = 1$.

A	B	AB	$\overline{AB}$
1	1	1	0
1	0	0	1
0	1	0	1
0	0	0	1

(a)

A	B	$\overline{A}$	$\overline{B}$	$\overline{A}+\overline{B}$
1	1	0	0	0
1	0	0	1	1
0	1	1	0	1
0	0	1	1	1

(b)

FIG. G-15. Truth tables to prove that $\overline{AB} = \overline{A} + \overline{B}$.

The truth table of Fig. G-15b is constructed in a similar manner. A comparison of the last columns of both tables reveals that the entries are the same, thereby proving Theorem 16.

As an illustrative example of circuit simplification, consider the circuit shown in Fig. G-16. A little study will reveal that the overall

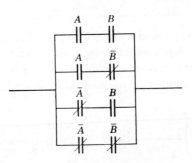

FIG. G-16. Contact configuration
to be simplified.

circuit is always closed despite the states of coils A and B. However, the same result can be obtained in a more elegant fashion through the use of Boolean Algebra. The contact configuration can be expressed as

$$AB + A\overline{B} + \overline{A}B + \overline{A}\overline{B}$$

With Theorems 11 and 9,

$$AB + A\bar{B} + \bar{A}B + \bar{A}\bar{B} = A(B + \bar{B}) + \bar{A}(B + \bar{B})$$
$$= A(1) + \bar{A}(1)$$
$$= A + \bar{A}$$
$$= 1 = \text{closed circuit}$$

Static Switching

Static switching refers to the use of a broad class of switching devices which have neither contacts nor moving parts. The idea is to provide greater reliability and longer life. Early static switching devices were electrical in nature and utilized magnetic-amplifier principles. More recently, electronic circuits involving diodes and transistors have been used with success. One could logically include devices from the newer field of *fluidics* in static control. These are typically pneumatic switching devices which utilize two levels of pressure as input and output signals.

The reader may be interested in seeing examples of how electronic circuits can be used in switching. First of all, one must realize that we can no longer think in terms of open and closed circuits. Instead, we must consider voltage in and voltage out. Input and output signals are, however, limited to two levels of voltage which can be represented as 0 and 1. Therefore, making the transition from contact circuits to electronic circuits presents little difficulty.

A circuit for an AND function is shown in Fig. G-17a. A positive

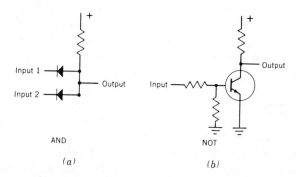

AND   NOT

(a)   (b)

FIG. G-17. Logic functions using electronic components.

voltage is applied at the top of the resistor. If both of the inputs are connected to ground potential (= 0), the output will be at essentially ground potential (= 0) because the voltage drop through the diodes is small. However, if the positive voltage (= 1) is applied to both inputs, current flow through the resistor will cease and the output will be at the level of the positive voltage (= 1). Further, the output will be 0 with a 0–1 combination at the inputs. The result is an AND function which can be expressed as

$$\text{Output} = (\text{Input 1})(\text{Input 2})$$

The output is 0 if either input is 0. If one chooses to think of 0 as no input (or output) and 1 as an input (or output), the earlier definition for an AND function prevails. That is, an AND function provides an output only if all inputs are present.

A transistor circuit which provides a NOT function is shown in Fig. G-17*b*. A positive voltage is applied as indicated. The transistor is cut off with the input at ground potential, and the positive voltage appears at the output. With the input positive, the transistor conducts with the result that the output is at ground potential. The behavior can be expressed as

$$\text{Output} = \overline{\text{Input}}$$

A schematic drawing of a turbulence amplifier is given in Fig. G-18

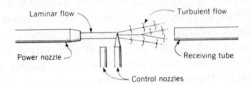

FIG. G-18. A NOR function using a
turbulence amplifier.

to illustrate how a logic function can be obtained with a fluidic device. The design is such that the jet from the power nozzle can remain laminar until it reaches the receiving tube. With this condition, pressure recovery (output) at the receiving tube is good. However, a jet from one of the control nozzles (input) causes the power jet to become turbulent, resulting in essentially no pressure recovery at the receiving tube. The turbulence amplifier is a NOR element because one or more inputs results in no output, while no inputs provide an output.

NEMA Function Symbols. The National Electrical Manufacturers Association (NEMA) has devised symbols for various functions required in static switching involving electrical devices. Some of these symbols are given in Table G-2. The letters *A* and *B* denote inputs while *E* represents the output.

The first four items in Table G-2 should be self-explanatory. The fifth is Off Return Memory. This function is useful where a holding contact would be required in an equivalent relay circuit. A momentary input *A* (and no input *B*) will cause an output *E*. The output will remain until there is an input *B*. An example of the use of the Off Return Memory function will follow shortly.

TABLE G-2
NEMA FUNCTION SYMBOLS

Function	Symbol	Definition
1. AND	A, B → E	$E = AB$
2. OR	A, B → E	$E = A + B$
3. NOT	A → E	$E = \overline{A}$
4. NOR	A, B → E	$E = \overline{A}\,\overline{B}$
5. Off Return Memory	A, B → E	$E = \overline{B}(A + E)$
6. Signal Converter		Converts input device signal to a logic input signal
7. Amplifier		Amplifies logic output signal to the power level required by output device

The Signal Converter in Table G-2 converts the signal from some input device (e.g., a limit switch) to one acceptable to logic elements. The output signal from a logic element must be amplified to the power level required to drive an output device (e.g., a solenoid). Item 7, an Amplifier, serves this purpose.

To illustrate the simplification of circuits formulated in logic functions, consider the circuit of Fig. G-19a. It can be expressed as

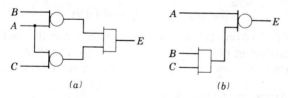

(a)   (b)

FIG. G-19. Equivalent circuits.

$$E = (A + B)(A + C)$$

With Theorem 12 (Table G-1),

$$E = A + BC$$

The simplified circuit is shown in Fig. G-19b. Note that one OR element has been eliminated.

Figure G-20 shows a static switching diagram which is equivalent to the diagram of Fig. G-9. Momentary closure of the Start pushbutton

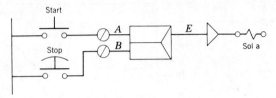

FIG. G-20. Static switching diagram equivalent to that of Fig. G-9.

produces a momentary input A. The Off Return Memory element responds with an output E, with the result that Sol a is energized. The solenoid remains energized until there is an input B produced by depressing the Stop button. The Stop button can be a normally-open switch because of the nature of an Off Return Memory element.

Circuit Design Example. An example will now be presented to illustrate the fact that circuits can be designed from a careful statement of the function to be performed. Consider a forming-press application wherein the operator is required to depress two widely-separated pushbuttons to initiate the press cycle. The objective, of course, is to be sure that both of the operator's hands are out of the press at the time of press closure. However, a lazy operator may tape down one switch so that only one hand is required to run the press. Let us assume that we wish to design a circuit which will sound an alarm should the operator tape down one of the pushbutton switches.

The two pushbutton switches are of the normally-open type and are designated as A and B. The alarm is to sound if one, but not both, of the switches is closed. Thus, the function to be performed can be expressed as

$$\text{Alarm} = E = (A + B)(\overline{AB})$$

The factor $A + B$ covers the taping of a switch, while the factor $\overline{AB}$ disables the alarm during normal press operation (which requires AB). The resulting diagram is shown in Fig. G-21. As a practical matter, a slight

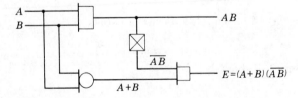

FIG. G-21. Press alarm diagram.

time delay would be placed between output E and the actual alarm so that, in normal operation, both buttons would not have to be pushed at precisely the same time.

Problems

G-1. Design the electrical circuits specified below:

a) A motor starter 1M is to be energized by a momentary contact Start pushbutton switch and can be deenergized only by depressing the Stop pushbutton switch. b) A motor starter 1M is to be energized by a momentary contact Start pushbutton switch but only if both normally-closed limit switches 1LS and 2LS are closed. The motor starter is to remain energized until either the Stop button is depressed or limit switch 2LS is opened. Once the motor starter is energized, it must remain so independent of the state of limit switch 1LS.

G-2. The sump shown in part *a* of the accompanying illustration receives small quantities of water on an intermittent basis. Water level is indicated through the use of a float and scale arrangement. When the water rises to a high level, one of the operating personnel starts the sump pump and brings the water level down to some lower limit. The electrical circuit is shown in part *b* of the figure. You are to modify the circuit such that pumping is automatic. Pumping is to begin when the water level reaches a certain upper limit and is to continue until the level reaches a certain lower limit. Pumping will then cease until the upper limit is again reached.

HINT: The float rod is sufficiently strong to support a cam which can be used to actuate one or more limit switches.

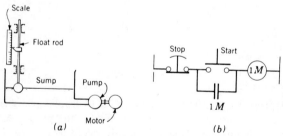

PROB. G-2

G-3. The electropneumatic system shown in the figure is to operate as follows: Start button depressed; piston rod advances to limit switch 1LS; piston

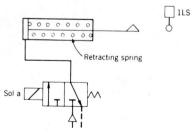

PROB. G-3

rod immediately returns to the retracted position and will remain there until the Start button is again pushed. A Retract button is to be provided which will cause retraction at any time during the operating cycle.

Design an electrical circuit so that the specifications will be met.

G-4. You are to design an electrical circuit to provide clamping for a pair of parts which are then manually welded. The schematic drawing is given in the accompanying illustration. Requirements are as follows:

a) Two pushbuttons are required to clamp the parts so that the operator must use both hands (safety measure). However, the clamps shall not engage unless the parts are properly located in the fixture as detected by the actuation of limit switches 1LS, 2LS, and 3LS. b) The operator must hold the two pushbuttons until clamping is accomplished as detected by the actuation of limit switches 4LS and 5LS. Until clamping is completed, release of one or both pushbuttons shall withdraw the clamps. c) After clamping is accomplished, the operator releases the pushbuttons and performs the welding operation. d) The actuation of a single pushbutton is to release the parts.

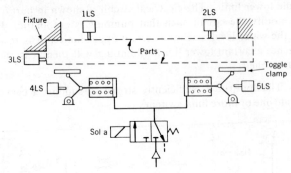

PROB. G-4

G-5. A proposed hydraulic circuit for a hydraulic press is shown in the figure. You are to design an electrical circuit for the press. A Start button is to initiate the cycle and cause the ram to advance thus closing the press. After the press is closed, full pressure is to be applied for several seconds after which the ram is to retract, completing the cycle; that is, the ram will remain at rest

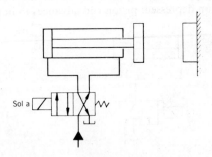

PROB. G-5

until the Start button is again depressed. A Retract button is to be provided which will cause the ram to retract during any portion of the cycle.

HINT: You may wish to use a limit switch to detect press closure.

G-6. This problem is a continuation of Prob. G-5.

a) A double-solenoid spring-centered valve is to be used in the hydraulic circuit. Design a new hydraulic circuit to accomodate this valve. b) Design a new electrical circuit. The cycle is to be the same as stated in Prob. G-5 except that depressing a Stop button is to stop the ram at any point in the cycle. (The Retract button may be omitted.) Both solenoids should be deenergized at the completion of the cycle.

G-7. A hydraulic circuit is shown in the accompanying illustration. Design an electrical circuit to provide the following cycle: Start button depressed; piston rod advances from the retracted position to the fully extended position; rod immediately reverses; rod stops when limit switch 1LS is encountered, dwells for several seconds, and then continues to the retracted position at which time the cycle is complete.

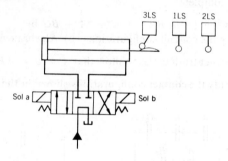

PROB. G-7

G-8. The object of this problem is to design two different electrical circuits for the control of the hydraulic press shown in the figure. One requirement for both circuits is that when the press is open, the ram is to be held in the retracted position under pressure by the constant energization of Sol b. As a safety feature,

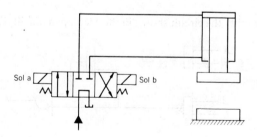

PROB. G-8

press actuation downward is to require that the operator use both hands to depress two pushbuttons. Press actuation upward is accomplished by releasing both pushbuttons. A possibility that must not be overlooked is that an operator may attempt to circumvent the safety feature by taping down one of the pushbuttons.

a) Design an electrical circuit such that if a switch is taped down, the press will close (when the other switch is actuated) but will remain so until the taped switch is released. b) Design an electrical circuit such that the press cannot be actuated at all following the taping of one pushbutton.

G-9. Pneumatic, hydraulic, and electrical circuits are to be designed according to the following specifications:

a) The operator places a part in a fixture and depresses a Start pushbutton switch. This activates a pneumatic cylinder which clamps the part in the fixture. b) There is a limit switch in the fixture which detects whether or not the part is properly clamped. The cycle is to continue if the limit switch is actuated. The cycle is to terminate and the clamping cylinder is to release if the limit switch is not actuated. c) With proper clamping, a hydraulic cylinder advances a tool through a controlled distance at the end of which, a limit switch is actuated. d) The hydraulic cylinder is to retract automatically. e) When the hydraulic cylinder is fully retracted, the pneumatic clamping cylinder is to be retracted. The cycle is then complete.

G-10. Show that $(A + B)(A + C) = A + BC$ by:
a) Sketching the contact configurations. b) Algebraic manipulation.

G-11. Construct truth tables to show that $\overline{A + B} = \overline{A}\,\overline{B}$.

G-12. Simplify the contact configurations shown in the figure. Use Boolean Algebra.

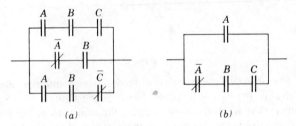

PROB. G-12

G-13. Simplify the circuit shown in the accompanying illustration.

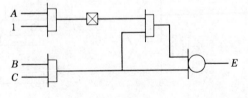

PROB. G-13

G-14. Convert your solution to Prob. G-3 to a static switching diagram.

MAGNITUDE RATIO AND PHASE ANGLE FOR $\dfrac{O}{I} = \dfrac{1}{\dfrac{s^2}{\omega_n^2} + \dfrac{2\zeta s}{\omega_n} + 1}$

M(W)(decibels)   W·N (radians/sec)   ANG (degrees)

	ZETA = 0.05			ZETA = 0.10	
W/WN	M(W)	ANG	W/WN	M(W)	ANG
0.100	0.0	-0.6	0.100	0.0	-1.2
0.200	0.4	-1.2	0.200	0.3	-2.4
0.300	0.8	-1.9	0.300	0.8	-3.8
0.400	1.5	-2.7	0.400	1.5	-5.4
0.500	2.5	-3.8	0.500	2.4	-7.6
0.600	3.8	-5.4	0.600	3.7	-10.6
0.700	5.8	-7.8	0.700	5.5	-15.4
0.800	8.7	-12.5	0.800	8.1	-24.0
0.850	10.7	-17.0	0.850	9.8	-31.5
0.900	13.5	-25.3	0.900	11.6	-43.5
0.925	15.3	-32.6	0.925	12.6	-52.0
0.950	17.3	-44.3	0.950	13.4	-62.8
0.975	19.2	-63.1	0.975	13.9	-75.8
1.000	20.0	-90.0	1.000	14.0	-90.0
1.050	16.7	-134.3	1.050	12.6	-116.0
1.100	12.5	-152.4	1.100	10.3	-133.7
1.150	9.3	-160.4	1.150	8.0	-144.5
1.200	6.8	-164.7	1.200	6.0	-151.4
1.400	0.3	-171.7	1.400	-0.0	-163.7
1.600	-3.9	-174.1	1.600	-4.0	-168.4
2.000	-9.6	-176.2	2.000	-9.6	-172.4
2.500	-14.4	-177.3	2.500	-14.4	-174.6
3.000	-18.1	-177.9	3.000	-18.1	-175.7
6.000	-30.9	-179.0	6.000	-30.9	-178.0
10.000	-39.9	-179.4	10.000	-39.9	-178.8

MAGNITUDE RATIO AND PHASE ANGLE FOR $\dfrac{O}{I} = \dfrac{1}{\dfrac{s^2}{\omega_n^2} + \dfrac{2\zeta s}{\omega_n} + 1}$ (continued)

ZETA = 0.15

W/WN	M(W)	ANG
0.100	0.0	−1.7
0.200	0.3	−3.6
0.300	0.8	−5.6
0.400	1.4	−8.1
0.500	2.3	−11.3
0.600	3.5	−15.7
0.700	5.2	−22.4
0.800	7.3	−33.7
0.850	8.5	−42.6
0.900	9.6	−54.9
0.925	10.1	−62.5
0.950	10.4	−71.1
0.975	10.6	−80.4
1.000	10.5	−90.0
1.050	9.6	−108.0
1.100	8.2	−122.5
1.150	6.5	−133.1
1.200	4.9	−140.7
1.400	−0.4	−156.4
1.600	−4.3	−162.9
2.000	−9.7	−168.7
2.500	−14.5	−171.9
3.000	−18.1	−173.6
6.000	−30.9	−177.1
10.000	−39.9	−178.3

ZETA = 0.20

W/WN	M(W)	ANG
0.100	0.0	−2.3
0.200	0.3	−4.8
0.300	0.7	−7.5
0.400	1.4	−10.8
0.500	2.2	−14.9
0.600	3.3	−20.6
0.700	4.7	−28.8
0.800	6.3	−41.6
0.850	7.2	−50.8
0.900	7.8	−62.2
0.925	8.0	−68.7
0.950	8.1	−75.6
0.975	8.1	−82.8
1.000	8.0	−90.0
1.050	7.3	−103.7
1.100	6.2	−115.5
1.150	5.0	−125.0
1.200	3.7	−132.5
1.400	−0.9	−149.7
1.600	−4.5	−157.7
2.000	−9.8	−165.1
2.500	−14.6	−169.2
3.000	−18.2	−171.5
6.000	−30.9	−176.1
10.000	−39.9	−177.7

ZETA = 0.25

W/WN		M(W)		ANG	
W/WN =	0.100	M(W) =	0.0	ANG =	−2.9
W/WN =	0.200	M(W) =	0.3	ANG =	−5.9
W/WN =	0.300	M(W) =	0.7	ANG =	−9.4
W/WN =	0.400	M(W) =	1.3	ANG =	−13.4
W/WN =	0.500	M(W) =	2.0	ANG =	−18.4
W/WN =	0.600	M(W) =	3.0	ANG =	−25.1
W/WN =	0.700	M(W) =	4.2	ANG =	−34.5
W/WN =	0.800	M(W) =	5.4	ANG =	−48.0
W/WN =	0.850	M(W) =	5.9	ANG =	−56.9
W/WN =	0.900	M(W) =	6.2	ANG =	−67.1
W/WN =	0.925	M(W) =	6.3	ANG =	−72.7
W/WN =	0.950	M(W) =	6.3	ANG =	−78.4
W/WN =	0.975	M(W) =	6.2	ANG =	−84.2
W/WN =	1.000	M(W) =	6.0	ANG =	−90.0
W/WN =	1.050	M(W) =	5.4	ANG =	−101.0
W/WN =	1.100	M(W) =	4.6	ANG =	−110.9
W/WN =	1.150	M(W) =	3.6	ANG =	−119.3
W/WN =	1.200	M(W) =	2.6	ANG =	−126.3
W/WN =	1.400	M(W) =	−1.5	ANG =	−143.9
W/WN =	1.600	M(W) =	−4.9	ANG =	−152.9
W/WN =	2.000	M(W) =	−10.0	ANG =	−161.6
W/WN =	2.500	M(W) =	−14.6	ANG =	−166.6
W/WN =	3.000	M(W) =	−18.2	ANG =	−169.4
W/WN =	6.000	M(W) =	−30.9	ANG =	−175.1
W/WN =	10.000	M(W) =	−39.9	ANG =	−177.1

ZETA = 0.30

W/WN		M(W)		ANG	
W/WN =	0.100	M(W) =	0.0	ANG =	−3.5
W/WN =	0.200	M(W) =	0.3	ANG =	−7.1
W/WN =	0.300	M(W) =	0.7	ANG =	−11.2
W/WN =	0.400	M(W) =	1.2	ANG =	−15.9
W/WN =	0.500	M(W) =	1.9	ANG =	−21.8
W/WN =	0.600	M(W) =	2.7	ANG =	−29.4
W/WN =	0.700	M(W) =	3.6	ANG =	−39.5
W/WN =	0.800	M(W) =	4.4	ANG =	−53.1
W/WN =	0.850	M(W) =	4.7	ANG =	−61.4
W/WN =	0.900	M(W) =	4.8	ANG =	−70.6
W/WN =	0.925	M(W) =	4.8	ANG =	−75.4
W/WN =	0.950	M(W) =	4.8	ANG =	−80.3
W/WN =	0.975	M(W) =	4.6	ANG =	−85.2
W/WN =	1.000	M(W) =	4.4	ANG =	−90.0
W/WN =	1.050	M(W) =	3.9	ANG =	−99.2
W/WN =	1.100	M(W) =	3.2	ANG =	−107.7
W/WN =	1.150	M(W) =	2.4	ANG =	−115.1
W/WN =	1.200	M(W) =	1.5	ANG =	−121.4
W/WN =	1.400	M(W) =	−2.1	ANG =	−138.8
W/WN =	1.600	M(W) =	−5.3	ANG =	−148.4
W/WN =	2.000	M(W) =	−10.2	ANG =	−158.2
W/WN =	2.500	M(W) =	−14.7	ANG =	−164.1
W/WN =	3.000	M(W) =	−18.3	ANG =	−167.3
W/WN =	6.000	M(W) =	−30.9	ANG =	−174.1
W/WN =	10.000	M(W) =	−39.9	ANG =	−176.5

MAGNITUDE RATIO AND PHASE ANGLE FOR $\dfrac{O}{I} = \dfrac{1}{\dfrac{s^2}{\omega_n^2} + \dfrac{2\zeta s}{\omega_n} + 1}$ *(continued)*

ZETA = 0.40

W/WN	M(W)	ANG
0.100	0.0	−4.6
0.200	0.2	−9.5
0.300	0.5	−14.8
0.400	0.9	−20.9
0.500	1.4	−28.1
0.600	1.9	−36.9
0.700	2.4	−47.7
0.800	2.7	−60.6
0.850	2.7	−67.8
0.900	2.6	−75.2
0.925	2.5	−79.0
0.950	2.3	−82.7
0.975	2.1	−86.4
1.000	1.9	−90.0
1.050	1.5	−97.0
1.100	0.9	−103.4
1.150	0.2	−109.3
1.200	−0.5	−114.6
1.400	−3.4	−130.6
1.600	−6.1	−140.6
2.000	−10.6	−151.9
2.500	−15.0	−159.1
3.000	−18.4	−163.3
6.000	−31.0	−172.2
10.000	−39.9	−175.4

ZETA = 0.50

W/WN	M(W)	ANG
0.100	0.0	−5.8
0.200	0.2	−11.8
0.300	0.4	−18.2
0.400	0.6	−25.5
0.500	0.9	−33.7
0.600	1.1	−43.2
0.700	1.2	−53.9
0.800	1.1	−65.8
0.850	1.0	−71.9
0.900	0.7	−78.1
0.925	0.6	−81.1
0.950	0.4	−84.1
0.975	0.2	−87.1
1.000	−0.0	−90.0
1.050	−0.5	−95.6
1.100	−1.0	−100.8
1.150	−1.5	−105.7
1.200	−2.1	−110.1
1.400	−4.6	−124.4
1.600	−7.0	−134.3
2.000	−11.1	−146.3
2.500	−15.3	−154.5
3.000	−18.6	−159.4
6.000	−31.0	−170.3
10.000	−40.0	−174.2

ZETA = 0.60

W/WN	M(W)	ANG
0.100	0.0	-6.9
0.200	0.0	-14.0
0.300	0.2	-21.6
0.400	0.3	-29.7
0.500	0.4	-38.7
0.600	0.3	-48.4
0.700	0.2	-58.7
0.800	-0.2	-69.4
0.850	-0.5	-74.8
0.900	-0.8	-80.0
0.925	-1.0	-82.6
0.950	-1.2	-85.1
0.975	-1.4	-87.6
1.000	-1.6	-90.0
1.050	-2.0	-94.7
1.100	-2.5	-99.0
1.150	-3.0	-103.2
1.200	-3.6	-107.0
1.400	-5.7	-119.7
1.600	-7.9	-129.1
2.000	-11.7	-141.3
2.500	-15.6	-150.3
3.000	-18.9	-155.8
6.000	-31.1	-168.4
10.000	-40.0	-173.1

ZETA = 0.70

W/WN	M(W)	ANG
0.100	0.0	-8.0
0.200	-0.0	-16.3
0.300	-0.0	-24.8
0.400	-0.0	-33.7
0.500	-0.2	-43.0
0.600	-0.5	-52.7
0.700	-0.9	-62.5
0.800	-1.4	-72.2
0.850	-1.7	-76.9
0.900	-2.1	-81.4
0.925	-2.3	-83.6
0.950	-2.5	-85.8
0.975	-2.7	-87.9
1.000	-2.9	-90.0
1.050	-3.4	-94.0
1.100	-3.8	-97.8
1.150	-4.3	-101.3
1.200	-4.8	-104.7
1.400	-6.8	-116.1
1.600	-8.7	-124.9
2.000	-12.3	-137.0
2.500	-16.0	-146.3
3.000	-19.1	-152.3
6.000	-31.1	-166.5
10.000	-40.0	-172.0

MAGNITUDE RATIO AND PHASE ANGLE FOR $\dfrac{0}{I} = \dfrac{1}{\dfrac{s^2}{\omega_n^2} + \dfrac{2\zeta s}{\omega_n} + 1}$ (continued)

ZETA = 0.80

W/WN	M(W)	ANG
0.100	-0.0	-9.2
0.200	-0.1	-18.4
0.300	-0.2	-27.8
0.400	-0.5	-37.3
0.500	-0.8	-46.8
0.600	-1.2	-56.3
0.700	-1.8	-65.5
0.800	-2.5	-74.3
0.850	-2.8	-78.5
0.900	-3.2	-82.5
0.925	-3.4	-84.4
0.950	-3.7	-86.3
0.975	-3.9	-88.2
1.000	-4.1	-90.0
1.050	-4.5	-93.5
1.100	-5.0	-96.8
1.150	-5.4	-99.9
1.200	-5.9	-102.9
1.400	-7.7	-113.2
1.600	-9.5	-121.4
2.000	-12.8	-133.2
2.500	-16.4	-142.7
3.000	-19.4	-149.0
6.000	-31.2	-164.7
10.000	-40.0	-170.8

ZETA = 0.90

W/WN	M(W)	ANG
0.100	-0.0	-10.3
0.200	-0.2	-20.6
0.300	-0.5	-30.7
0.400	-0.9	-40.6
0.500	-1.4	-50.2
0.600	-2.0	-59.3
0.700	-2.7	-68.0
0.800	-3.4	-76.0
0.850	-3.8	-79.7
0.900	-4.2	-83.3
0.925	-4.5	-85.0
0.950	-4.7	-86.7
0.975	-4.9	-88.4
1.000	-5.1	-90.0
1.050	-5.5	-93.1
1.100	-6.0	-96.1
1.150	-6.4	-98.9
1.200	-6.9	-101.5
1.400	-8.6	-110.9
1.600	-10.3	-118.4
2.000	-13.4	-129.8
2.500	-16.8	-139.4
3.000	-19.7	-146.0
6.000	-31.3	-162.9
10.000	-40.1	-169.7

Bibliography

1. AHRENDT, W. R. *Servomechanism Practice.* New York: McGraw-Hill Book Company, Inc., 1954.
2. BLACKBURN, J. F., G. REETHOF, and J. L. SHEARER. *Fluid Power Control.* New York: John Wiley & Sons, Inc., 1960.
3. BROWN, G. S., and D. P. CAMPBELL. *Principles of Servomechanisms.* New York: John Wiley & Sons, Inc., 1955.
4. CHESTNUT, H., and R. W. MAYER. *Servomechanisms and Regulating System Design,* vol. 1. New York: John Wiley & Sons, Inc., 1959.
5. CHURCHILL, R. V. *Operational Mathematics.* New York: McGraw-Hill Book Company, Inc., 1958.
6. CONSIDINE, D. M. *Process Instruments and Control Handbook.* New York: McGraw-Hill Book Company, Inc., 1957.
7. *Control Engineering.* A Special Report on Digital Computers in Industry, September 1966.
8. DORF, R. C. *Time-Domain Analysis and Design of Control Systems.* Reading, Mass.: Addison-Wesley Publishing Company, Inc., 1965.
9. DWIGHT, H. B. *Tables of Integrals and Other Mathematical Data.* New York: The Macmillian Company, 1947.
10. ECKMAN, D. P. *Automatic Process Control.* New York: John Wiley & Sons, Inc., 1958.
11. EVANS, W. R. "Control System Synthesis by Root Locus Method," *Trans. AIEE,* vol. 69, pp. 66–69, 1950.
12. EVANS, W. R. "The Use of Zeros and Poles for Frequency Response or Transient Response," *Trans. ASME,* vol. 76, pp. 1335–1344, 1954.
13. GIBSON, J. E. *Nonlinear Automatic Control.* New York: McGraw-Hill Book Company, Inc., 1963.
14. GRABBE, E. M., S. RAMO, and D. E. WOOLDRIDGE. *Handbook of Automation, Computation, and Control,* vol. 1. New York: John Wiley & Sons, Inc., 1958.
15. GRAHAM, D., and D. McRUER. *Analysis of Nonlinear Control Systems.* New York: John Wiley & Sons, Inc., 1961.
16. HILDEBRAND, F. B. *Advanced Calculus for Engineers.* New Jersey: Prentice-Hall, Inc., 1949.

453

17. JOHNSON, C. L. *Analog Computer Techniques*. New York: McGraw-Hill Book Company, Inc., 1956.
18. KARPLUS, W. J., and W. W. SOROKA. *Analog Methods Computation and Simulation*. New York: McGraw-Hill Book Company, Inc., 1959.
19. KUO, B. C. *Automatic Control Systems*. New Jersey: Prentice-Hall, Inc., 1962.
20. KUO, BENJAMIN C. *Analysis and Synthesis of Sampled-Data Control Systems*. New Jersey: Prentice-Hall, Inc., 1963.
21. LEVY, E. C. *A New Table of Laplace Transform Pairs*. Copyright by Ezra C. Levy, 520 N. Highland Avenue, Los Angeles 36, California, 1959.
22. LIN, Shih-Nge. "A Method of Successive Approximations of Evaluating the Real and Complex Roots of Cubic and Higher Order Equations," *J. of Math. and Physics*, vol. 20, pp. 231–242, 1941.
23. LANGILL, A. W., JR. *Automatic Control Systems Engineering, Volume II*. New Jersey: Prentice-Hall, Inc., 1965.
24. NOTON, A. R. M. *Variational Methods in Control Engineering*. Pergamon Press, 1965.
25. RAZAZINI, J. R. and FRANKLIN, G. F. *Sampled Data Control Systems*. New York: McGraw-Hill Book Company, 1958.
26. SEIREG, A. A. H. "Static and Dynamic Behavior of Flexible Torsional Couplings with Nonlinear Characteristics." Ph.D. thesis, University of Wisconsin, 1954.
27. SMITH, G. W. *Principles of Analog Computation*. New York: McGraw-Hill Book Company, Inc., 1959.
28. SMITH, JAMES and WOLFORD. *Analog and Digital Computer Methods in Engineering*. Scranton, Pa.: International Textbook Co., 1965.
29. STANTON, R. G. *Numerical Methods for Science and Engineering*. New Jersey: Prentice-Hall, Inc., 1961.
30. THOMPSON, W. T. *Laplace Transformation*. New Jersey: Prentice-Hall, Inc., 1950.
31. TOU, J. T. *Modern Control Theory*. New York: McGraw-Hill Book Company, Inc., 1964.
32. TRUXAL, J. G. *Automatic Feedback Control System Synthesis*. New York: McGraw-Hill Book Company, Inc., 1955.
33. WYLIE, C. R., JR. *Advanced Engineering Mathematics*. New York: McGraw-Hill Book Company, Inc., 1960.

Index